TRANS *FATTY ACIDS IN HUMAN NUTRITION*

edited by

Jean Louis Sébédio

INRA, Unitè de Nutrition Lipidique, 17 rue Sully, B.V. 1540, 21034 Dijon Cedex, France

and

William W. Christie

Scottish Crop Research Institute, Invergowrie, Dundee (DD2 5DA), Scotland

THE OILY PRESS

DUNDEE

ISBN 0 9514171 8 5

British Library Cataloguing-in-Publication Data. A catalogue record for this book is available from the British Library.

This is - Volume 9 in the Oily Press Lipid Library

(Volume 1 - "Gas Chromatography and Lipids" by William W. Christie; Volume 2 - Advances in Lipid Methodology - One" edited by William W. Christie; Volume 3 - "A Lipid Glossary" by Frank D. Gunstone and Bengt G. Herslöf; Volume 4 - "Advances in Lipid Methodology - Two" edited by William W. Christie; Volume 5, "Lipids: Molecular Organization, Physical Functions and Technical Applications" by Kåre Larsson; Volume 6 - "Waxes: Chemistry, Molecular Biology And Functions" edited by Richard J. Hamilton; Volume 7 - "Advances in Lipid Methodology - Three" edited by William W. Christie; Volume 8 - "Advances in Lipid Methodology - Four" edited by William W. Christie).

Printed in Great Britain by Bell and Bain Ltd., Glasgow

PREFACE

While numerous studies on the health effect of partially hydrogenated vegetable oils and of other sources of *trans* fatty acids were carried out more than twenty years ago, there has been increasing concern in the 1990s about the health effects of the *trans* mono-ethylenic isomers formed during commercial hydrogenation processes. In this book, authors who are recognized international authorities in their fields have covered, the occurrence, consumption, analysis, biochemistry and health effects of the different types of *trans* fatty acids, also including conjugated linoleic acid (CLA), which has many beneficial effects apparently.

In the first chapter, Robert Wolff *et al.* have described the distribution profiles of the *trans* 18:1 acids of natural origin, resulting from biohydrogenation in the rumen of cows and other animals. An appendix by Robert Ackman lists other natural *trans* fatty acids. Robert Ackman and Ted Mag show in the second chapter, how industrial processing (hydrogenation and deodorization) of vegetable oils (and fish oil) can lead to the formation of *trans* mono- and polyunsaturated fatty acids, that enter the food chain. In the third chapter, Margaret Craig-Schmidt describes the methods used to estimate the consumption of *trans* fatty acids and critically reviews the data on consumption in different countries and regions of the world. In the next chapter, Nimal Ratnayake shows how important is the choice of an analytical method in order to obtain reliable data for quantification of *trans* fatty acids, a feature stressed in many of the other chapters. Faulty methodology has lead to many erroneous conclusions in the past.

The biochemistry of *trans* mono- and polyunsaturated fatty acid is covered in Chapters 5 (Hölmer) and 6 (Sèbèdio and Chardigny), respectively. A variety of different biological effects have been observed, and many of these are dependent on specific isomers and thence on the dietary origin. Interestingly, metabolism of the *trans* n-6 or the n-3 polyunsaturated fatty acids follows that of the corresponding cis isomers. The health effects of *trans* monounsaturated fatty acids (in relation to lipoprotein metabolism and coronary heart disease, especially) are then described by Ronald Mensink and Peter Zock and by Antti Aro, respectively. It seems apparent that many of the hazards associated with *trans* fatty acids may have been exaggerated, although they may not be negligible.

Finally, there has been a remarkable growth of interest in the health benefits of conjugated linoleic acid, which also has a *trans* double bond. In particular, anticarcinogenic and antiatherogenic properties are ascribed to it and are discussed in the last chapter by Banni and Martin. Here also, problems associated with the analysis are described in some detail, as again biological effects may be due to specific isomers.

We hope that the availability of so much information in a single volume will serve to clarify the general effects of *trans* fatty acids in human nutrition. It may dispel some myths and hopefully stimulate more and better research.

Jean Louis Sébédio
William W. Christie

CONTENTS

CHAPTER 1

OCCURRENCE AND DISTRIBUTION PROFILES OF TRANS-*18:1 ACIDS* IN EDIBLE FATS OF NATURAL ORIGIN

Robert L. Wolff[a], Dietz Precht[b] and Joachim Molkentin[b]

[a]*Institut des Sciences et Techniques des Aliments de Bordeaux, Université Bordeaux 1, Allée des Facultés, 33405, Talence Cedex, France,* [b]*Bundesanstalt für Milchforschung, Institut für Chemie und Physik, Postfach 6069, D-24121 Kiel, Germany*

A. INTRODUCTION

This introductory chapter is devoted to the survey and critical evaluation of the present literature knowledge about the *trans*-18:1 acid content and profile in natural edible fats. Straightaway, it should be made clear that such fats are essentially those originating from ruminants, and it should be kept in mind that, from time immemorial, human beings, either as hunters or breeders, have consumed ruminant fats, and thus *trans*-18:1 acids. Despite the great number of reviews that appeared on potential health implications of *trans* fatty acids (for the period 1992-1995, see for example [11,15,20,34,59,69,83], ruminant fats do not seem to have been approached in depth in this regard. Apparently, these acids, though having an uncommon structure (uncommon as compared to unsaturations of the *cis* configuration that are introduced in fatty acids of mammalian, and hence, of human cells), seem to be "protected" against criticism by their natural origin, in contrast to the same acids present in partially hydrogenated vegetable oils, which are man made. *Trans*-18:1 acids from this latter origin are questioned as to their potential harmfulness to health especially, though their structures are similar in essence to those found in natural sources. The main differences are in the relative proportions of individual *trans*-18:1 isomers and in the relative contribution of each source to the total dietary intake. However, ruminant fats may contribute a significant part, and sometimes the major one, to the overall *trans*-18:1 acid intake by humans in several countries [79]. It is thus necessary to have available reliable and accurate values about their content in foods. One aim of this chapter is to try to point out sound data that should be used with confidence when evaluating *trans*-18:1 acid consumption from natural sources.

Several *trans* unsaturated fatty acids with odd structures occur naturally in some plants [36,67], but these have generally no food applications and they are not consumed by humans. Consequently, we will not include these exceptions in the present chapter. As outlined above, natural edible sources of *trans* acids, mainly *trans*-18:1 acids, are restricted in practice to ruminant fats. We thus discuss in detail the present knowledge of these isomers, which are found mainly in beef meat lipids, fatty tissues, or milk fat. However, the analysis of *trans* acids has not always been an easy task, and a short critical description of the more common analytical methodologies will be given to familiarize the reader with their advantages and limitations. The deficiencies of some of them should be kept in mind when critically evaluating quantitative data published in the literature. More details on updated methodologies for *trans* acid analysis are given in Chapter 4 of this book.

Non-ruminant animals that have themselves eaten some ruminant fats or partially hydrogenated oils and that are dietary sources for humans may also contain *trans*-18:1 isomers. This is also the case for pig meat and fat and is the reason why we include the study of these edible sources in this chapter.

B. DIVERSITY OF *TRANS*-18:1 ACIDS IN RUMINANT FATS

Though elaidic (*trans*-9 18:1) acid was discovered as a solid form of oleic (*cis*-9 18:1) acid as early as 1832 by Boudet (cited by Moore [48]), the history of naturally occurring *trans* acids begins apparently in 1921 with the detection by Twitchell [73] of a small amount of an "unsaturated solid fatty acid" in tallow. Following this first observation, Bertram [6], in 1928, isolated and characterized a new C18 "elaidic" acid in beef and sheep fats, and also in butterfat. He established that its ethylenic bond was in position $\Delta 11$, and he proposed the name vaccenic acid, from the Latin *vacca*, cow. The existence of vaccenic acid was subsequently confirmed by other investigators, but it was not until 1948 that the *trans* ("elaidic") configuration of this acid was confirmed by Rao and Daubert [62], in a comparative study of its infrared spectrum with that of authentic elaidic acid. Later, Backderf and Brown [3] established that in addition to vaccenic acid, butterfat contained a *trans*-16-octadecenoic acid present to the extent of 1 - 2% of total fatty acids. In the mean time, elaidic acid had been characterized in beef fat by Swern *et al.* [72].

It was not until 1966 that Katz and Keeney [35] could establish the complete profile and the wide range of *trans*-18:1 acids that occur in cow rumen digesta and bacteria. Isomeric fatty acids with ethylenic bonds spanning from position $\Delta 5$ to position $\Delta 16$ were identified. Later, Bickerstaff *et al.* [7] followed in the goat the migration of *trans*-18:1 isomers from digesta to milk through lymph, arterial plasma and mammary veinous plasma. The complexity of these isomers was found later in cow milk fat by Hay and Morrison [27] and by Parodi [53]. A similar range of isomers was found in cow adipose tissues [28,53]. Thus, *trans*-18:1 isomers that result from the biohydrogenation of polyunsaturated fatty acids by rumen bacteria retain much of their distribution profile when they are absorbed and transferred into lipids from the milk [27,53] or the fatty tissues [28,53] of the animal.

C. METHODOLOGICAL LIMITATIONS IN THE QUANTIFICATION OF *TRANS*-18:1 ACIDS

The determination of *trans* fatty acid contents in edible fats and oils has been performed by several analytical procedures. Methods applied were infrared (IR)-spectroscopy, gas-liquid chromatography (GLC), silver-ion impregnated (argentation) thin-layer chromatography (Ag-TLC) along with GLC, and ozonolysis in combination with GLC. It is the aim of the present section to try to determine the limitations of these methodologies with regard to the accurate determination of total and individual *trans*-18:1 acids, in order to interpret literature data.

1. Infrared Spectroscopy

In the past, quantification of *trans* fatty acids was frequently by IR-spectroscopy and measurements of "specific" absorption at about 970 cm^{-1} of *trans* ethylenic bonds. It is clear that IR-spectroscopy was, and is still, inappropriate to study individual *trans*-unsaturated fatty acids, because it does not give information on the type of fatty acids, i.e. chain-length, number and position of ethylenic bonds. Generally, IR results were at best semi-quantitative for samples with low *trans* contents (less than ca. 5%), which is generally the case for ruminant fats. For example, DeMan and DeMan [16] have observed that values of ca. 2 or 7% could be obtained for the same sample, depending only on the method employed (uncorrected IR absorption of methyl esters and triacylglycerols, respectively). With a few samples, Smith *et al.* [66] also observed that results obtained by IR-spectroscopy could be twice those determined by GLC coupled with Ag-TLC.

Although results can be improved by using Fourier-transform IR-spectroscopy, the problem of exact determinations of low *trans* fatty acid content (less than 5%) remains [74-76)]. Moreover, IR analyses alone still do not give information on the type of fatty acid. Undoubtedly, GLC is a more convenient means to study individual *trans*-unsaturated fatty acids, though it has its own limitations.

2. Gas-Liquid Chromatography on Packed Columns

In the seventies, columns packed with Silar-10C, Silar-12C, SP-2340, and especially OV-275, had general acceptance for the separation of *trans*-18:1 isomers. Ottenstein *et al.* [51] could obtain a base-line resolution (R = 1.5) of *trans*-9 and *cis*-9 18:1 isomers in 30 min or less. Though *trans*-9-18:1 acid is easily commercially available, it is a very minor component in ruminant fats. In these fats, the double bond spans from position $\Delta 5$ to $\Delta 16$, and isomers having their ethylenic bonds close to the methyl end have retention times slightly longer than those having their ethylenic bond close to the carboxylic end. This leads to an appreciable enlargement of the *trans*-18:1 acid peak, and the resolution from the generally dominating *cis*-9-18:1 acid is somewhat decreased. According to Conacher and Iyengar [14], direct GLC on a Silar-10C packed column gave quantitative results for *trans*-18:1 acids that were 10-20% lower than those obtained by combining GLC and Ag-TLC. Heckers *et al.* [29] found also that quantitative results were 16 - 38% lower than by Ag-TLC/GLC, depending on the total content of *trans*-18:1 acids. Using stationary phases of medium polarity, such as FFAP, SP-1000, SP-2100 or CPWax-58, leads to a reversal of retention times concerning *cis*- and *trans*-18:1 acids. Thus, *trans*-18:1 isomers elute after oleic acid, but are partly overlapped by *cis* isomers as well [44].

3. Gas-Liquid Chromatography on Capillary Columns

The main progress of capillary columns as compared to packed columns was a considerable reduction in the peak widths, and a partial separation of individual *trans*- and *cis*-18:1 isomers could be achieved. However, considering each of the *trans*-18:1 and *cis*-18:1 acid groups as a whole, the resolution was not improved greatly, particularly with short columns (10 - 15 m). Here too, isomers having their ethylenic bonds closest to the methyl end elute under the *cis*-18:1 acids.

Because the efficiency is proportional to the length of the column, an increase in the column length will increase the resolution. This was illustrated in 1980 by Lanza *et al.* [39] for capillary columns coated with SP-2340. Modifying the selectivity may also profoundly influence the resolution. When the length of the column and the selectivity are both increased, the resolution is greatly improved [52].

Even with the longest and most efficient columns (50 to 100-m columns coated with cyanoalkyl polysiloxane phases), a complete separation of *trans-* from *cis-*18:1 isomers has apparently not been achieved. Using a 100-m CPSil-88 capillary column, in ruminant fats at least, the group *trans-*Δ12 to *trans-*Δ14 isomers can be resolved from oleic acid, but this still includes *cis-*Δ6 - Δ8 isomers [61]. Moreover, with optimal GLC conditions, the peak of *trans-*Δ15 is separated from *cis-*Δ9 in a large number of milk fats as well. In contrast to partially hydrogenated vegetable oils, the overlapping peak of *cis-*Δ10 is present only in trace amounts in bovine milk fat [59].

According to Chen *et al.* [12], who analysed the fatty acids prepared from human milk on a 100-m SP-2560 capillary column, and who compared these results with Ag-TLC/GLC data, the overlap of *trans-*18:1 acids with *cis-*18:1 acids ranged from 9 - 30% (mean, 21%; $n = 198$). Compared to that, use of 30 - 50-m columns allows only the group of isomers *trans-*Δ6 to -Δ11 to be separated without overlaps by *cis-*isomers, which for bovine milk fat is equivalent to *ca.* 60% of the total *trans-*18:1 acid content. Overlaps are minimized, but not suppressed, when using a 100-m CPSil-88 capillary column for the study of ruminant fats [47,80].

Single GLC analyses can be performed to quantify *trans-*18:1 acids exclusively on the condition that results are corrected by comparison with data obtained with the combination Ag-TLC/GLC. For the specific case of cow milk fats, the total *trans-*18:1 acid content can be estimated from direct GLC data with the help of statistically derived formulae, which are based upon data from direct GLC and from Ag-TLC/GLC [30,47,61]. Precht and Molkentin [55,61] could also derive *trans* fatty acid formulae from triacylglycerol compositions and Ag-TLC/GLC of fatty acid methyl esters data.

4. Argentation Thin-Layer Chromatography

Ag-TLC was introduced in the sixties [17,49] as a means to separate fatty acid derivatives according to their degree of unsaturation and to the geometry of ethylenic bonds. Depending on the position of the ethylenic bond along the hydrocarbon chain, individual isomers have slightly different R_f [23]. Despite these differences in mobility, isomers with double bonds between positions Δ5 and Δ16 are generally recovered in a single, though heterogeneous band. An example of the reproducibility of results obtained by Ag-TLC/GLC on a capillary column (50-m CPSil-88), and applied to meat lipids, is given in Table 1.1. Similar observations have been published for butterfat [81], which would suggest that this methodology might be of general application.

5. Separation of Individual Trans-18:1 Isomers by GLC

To get an insight into the distribution of individual *trans-*18:1 isomers after their isolation by Ag-TLC, it is necessary to perform analyses with a capillary column, preferentially of great length (100 m), though some shorter columns (50 m) can give valuable information on this profile as well [80] (Figure 1.1).

Table 1.1

Reproducibility of results obtained by Ag-TLC coupled with capillary GLC (50-m CPSil-88 column). Data are from five independent fractionations by Ag-TLC of fatty acid isopropyl esters prepared from lipids of one beef cut purchased in September. Total *trans*-18:1 acid percentages were established with 16:0 and 18:0 acids as standards (these acids were from the saturated fraction isolated by Ag-TLC and pooled with the *trans*-18:1 acid fraction).

Fractionation number	Total *trans*-18:1 acids[a]	*Trans*-18:1 isomers[b]					
		Δ6–Δ9	Δ10+Δ11	Δ12	Δ13+Δ14	Δ15	Δ16
1	1.57	11.90	65.10	6.41	8.68	3.14	4.76
2	1.55	11.97	63.52	6.24	9.13	4.46	4.67
3	1.63	12.38	63.31	6.37	8.78	4.60	4.56
4	1.62	11.36	59.34	9.41	9.19	6.62	4.08
5	1.63	9.93	58.94	11.10	8.43	7.19	4.40
Mean ± S.D.	1.60 ± 0.04	11.51 ± 0.95	62.04 ± 2.74	7.91 ± 2.23	8.84 ± 0.32	5.20 ± 1.67	4.49 ± 0.27

[a]As weight percent of total fatty acids.
[b]As weight percent of total *trans*-18:1 isomers.

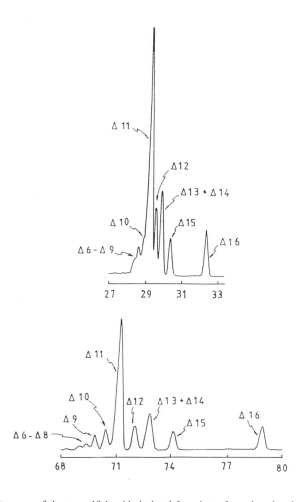

Fig.1.1. Chromatograms of the *trans*-18:1 acids isolated from butterfat and analysed as isopropyl esters. Upper chromatogram: analysis on a 50 m × 0.25 mm i.d. CPSil-88 fused-silica capillary column (Chrompack, Middelburg, The Netherlands) operated at 160°C with helium as the carrier gas (inlet pressure, 120 kPa). Lower chromatogram: analysis on a 100 m × 0.25 mm i.d. CPSil-88 fused-silica capillary column (same commercial source) with helium as the carrier gas (inlet pressure, 180 kPa) at 160°C. Identification of individual isomers was by comparison with synthetic compounds, except for the *trans*-16 18:1 isomer (from ref. [80], reprinted by permission of the publisher).

Precht and Molkentin [47,59] and Wolff and Bayard [80] have shown independently that 100-m CPSil-88 columns were able to separate most individual *trans*-18:1 isomers, after their fractionation by Ag-TLC. However, the Δ6 to Δ8 isomers on the one hand, and the Δ13 and Δ14 isomers on the other hand, remain poorly resolved or completely unresolved, respectively. When working at relatively low temperature [80], other isomers are almost base-line resolved (Figure 1.1). The use of such columns might usefully replace ozonolysis-based procedures in most instances.

Chromatograms obtained with 100-m SP-2560 columns [1,12,32] show *trans*-18:1 isomers emerging under an uneven peak that precedes oleic acid. Only summits or shoulders corresponding to individual isomers are distinguishable, with poor resolution, which makes the quantification questionable.

Consequently, we will give particular credence to quantifications of *trans*-18:1 acids by combined Ag-TLC/GLC on very long (50-100m) capillary columns. For results obtained by IR-spectroscopy or by GLC alone, it is evident that correction factors, if available, will be needed; raw data obtained in such a way should be considered with caution.

D. *TRANS*-18:1 ACIDS IN RUMINANT MEAT LIPIDS AND FATTY TISSUES

1. Beef Lipids

In 1983, it was estimated by the USDA that beef could contribute *ca.* 11% of the food fat available in the U.S. [65], which would correspond to a fat consumption of 18 g/person/day, though some authors have made higher estimates for the same country (up to 32 g/person/day) [18]. Significantly lower values (less than 5 g/person/day) have been proposed for European populations [78], even if in some countries, the consumption of beef does not significantly differ from that in the U.S. (i.e. France and the U.S., *ca.* 50 g/person/day) [65,78]. For Scotland [10], it has been suggested that red meat could be a major natural dietary source of *trans*-18:1 acids, and more generally in the U.K., that the daily consumption of *trans* isomers from meat would be approximately 1.5 g/person [24]. It is thus appropriate to review literature data on the *trans*-18:1 acid content in beef fats. Unfortunately, these data are scarce and scattered, and seldom evaluated for their accuracy. Moreover, one has to distinguish between three possible origins of *trans*-18:1 acids from beef: lipids from the lean part of meat (including marbling), the fat surrounding the cuts, and other fatty tissues that are generally used for the production of tallow.

i. Meat (lean part) lipids. Lanza and Slover [38], using a combination of Ag-TLC and capillary GLC (100-m SP-2340 column), found 4.5% of *trans*-18:1 acids in the separable lean of one sample of beef rib roast. With the same methodology, Lin *et al.* [40] established the *trans*-18:1 acid content in six samples of beef *longissimus* muscle, and obtained a mean of 1.7 ± 0.9% relative to total fatty acids. Lanza *et al.* [39] found 4.6% of *trans*-18:1 acids by simple capillary GLC in

Table 1.2

Distribution of *trans*-18:1 isomers in beef meat (lean part) lipids as a function of the period of collection of the cuts and of the lipid fraction.

Fraction	Period[a]	n[b]	*Trans*-18:1 acid content	*Trans*-18:1 isomers[c]							Ref.[d]
				$\Delta 6$-$\Delta 9$	$\Delta 10 + \Delta 11$	$\Delta 12$	$\Delta 13 + \Delta 14$	$\Delta 15$	$\Delta 16$		
Total lipids	January	10	1.95 ± 0.94	8.7 ± 2.4	66.9 ± 7.2	5.9 ± 2.2	9.5 ± 2.4	4.1 ± 1.2	4.9 ± 1.5	[79]
Total lipids	September	11	1.91 ± 0.80	10.2 ± 3.6	56.4 ± 10.9	9.5 ± 3.0	11.9 ± 3.4	6.7 ± 1.7	5.3 ± 1.7	U.R.
Triacylglycerols	September	2	-	11.3	59.6	5.9	11.7	5.3	6.2	U.R.
Polar lipids	September	2	-	13.7	55.1	6.2	13.2	5.5	6.3	U.R.

[a]Period of collection of the cuts.
[b]Number of samples.
[c]Values are the percentages of the indicated isomers relative to total *trans*-18:1 acids.
[d]References; U.R., unpublished results (Wolff and Bayard).

the fat from a single sample of raw beef (origin and preparation of the cut not stated), whereas the corresponding value for a cooked sample was 5.1%. In a more comprehensive study, Slover *et al.* [65] analysed the fatty acids from 269 raw beef cuts (animals reared in the U.S.) by capillary GLC on a 100-m SP-2340 column. The lean part of the cuts was prepared by carefully removing all apparent fatty tissues and bone plus connective tissue. This was intended to represent the edible portion of the cuts. The minimum *trans*-18:1 acid level was 1.3%, whereas the maximum was 4.4%, with a mean of 3.2%. Data for bovines (calves, age not stated) reared in Spain, and reported by Lluch *et al.* [42], were not in agreement with the preceding results, being significantly higher. Here too, fatty acids were analysed by capillary GLC (50-m CPSil-88 column). For the muscle tissue of 77 cuts, the mean *trans*-18:1 acid content was 5.9%, with a low of 0.2% and a high of 12%. An even higher mean value, 8.5 ± 2.7%, was reported by the same laboratory [8] for total *trans* acids in 45 cuts obtained from Spanish slaughterhouses. Taking into account *trans*-polyunsaturated acids (mainly *cis,trans*-conjugated 18:2 acids) that also occur in beef meat, this should correspond to *ca.* 7 - 7.5% *trans*-18:1 acids.

Combining Ag-TLC and capillary GLC on a 50-m CPSil-88 column, Wolff [79] analysed the carefully dissected lean part of ten different beef retail cuts from ten different animals. The cuts were purchased in winter. A mean *trans*-18:1 acid content of 2.0 ± 0.9% was established (minimum: 0.8%; maximum: 3.5%) (Table 1.2). The analysis of eleven supplementary cuts obtained from cows slaughtered at the end of summer and analysed in the same conditions (Wolff and Bayard, unpublished data) led to exactly the same results (Table 1.2). It was also determined, in the first study, that beef meat triacylglycerols had a significantly higher level of *trans*-18:1 acids (2.5 ± 0.9; n = 10) than beef meat phospholipids (0.8 ± 0.3; n = 10). This observation was in qualitative agreement with that of Wood [82], who found 2.4 - 5.7% of *trans*-18:1 acids in neutral lipids, and 1.1 - 1.2% in phospholipids from three samples of beef *Sterno mandibularis* muscle.

Wood [82] was apparently the first to study individual *trans*-18:1 isomers in beef meat lipids. He reported that the double bonds in these components span from position $\Delta6$ to position $\Delta14$, with the $\Delta10$ isomer being the major constituent. Isomers were identified by capillary GLC on a 100-m SP-2340 column by comparison with authentic standards. Wolff [79] also analysed individual *trans*-18:1 isomers from beef meat lipids with a 50-m CPSil-88 capillary column. However, his observations did not fully agree with those of Wood [82]. Isomers having their ethylenic bonds between positions $\Delta6$ and $\Delta16$ were identified, here too with authentic standards, and the major *trans*-18:1 isomer was shown to be *trans*-11 18:1 acid, as in beef perirenal [28,53] or subcutaneous fat [53], beef tallow [5], and cow, goat or ewe milk fats [59,81]. This was unambiguously demonstrated by adding to the *trans*-18:1 acid fraction, isolated by Ag-TLC, authentic *trans*-10 or *trans*-11 18:1 acids, which were partly but sufficiently resolved during GLC. To explain his observations, Wood [82] speculated that the animals from which the cuts were obtained might have eaten some partially hydrogenated fat at one time, an explanation which is unlikely.

Table 1.3
Content and distribution of *trans*-18:1 isomers in bovine adipose tissues.

Origin		Total *trans*-18:1 isomers[a]	n[b]	Individual *trans*-18:1 isomers[c]											Ref.[d]
				Δ6	Δ7	Δ8	Δ9	Δ10	Δ11	Δ12	Δ13	Δ14	Δ15	Δ16	
Tallow[c]	Mean	4.9 ± 0.9	10	7.6 ± 1.8				66.5 ± 2.8		5.1 ± 0.7	11.0 ± 0.9		4.4 ± 0.4	5.5 ± 0.4	[5]
	Max.	6.2	10	10.9				71.7		6.7	11.8		5.2	6.3	[5]
	Min.	3.4	10	4.5				63.4		4.1	9.8		4.1	4.8	[5]
Perirenal		3.6	1	-	1.0	2.1	5.0	11.9	46.9	6.0	6.6	7.4	5.5	7.6	[28]
Perirenal		-	1	0.3	0.5	1.6	8.9	5.4	68.6	2.7	2.6	3.6	2.6	3.2	[53]
Subcutaneous		-	1	0.2	0.3	1.5	13.6	6.4	64.4	2.4	2.3	3.6	2.3	3.0	[53]

[a] Weight percent of total fatty acids.
[b] Number of samples.
[c] Weight percent of total *trans*-18:1 isomers.
[d] References.
[e] For tallow, the Δ6 to Δ9, Δ10 + Δ11 and Δ13 + Δ14 isomers were not or poorly resolved by capillary GLC.

Differences between Wood's and Wolff's results cannot be explained by the fact that the first author analysed neutral lipids whereas the second analysed total meat lipids. In a complementary study, Wolff and Bayard (unpublished observations) have separated total polar lipids and triacylglycerols by TLC. The *trans*-18:1 acid fractions from the two categories of lipids were isolated by Ag-TLC and analysed by capillary GLC. No gross differences could be observed in the distribution profiles of *trans*-18:1 acids from both origins (Table 1.3). Consequently, the observation by Wood [82] of the *trans*-10 18:1 acid being the major *trans*-18:1 isomer in beef meat lipids, remains atypical and unexplained. One may also add that the feed is apparently not able to alter greatly the *trans*-18:1 isomer profile of beef meat lipids, at least for a given country. No major differences were observed between animals slaughtered in summer (Wolff and Bayard, unpublished observations) or in winter [79] (Table 1.3).

ii. Fatty tissues and tallow. Another source of *trans*-18:1 is constituted by the fat surrounding the lean part of beef cuts. Depending on the cooking method and on individual habits, a more or less important part of this fat may be ingested, and this will contribute to the overall *trans*-18:1 acid intake from beef. Slover et al. [65] analysed such fatty tissues prepared from cows reared in the U.S., and reported mean values ranging from 6.2 to 6.8%, which are approximately twice the corresponding values for *trans*-18:1 acids in the lean part.

In a few other studies, samples of depot fats were analysed for their trans-18:1 acid content. These fats are generally not used as such for edible purposes, but rather for tallow production. Lin et al. [40] have established a mean trans-18:1 acid content of 3.1 ± 1.4% in six samples of beef subcutaneous fat. Other isolated data are from Enig et al. [19] (one sample of tallow, 1.8% of trans-18:1 acids), Hyvönen et al. [33] (one sample of animal household fat, presumably tallow, 2.3%), Hay and Morrison [27] (one sample of ox perirenal fat, 3.6% of trans-18:1 acids plus 0.3% of trans-16:1 acids), Mansour and Sinclair [45] (one sample of dripping, 3.10%), and from Precht and Molkentin [58] (one sample of tallow, 4.39% of trans-18:1 acids). Kochbar and Matsui [37] found 4.1 - 5.6% total trans content (by IR-spectroscopy) in three margarines that were possibly made with 95-97% tallow. Older data obtained with IR-spectroscopy by Kaufman and Mankel [36] indicated a wide range of trans acids (chain-length not specified) in beef (2 - 12%) and calf (1 - 17.3%) depot fats (subcutaneous, perirenal, pericardial, and hepatic), but smaller and lower intervals for beef (4.7 -6.5%) and calf (1.0 - 4.5%) organ fats (kidney, heart, and liver). Due to the methodology used, the accuracy of these last data is questionable, however. An even wider range (2.1 - 20.0%; mean, 9.6%) was recently published by Lluch et al. [42] for 77 samples of perirenal fat from Spanish bovines. Although established by capillary GLC (50-m CPSil-88 column), the significance of these data as compared to most other values in the literature (see above) is not clear. Apparently, the animals were calves, not cows. Moreover, one may speculate that other parameters than age may influence the trans-18:1 acid content of adipose tissues, and that some major differences in cattle feeding exist between Spain and other countries.

More recently, Bayard and Wolff [5] more specifically addressed the *trans*-18:1 acid content and profile in edible refined beef tallow. For this purpose, GLC on a 50-m CPSil-88 capillary column was combined with Ag-TLC and samples obtained monthly throughout the year were analysed. The mean content of *trans*-18:1 acids relative to total fatty acids was 4.9 ± 0.9% (n = 10) of total fatty acids with a minimum of 3.4% and a maximum of 6.2% (Table 1.3). Seasonal variations were not clearly apparent. Because the tallow samples analysed were prepared with different fatty tissues from a great number of bovines, of all ages, and reared on different feeds, the value of 4.9% appears to be representative of the *trans*-18:1 acid content of an average bovine fatty tissue. The distribution profile of individual isomers was also established by capillary GLC, and it was observed, qualitatively at least, that the distribution profile did not differ largely from previous patterns established through ozonolysis-based procedures with *trans*-18:1 acids prepared from perirenal and subcutaneous fat [28,53] (Table 1.3). The main isomer was vaccenic acid, with other isomers having their ethylenic bonds between positions Δ6 and Δ16 in lesser amounts. Some small but definite differences were noted between the *trans*-18:1 acid profile in tallow as compared to that in milk fat, but not with that of *trans*-18:1 isomers in meat lipids. This was interpreted as indicative of some slight differences in the specificity of the tissues (mammary gland *versus* adipose tissues) for the acylation of individual *trans*-18:1 isomers in lipids [5].

From the preceding data, a few conclusions can be drawn concerning beef tissues. In general, *trans*-18:1 acids tend to be present in higher amounts in adipose tissues (5 - 6%) than in muscles (2 - 3%). An explanation for this phenomenon would be that muscles contain phospholipids in a higher proportion than adipose tissue, and because the level of *trans*-18:1 acids is less in phospholipids than in triacylglycerols, total lipids from muscles will have a lower overall content of *trans*-18:1 acids than adipose tissue lipids, almost exclusively composed of triacylglycerols. However, the distribution profiles of individual *trans*-18:1 isomers in meat lipids and tallow are practically identical, with vaccenic acid accounting for approximately one-half of total isomers, and with other isomers from Δ5 to Δ16 in lesser amounts. Concerning the contribution of beef to the intake of *trans*-18:1 acids, a simple calculation indicates that 100 g of beef meat, containing 10 g of fat, will provide *trans*-18:1 acids in the range 200 - 350 mg, approximately half of which is vaccenic acid. In comparison, a 40-g piece of cow cheese (30% fat, containing 3.5% *trans*-18:1 acids) provides approximately 400 mg of *trans*-18:1 acids. Thus, beef meat can contribute only minimally to the daily intake of *trans*-18:1 acids, and claims about its so-called importance as a natural source of *trans*-18:1 acids [10] are groundless.

2. Other Ruminant Lipids

Beef is generally the major source of ruminant meat. However, in some countries (United Kingdom, Ireland, Spain, Greece), sheep may contribute 25 - 40% of the production of total ruminant meat for edible purposes, and it cannot be neglected as a natural source of *trans*-18:1 acids. Unfortunately, our knowledge of the *trans*-18:1 acid content of sheep meat is limited. Most available data have been obtained by IR-spectroscopy, and they generally relate to adipose or total tissues, and practically never to meat.

In the early work of Hartman *et al.* [25], the perinephric fat of one pasture-fed sheep was shown to contain 11.2% of *trans* acids (by IR-spectroscopy). Similar values, determined with the same methodology, were later reported by Kaufman and Mankel [36] for fatty tissues from different anatomical locations (range, 10.7 - 15.8%). On the other hand, the fat from kidneys, heart, and liver, contained slightly less *trans* acids (7.0 - 8.2%). Comparing the *trans* acid content of neutral lipids (essentially triacylglycerols) from total maternal and fetal tissues, Shorland *et al.* [64] observed that *trans* acids were absent from fetuses, whereas they accounted for 12 - 13.5% in the maternal tissue neutral lipids. Similar observations were made by Body *et al.* [9] (0.6% in fetal lamb and 12.0% in maternal ewe total tissues). Garton and Duncan [21] could confirm the absence of *trans* acids in adipose tissues from neonatal lambs (either still-born or slaughtered shortly after birth). However, *trans* acids were present at similar levels in phospholipids from both maternal and fetal total tissues (12.5 and 9.3%, respectively) [21].

An interesting observation was that adipose tissues had different *trans* acid contents, dependent on the diet [21]. When lambs were fed from weaning at ten weeks of age on a diet of grass cubes, in which α-linolenic acid was the main fatty acid (65%), the *trans* acid content was 6.9 - 12.0% of adipose tissue triacylglycerols, whereas it was only 1.1 - 1.6% for lambs fed a pelleted ration in which linoleic acid accounted for 50% of total dietary fatty acids. This clearly demonstrated the preponderant influence of α-linolenic acid in the diet on the *trans*-18:1 acid content of fatty tissues.

Christie and Moore [13] were apparently the first to specifically study the trans-18:1 acid content of different sheep organs with a combination of Ag-TLC and GLC. With this methodology, the range for trans-18:1 acids in triacylglycerols (1.6 - 4.6 mole%; mean, 3.7 ± 1.0 mole%) was significantly lower than those previously reported by authors who used IR-spectroscopy. However, the animals under investigation were stall-fed on hay and concentrates. Consequently, the discrepancy between results by Christie and Moore [13] and older data cannot be exclusively attributed to methodological differences (i.e. Ag-TLC/GLC versus IR-spectroscopy).

In an unpublished study, Wolff has determined the *trans*-18:1 acid content of total lipids from the lean part of four sheep cuts with the methodological combination Ag-TLC and GLC on a 50-m CPSil-88 capillary column. The results (Table 1.4) essentially confirmed those from Christie and Moore [13], being considerably lower (range, 2.2 - 5.6%) than older data obtained by IR-spectroscopy. Because the cuts were purchased at random in different supermarkets, it is believed that the results are representative of sheep meat, irrespective of the diets of the animals. Though it is evident that the feed may influence the percentages of *trans*-18:1 acids in sheep organs, it would appear that

Table 1.4

Trans-18:1 acid content in lipids from meat (lean part) of some ruminants (weight% of total fatty acids). Data were obtained by combining Ag-TLC and capillary GLC on a 50-m CP-Sil 88 column for total lipids and by direct GLC and use of a correction factor (× 1.25) for triacylglycerols and polar lipids.

Fraction	Species			
	Sheep[a] (n = 4)[e]	Beef[b] (n = 10)	Buffalo[c] (n =3)	Reindeer[d] (n = 3)
Total lipids	4.0 ± 1.4[f]	1.9 ± 0.9	2.9 ± 0.5	0.7 ± 0.2
Triacylglycerols[g]	4.6 ± 1.7	2.5 ± 0.9	3.5 ± 0.6	1.0 ± 0.5
Polar lipids	1.3 ± 0.3	0.8 ± 0.3	1.2 ± 0.5	0.4 ± 0.1

[a]Four cuts from four different animals purchased in France.
[b]From ref. [79]. Animals slaughtered in winter.
[c]*Bison bison*, imported from Canada. Three cuts probably from a single animal.
[d]*Rangifer tarandus*, purchased in Finland. Three cuts from an unknown number of animals.
[e]Number of analyses.
[f]Mean ± standard deviation.
[g]Triacylglycerols and polar lipids separated by thin-layer chromatography.

IR-spectroscopy determinations gave systematically higher values than Ag-TLC coupled with GLC. It should also be noted that the *trans*-18:1 acid content in meat triacylglycerols is considerably higher than in polar lipids (Table 1.4). In this respect, sheep meat does not differ from beef meat, and more generally from other ruminants, for example buffalo or reindeer (Table 1.4).

It is clear from this brief survey that accurate and reliable data on the *trans*-18:1 acid content of sheep meat are scarce. However, it would appear that the *trans*-18:1 acid content in sheep meat is higher than in beef meat (*ca. 4% versus 2%*). The same holds true for milk fat of both species, ewe milk fat containing a higher level of *trans*-18:1 acids than cow milk fat (see next section). These differences may be related to some species differences, but we believe that the predominant parameter that influences the *trans*-18:1 acid content is of dietary origin (content of α-linolenic acid in the diet).

3. Non-Ruminant Lipids

As stated in the introduction, animals that have been fed with foods containing *trans*-18:1 acids may themselves present such isomers in their tissue lipids. This is apparently the case for pigs, for which it has been repeatedly reported that their tissue lipids contain small amounts of *trans* acids. This was first observed in lard as early as 1930 by Grossfeld and Simmer [22]. These authors used complex combinations of fractionations of lead soaps of fatty acids, mercury salt separations, and solvent crystallizations. At that time, *trans* fatty acids were reported as vaccenic acid, and its amount was estimated to be 0.2%.

The presence of *trans* acids in pig tissues (perinephric fat) was confirmed by Hartman *et al.* [26], by IR absorbtion measurements, at a level of 0.9%. More recent studies using a combination of Ag-TLC and capillary GLC [40] indicated that pork *longissimus* muscle lipids contained 0.12% *trans*-18:1 acids, whereas these acids accounted for 0.17% in subcutaneous fat. Also using capillary GLC (50-m CPSil-88 column), and analysing the subcutaneous and intramuscular fats from about eighty pigs reared in Spain, Lluch *et al.* [41] found 0.6% of *trans*-18:1 acids in lipids from both origins. These authors detected a clear influence of feeding on the *trans*-18:1 acid content. Mansour and Sinclair [45] have analysed one sample of Australian lard and found 0.34% of *trans*-18:1 acids (direct GLC on a 50-m BPX-70 column). Combe (personal communication) found 1.44% of *trans*-18:1 acids in one sample of household lard purchased in France (by Ag-TLC/GLC). On examination of the *trans*-18:1 isomers purified by Ag-TLC and analysed on a 50-m CPSil-88 capillary column, this author could conclude that their distribution profile was identical to that of ruminant fats, with vaccenic acid being the major isomer.

So, it would appear that pork meat fat and lard actually contain *trans*-18:1 acids in small and variable amounts (0 - 1.5%). These acids are derived from the feed fat (milk fat, tallow, or hydrogenated vegetable oils) and are supposed to vary qualitatively and quantitatively as a function of the amount and distribution of dietary *trans*-18:1 isomers. From data available, it is difficult to establish a precise mean value, but it is sure that pork is of minor importance, though not necessarily negligible, in the dietary consumption of *trans*-18:1 acids by humans. It should be added that some studies [11] also include in assessments of *trans* acid intakes the contribution of fats from poultry, eggs and fish. Though the presence of *trans* fatty acids in these foods is theoretically possible through fat supplementation, we are not aware of published data demonstrating their effective presence.

E. *TRANS*-18:1 ACIDS IN RUMINANT MILK FATS

1. Cow Milk Fat

Cow milk, under all its forms (milk, butter, cream, cheese...), may be an important dietary source of fat. Its consumption by European people, with the exception of Mediterranean countries, generally exceeds 280 kg/person/year, and it can even reach 400 kg/person/year (France) [78]. Based on a 3.7% content of fat, this represents a fat intake of about 40 g/person/day, which is almost one-half the total fat and oil intake by French people. Consequently, cow milk fat can contribute a non-negligible part to the daily intake of *trans*-18:1 acids. This emphasizes the need for a precise knowledge of the *trans*-18:1 acid content in cow milk fat, especially as this content is submitted to large variations, which are attributable to feeding conditions, milk output, genetic constitution (breed, single cow or bulk milk) as well as stage of lactation. Furthermore, superimposed on these natural variations are spurious variations of analytical origin [16,78].

For this reason, this section will principally consider studies in which *trans*-18:1 acids were quantified by GLC, with a particular attention for recent works employing Ag-TLC in combination with capillary GLC, and based on large collections of samples. Moreover, results based upon triacylglycerol formulae that

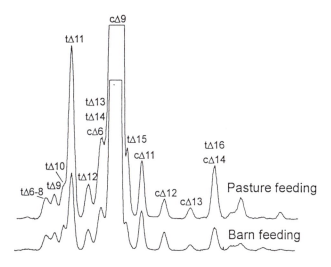

Fig. 1.2. Partial chromatograms of fatty acid methyl esters (18:1 acid regions) derived from milk fat of the pasture and barn feeding periods. Analyses on a 100 m × 0.25 mm i.d. CPSil-88 capillary column (same commercial source as in Fig. 1.1) at 175°C with hydrogen as the carrier gas (inlet pressure, 160 kPa) (from ref. [57], reprinted by permission of the publisher).

were derived from Ag-TLC/GLC of fatty acid methyl esters and GLC of triacylglycerols, are considered as well. Older data obtained by IR absorption measurements are out of the scope of this overview. Their accuracy was insufficient, and they included other *trans* acids than the 18:1 isomers.

When considering bovine milk fat, there are several pitfalls to avoid. As mentioned above, the feeding conditions have a direct and profound influence on the *trans*-18:1 acid content of milk fat, which results in definite seasonal variations [60,81]. Consequently, data obtained from isolated samples, and that do not take into account this variability, are meaningless. This variability may explain why some authors, when trying to assess the *trans*-18:1 acid consumption by humans, use different values, that were gathered at random in the literature, for the *trans*-18:1 acid content in milk, butter, cream, or cheese fats. *A priori*, if there exists a mean annual value for the *trans*-18:1 acid content of cow milk fat, this value should apply as well to all milk-derived products. Other limitations to the credibility of literature data are linked to experimental procedures and analytical conditions. Because the precise quantification of *trans*-18:1 acids by Ag-TLC/GLC also relies on the exact quantification of all other fatty acids, this implies that all fatty acids, including the shortest ones, are accurately taken into account. This can be achieved by using temperature-programming of the column, starting at 45-60°C, in order to elute butyric acid derivatives without interference with the solvent [78]. Isothermal analyses at a high temperature inevitably lead to a loss of short- and medium-chain fatty acids which coelute with the solvent. Moreover, fatty acid ester solutions should never be evaporated. Finally, and due to the unequal response of the flame-ionization detector, response factors should be used to correct rough area percentages displayed by integrators [78,81].

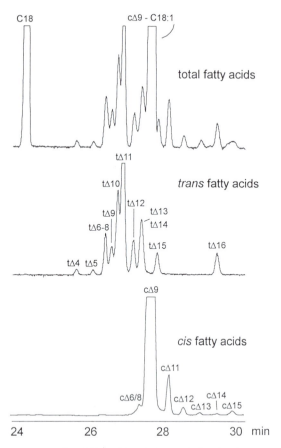

Fig. 1.3. Partial chromatograms of total 18:1 acid methyl esters, before fractionation, and of *trans*-18:1 and *cis*-18:1 acids after fractionation by silver-ion thin-layer chromatography. Analyses on a 100-m CPSil 88 capillary column (same operating conditions as in Fig.1.2; from ref. [61], reprinted by permission of the publisher).

Figure 1.2 is intended to illustrate the dependence of the chromatographic profile of a sample on the feeding conditions of the cattle. It shows partial gas-liquid chromatograms of unfractionated 18:1 acids from bovine milk fat [57]. The upper chromatogram originates from typical pasture feeding, the lower from barn feeding with high amounts of concentrates. The quantities of the individual *trans* fatty acids Δ6 to Δ14 indicate the considerably higher contents of total *trans*-18:1 with typical pasture feeding in summer compared to stall-feeding in winter. Thus, a representative mean value for the *trans*-18:1 acid content in milk fat cannot be derived from only a limited number of samples.

Moreover, even new studies applying highly polar 100-m capillary columns for GLC of fatty acids have shown that overlaps between *cis*- and *trans*-18:1 acids still occur [47,59,61,80]. These can in particular be recognized after Ag-TLC separation of fatty acid derivatives. Although the direct analysis with 100-m CPSil 88 columns already results in a good approximation of the *trans*-18:1 acid

Table 1.5

Mean values (M.V.), standard deviations (S.D.), minimal (Min.) and maximal (Max.) contents ot total
trans-18:1 acids in bovine milk fats (*n*, number of samples) from different countries (weight% of total
fatty acids). This table does not include results obtained by infrared spectroscopy.

M.V.	S.D.	Min.	Max.	*n*	Year	Country	Ref.[a]
1.8		1.8	1.8	5	1978	U.S.A.	[66]
3.4		3.1	3.8	3	1983	U.S.A.	[19]
4.0		3.1	4.9	2	1984	Finland	[31]
2.7				1	1981	Sweden	[2]
4.63		3.59	5.66	2	1982	Denmark	[43]
3.33	0.99	1.75	5.20	31	1994	Austria	[30]
3.62	1.22	1.29	6.75	1756	1995	Germany	[61]
3.83	1.34	1.91	6.34	100	1995	Germany	[61]
3.19				1	1995	Belgium	[61]
4.21	0.60			4	1995	Denmark	[61]
4.04	0.30			10	1995	Spain	[61]
3.8		2.46	5.10	24	1994	France	[78]
4.47	0.92			10	1995	France	[61]
3.3		2.4	4.3	60	1996	France	[81]
4.01	0.15			4	1995	Greece	[61]
5.34				1	1971	Italy	[70]
4.14	1.30			12	1995	Italy	[61]
5.91	0.92			22	1995	Ireland	[61]
3.51				1	1995	Luxemburg	[61]
4.09	0.91			24	1995	The Netherlands	[61]
4.78	0.91			23	1995	U.K.	[61]

[a]Reference

content, only the combination of GLC with Ag-TLC allows an exact
quantification of *trans*-18:1 acids. Figure 1.3 shows partial chromatograms of the
unfractionated FAME as well as the *trans* and *cis* fractions separated by Ag-TLC
from a bovine milk fat sample. *Cis* isomers overlap with *trans* isomers at *trans*-
Δ13/14 (*cis*-Δ6/8), and *trans*-Δ16 (*cis*-Δ14). Sometimes, the *trans*-Δ15 isomer
cannot be separated from oleic acid (*cis*-Δ9) as well. From the chromatograms, it
follows that vaccenic acid is the most prominent *trans* isomer in cow milk fat.

Chromatograms with the resolution shown in Figure 1.3 were published by
Wolff and Bayard [80], as well as by Molkentin and Precht [47,59], in 1995 for
the first time. In studies from both laboratories, 100-m CPSil-88 capillary
columns were used. With shorter columns, the 18:1 acids *trans*-Δ12 to *trans*-Δ15
coelute with *cis*-Δ6 to *cis*-Δ11. It is thus expected that data obtained in such a way
are erroneous. Without pre-separation by Ag-TLC, total *trans*-18:1 acid contents
determined this way are undoubtedly underestimates. Moreover, the isomer *trans*-
Δ16 is frequently not considered, but it should be noted that its quantification is
complicated by an overlap with *cis*-Δ14 (Figure 1.3).

Although many reports have dealt with the *trans* fatty acid content of bovine
milk fat (for references of studies using IR-spectroscopy, see [78]), surprisingly
few data obtained by combined Ag-TLC/GLC are available in the literature. In
Table 1.5, analytical results on the content of total *trans*-18:1 acids in bovine milk
fats from different countries are listed. Due to the large variations associated with

Table 1.6

Isomeric distribution of *trans*-octadecenoic acids (weight % of total fatty acids) in German milk fats comprising most different conditions of feeding and lactation (n = 1756; M.V., mean value; Median, median value; S.D., standard deviation; Min., minimal content; Max., maximal content) (from ref. [61]).

Trans-position	M.V.	S.D.	Min.	Max.	Median
Δ4	0.05	0.007	0.02	0.08	0.05
Δ5	0.05	0.008	0.00	0.11	0.05
Δ6-8	0.16	0.040	0.07	0.27	0.15
Δ9	0.23	0.026	0.16	0.30	0.23
Δ10	0.17[a]	0.031	0.03	0.30	0.17
Δ11	1.72[a]	0.976	0.35	4.43	1.42
Δ12	0.21	0.031	0.10	0.31	0.21
Δ13/14	0.49	0.088	0.00	0.85	0.48
Δ15	0.28	0.081	0.04	0.48	0.27
Δ16	0.33	0.060	0.11	0.52	0.33

[a]Based upon 1707 milk fats

different conditions of feeding and lactation, in the case of only small numbers of samples, the data cannot be regarded as representative for the respective country. The data from Smith *et al.* [66], who analysed five brands of U.S. butter, and found the same value of 1.8% for each sample, are likely to be imprecise, as separation by Ag-TLC was not quite complete.

In extensive new studies by Precht and Molkentin [61], the total content of *trans*-18:1 acids as well as the content of the individual isomers, were determined in 1756 samples of German milk. The isomeric distribution is of particular interest in regard to a comparison with partially hydrogenated fats [77]. To that end, Ag-TLC combined with GLC as well as triacylglycerol formulae, were applied. In addition to mean and median values, Table 1.6 gives the standard deviations and ranges of variation for the individual *trans*-18:1 isomers.

The main positional isomer of *trans*-18:1 acids is Δ11 (vaccenic acid), which represents about one-half of total *trans*-octadecenoic acids. Similar relations were found in earlier [27,44,53] and recent studies [78] as well. The mean contents of individual *trans*-18:1 positional isomers *trans*-Δ4 to *trans*-Δ16 [61] in these milk fats are presented graphically in the upper part of Figure 1.4. Moreover, the lower part of Figure 1.4 shows the isomeric distribution of *trans*-18:1 isomers subdivided into samples from barn feeding, transition period (from barn to pasture feeding in spring and *vice versa* in late fall) and pasture feeding [60]. Great differences are particularly obvious regarding vaccenic acid (*trans*-Δ11). These differences among the three distributions, that are based upon the mean values from the analysis of hundreds of different milk fats, clearly confirm that data on isomeric distributions derived from only a few samples cannot be representative.

Further isomeric distributions of *trans*-18:1 acids in bovine milk fat from ten E.U. countries are presented in Table 1.8. In this table, the mean C-54 contents include all triacylglycerols with this total acyl-C number. They characterize the feeding conditions, being found in high amounts during pasture feeding and in

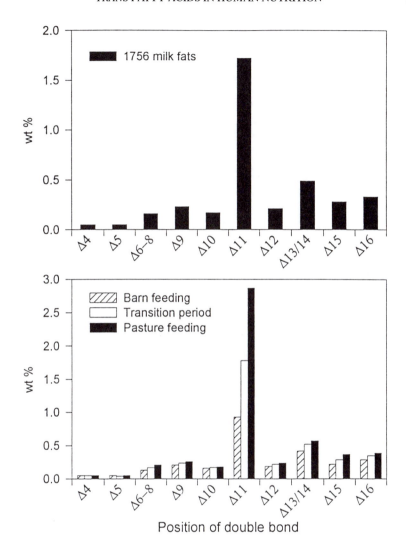

Fig. 1.4. Upper panel: mean isomeric distribution of *trans*-18:1 positional isomers in German milk fats (*n* = 1756) (weight % of total fatty acids). Lower panel: mean isomeric distribution in German milk fats subdivided in samples from the barn feeding period in winter (*n* = 927), the transition period (*n* = 236), and the pasture feeding period in summer (*n* = 593) (from ref. [60], reprinted by permission of the publisher).

low amounts during barn feeding [54]. Values exceeding *ca.* 5.5 wt% result from pasture feeding, so that most of the mean C-54 contents in Table 1.7 are indicative of pasture feeding. Thus, except for Ireland, where pasture feeding is prevailing, the contents of *trans*-18:1 acids cannot be regarded as representative for the respective country. In comparison, the isomeric distribution in Table 1.6 is based upon a mean C-54 content of 4.73 wt%, being representative for the whole range of variations from a low of 1.42 to a high of 9.02 wt% [61].

Table 1.7

Mean distributions of positional isomers of *trans*-octadecenoic acids (weight % of total fatty acids) and mean C-54 triacylglycerol contents (weight %) in milk fats (*n*, number of samples) from ten E.U. countries [61].

	B	DK		S		F		GR		I		IR		LUX	NL		UK	
C-54 (total)	4.21	6.07	1.16	6.43	0.60	5.68	0.98	6.22	0.07	5.26	2.04	7.17	0.80	4.98	4.87	1.19	6.58	0.80
n	1	4		10		10		4		12		22		1	24		23	
	mean	mean	s.d.	mean	s.d.	mean	s.d.	mean	s.d.	mean	s.d.	mean	s.d.	mean	mean	s.d.	mean	s.d.
trans (total)	3.19	4.21	0.60	4.04	0.30	4.47	0.92	4.01	0.15	4.14	1.30	5.91	0.92	3.51	4.09	0.91	4.78	0.91
Δ4	0.04	0.05	0.01	0.04	0.00	0.06	0.01	0.05	0.00	0.05	0.01	0.06	0.00	0.06	0.05	0.01	0.06	0.01
Δ5	0.04	0.04	0.00	0.03	0.00	0.04	0.01	0.04	0.00	0.04	0.01	0.05	0.01	0.05	0.05	0.01	0.04	0.01
Δ6-8	0.16	0.19	0.02	0.20	0.01	0.20	0.03	0.20	0.00	0.20	0.03	0.23	0.02	0.16	0.17	0.03	0.21	0.03
Δ9	0.22	0.25	0.02	0.25	0.01	0.24	0.01	0.24	0.00	0.24	0.02	0.26	0.01	0.23	0.25	0.02	0.25	0.01
Δ10	0.17	0.17	0.01	0.20	0.01	0.19	0.02	0.19	0.02	0.21	0.04	0.16	0.02	0.16	0.17	0.03	0.18	0.02
Δ11	1.47	2.00	0.44	2.11	0.20	2.36	0.75	1.91	0.09	2.10	0.90	3.54	0.80	1.48	1.98	0.82	2.51	0.79
Δ12	0.18	0.24	0.02	0.23	0.01	0.24	0.02	0.24	0.01	0.23	0.04	0.25	0.02	0.21	0.24	0.02	0.25	0.02
Δ13/14	0.38	0.57	0.07	0.56	0.04	0.54	0.04	0.55	0.03	0.52	0.12	0.58	0.05	0.49	0.57	0.06	0.61	0.05
Δ15	0.23	0.31	0.04	0.31	0.03	0.33	0.05	0.32	0.01	0.29	0.10	0.44	0.05	0.30	0.31	0.06	0.37	0.05
Δ16	0.29	0.37	0.01	0.35	0.01	0.38	0.04	0.34	0.01	0.34	0.07	0.43	0.03	0.32	0.39	0.03	0.40	0.04

Table 1.8

Distribution of positional *trans* 18:1 isomers in bovine milk fat (% of total *trans*-18:1 acids; *n*, number of samples; s.d., standard deviation).

Ref.[a]	[61]		[78]	[78]	[44]	[53]	[27]
Year	1996		1994	1994	1983	1976	1970
Country	Germany		France	France	Denmark	Australia	UK
Trans-position	$n = 1756$	Spring s.d.	Autumn $n = 12$	$n = 12$	$n = 5$	$n = 18$	$n = 1$
Δ4	1.6	0.6					
Δ5	1.5	0.6					
Δ6						0.3	1.0
Δ7	4.7	0.6	7.2	9.6		0.3	0.8
Δ8						1.5	3.2
Δ9	6.9	1.6				8.8	10.2
Δ10	4.7	1.2	58.2	50.4		5.5	10.5
Δ11	43.2	14.3			72.4	60.5	35.7
Δ12	6.3	1.0	6.2	7.2		4.1	4.1
Δ13	14.2	3.1	15.4	17.2	7.5	4.4	10.5
Δ14					7.8	5.2	9.0
Δ15	7.9	1.2	5.7	6.6	5.4	3.9	6.8
Δ16	9.8	2.0	7.3	9.0	7.0	5.5	7.5

[a]Reference.

Table 1.8 summarizes the mean proportions of the individual positional isomers of *trans*-18:1 in milk fat relative to the total content of *trans*-18:1 acids found in different studies. The recent results have been achieved either by Ag-TLC/GLC [44, 78] or by triacylglycerol formulae [61], which were derived from Ag-TLC/GLC of FAME and triacylglycerol analyses. Parodi [53], as well as Hay and Morrison [27], performed reductive ozonolysis or permanganate-periodate oxidation after isolation of the *trans*-18:1 acid fraction by preparative GLC and Ag-TLC. The data confirm that in contrast to vaccenic acid (*trans*-Δ11), elaidic acid (*trans*-Δ9) occurs in relatively low amounts in bovine milk fat, with the mean vaccenic/elaidic acid ratio being *ca.* 7:1 (range: 16:1 - 2:1) [61].

However, compared with earlier data on Australian butterfats by Parodi [53], Table 1.8 shows lower contents of vaccenic acid (43.2 instead of 60.5%) in the recent analyses of German milk fats by Precht and Molkentin [61]. On the other hand, these recent analyses led to higher contents of *trans* Δ4, Δ5 and Δ6-8. Considering the results of Hay and Morrison [27], relatively high amounts of *trans*-Δ9 and *trans*-Δ10 were found. However, these first data on the isomeric distribution of *trans*-18:1 in British butterfat (from cream) published in 1970 cannot be regarded as representative, since they were based on only one sample. Further data published in 1973 by Strocchi *et al.* [71] for Italian butterfat comprised the *trans*-isomers Δ7 to Δ16. The values published by Wolff [78,81], as well as those from Lund and Jensen [44], are in good agreement with the data by Precht and Molkentin [61], when the respective isomers from column 2 in Table 1.8 are summed up.

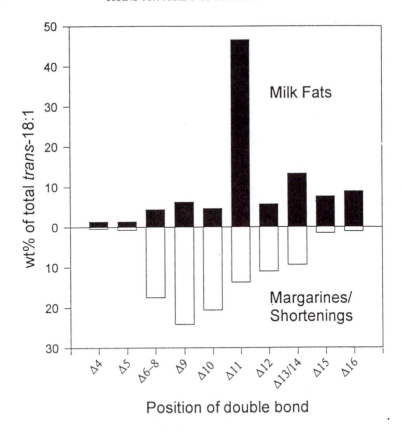

Fig. 1.5. Relative isomeric distribution of *trans*-18:1 positional isomers in 1756 German milk fats [61] and in 62 German margarines and shortenings [59] (weight % of total *trans*-18:1 isomers).

For a direct comparison with other fats, the upper part of Figure 1.5 shows the mean contents of the 18:1 positional isomers *trans*-Δ4 to *trans*-Δ16 from the representative collection of German milk fats [61] relative to the mean total *trans*-18:1 content (weight%). As can be seen from Figure 1.5, the isomeric distribution of *trans*-18:1 acids in milk fat differs considerably from the distribution in margarines and shortenings (lower part), that results from Ag-TLC/GLC analyses of 62 German samples by Precht and Molkentin [59]. In these partially hydrogenated vegetable fats, the content of *trans*-Δ11 is only 13.7% (milk fat, 46.6%), whereas the isomers Δ6 to Δ10 altogether account for 62.2% (milk fat, 15.2%).

With regard to studies comparing partially hydrogenated vegetable fats with ruminant milk fats, quantitative data on the contents of individual *trans*-18:1 isomers are of special interest. In the past, the isomeric distribution has been largely neglected, since a far reaching resolution or exact determination was impossible. However, it might be just the varying isomeric distribution in different products [59] that is of significance, as the differences in the risk to health between *trans* fatty acids from milk fats and from partially hydrogenated

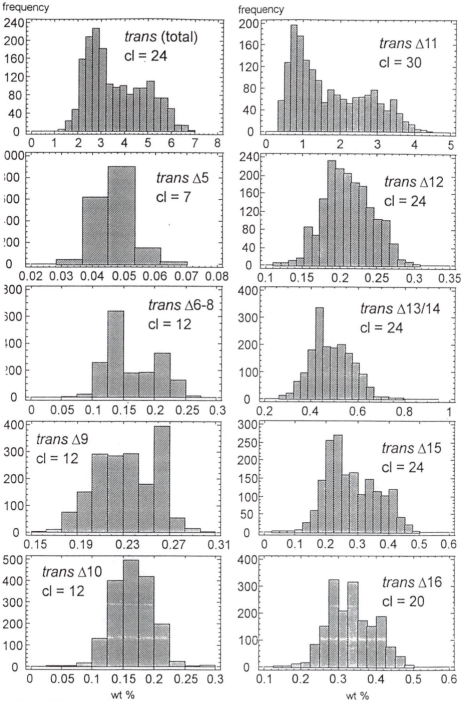

Fig. 1.6. Frequency distribution of total *trans*-18:1 contents as well as of the individual positional isomers of *trans*-18:1 in 1756 different German samples of milk fat (weight % of total fatty acids; cl, number of classes) (from ref. [61], reprinted by permission of the publisher).

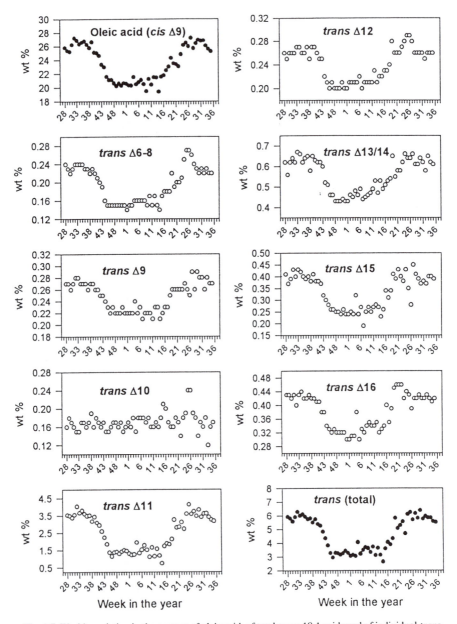

Fig. 1.7. Weekly variation in the content of oleic acid, of total *trans*-18:1 acids and of individual *trans*-18:1 positional isomers in butterfats from a large collection area in Germany (weight % of total fatty acids) (from ref. [60], reprinted by permission of the publisher).

Table 1.9

Content and distribution of trans-18:1 isomers in French butters as a function of the period of collection. Twelve different commercial samples analysed at each period.

Collection period	Total trans-18:1 acids[a]	Trans-18:1 isomers[b]						Ref.[c]
		Δ6-Δ9	Δ10 + Δ11	Δ12	Δ13 + Δ14	Δ15	Δ16	
January	2.37 ± 0.27	10.1 ± 1.2	48.4 ± 3.0	8.9 ± 0.7	17.0 ± 0.9	6.6 ± 0.6	9.0 ± 1.1	[81]
March	2.37 ± 0.23	6.7 ± 1.3	47.8 ± 1.9	10.2 ± 0.7	19.4 ± 1.1	7.4 ± 0.5	8.4 ± 0.7	[81]
May-June	4.28 ± 0.47	7.2 ± 0.8	58.2 ± 3.8	6.2 ± 0.7	15.4 ± 1.7	5.7 ± 0.6	7.3 ± 0.9	[79]
July-August	3.68 ± 0.68	8.0 ± 1.2	56.5 ± 3.8	7.3 ± 2.4	14.9 ± 1.4	5.7 ± 0.6	7.6 ± 1.1	[81]
October-November	3.22 ± 0.44	9.6 ± 1.3	50.4 ± 3.9	7.2 ± 0.7	17.2 ± 1.6	6.6 ± 1.6	9.0 ± 1.2	[79]

[a] As weight percent of total fatty acids (mean ± standard deviation).
[b] As weight percent of total trans-18:1 isomers (mean ± standard deviation).
[c] References.

vegetable fats found for instance by Willett *et al.* [77] might only originate from a few isomers. So, certain *trans*-18:1 isomers such as elaidic acid (*trans*-9 18:1 acid), which compared to ruminant fats is found in higher amounts in partially hydrogenated vegetable fats [59], might exert particular negative effects [77].

The variations in the contents of individual *trans*-18:1 acids in bovine milk fats can be gathered from Figure 1.6, that shows frequency distributions of the total content of *trans*-18:1 acids as well as of the individual positional isomers in German mikfats [61]. These studies exhibited considerably higher *trans* contents during pasture feeding. Such milk fats can be related to the right relative maximum in the frequency distribution of total *trans*-18:1 acids, whereas the left relative maximum is attributable to milk fats from barn feeding. This differentiation is to be seen in case of *trans*-Δ11 and *trans*-Δ15 by two relative maxima as well.

In recent studies published by Precht and Molkentin [55,57,60], the seasonal variations in *trans*-18:1 acid contents in milk fats from a large milk collection area were determined directly. Moreover, in another study, these authors analysed the contents of individual positional *trans*-18:1 isomers in butter samples made from bulk milks, that were collected weekly over a whole year. Figure 1.7 illustrates the values found weekly for oleic acid (*cis*-9-18:1), the different positional isomers of *trans*-18:1, as well the total content of *trans*-18:1 isomers. Except for the *trans*-Δ10 isomer, all other positional *trans* isomers as well as total *trans*-18:1 exhibit a seasonal course that is identical to oleic acid, showing considerably lower contents during winter (*ca.* 44th to 15th week of next year) compared to the summer feeding period. Similar results were found by Wolff *et al.* [81], who analysed sixty samples of French butter at five different periods of the year by Ag-TLC/GLC. Most of the samples were major national brands. Due to the sampling method, results from this study were independent of the breed, stage of lactation, age, or calving period, and applied to a considerable quantity of pooled milks from many regions. The highest levels of total *trans*-18:1 occurred from May to June (4.3 wt%), the lowest from January to March (2.4%), with a mean estimated annual value of 3.3% (Table 1.9). In the same way, the content of vaccenic acid (plus *trans*-10 18:1 acid) relative to total *trans*-18:1 was highest in spring (*ca.* 58%) and lowest in winter (*ca.* 48%) (Table 1.9). Thus, in agreement with Figure 1.4, the content of *trans*-Δ11 (plus Δ10) is already decreasing during summer time. The other isomers determined in the study by Wolff *et al.* [81] (Δ6-Δ9, Δ12, Δ13/Δ14, Δ15 and Δ16) partly show slightly different seasonal courses with highest relative contents in summer. In earlier studies on butter samples from the autumn (*n* = 12) and the spring period (*n* = 12) by Ag-TLC/GLC, Wolff [78] had found total contents of *trans*-18:1 acids of 3.22 % and 4.28 %, respectively. Highest and lowest values were 5.10% (spring) and 2.46% (autumn). Corresponding results of 1.97% (winter) and 4.37% (summer) were found by Precht [55] for milk fat samples from a single but large milk collection area. So, the *trans*-18:1 acid content may double throughout the year. However, both the German and French studies agreed on a mean annual value in the range 3.3 - 3.8% of total fatty acids.

All these seasonal effects are attributable to the high amount of up to 75%

polyunsaturated C18 fatty acids, mainly α-linolenic acid, in the pasture fodder and to the low amount of these fatty acids in the barn fodder. By the biohydrogenation in the rumen, the polyunsaturated fatty acids are transformed into stearic and oleic acids, but partly also into *trans*-octadecenoic acids.

In addition to seasonal variations, the food supply of the cows can influence the *trans* -18:1 acid content in bovine milk fat. The studies by Precht *et al.* [56,60] revealed a particular low *trans* content in milk fat from under-fed cows. Due to the limited ability of food intake, high-performing cows normally show a negative balance of energy at the top of lactation. The lower *trans* fatty acid content resulting from this energy deficiency is attributable to the fact that the C18 acids mobilized from the adipose tissue in this state of nutrition are not subjected to biohydrogenation like fatty acids taken up with the fodder. Under the influence of the Δ9-desaturase, stearic acid is converted only to oleic acid, without any formation of *trans* isomers.

All these numerous investigations show in particular that analysing only few milk fat samples, *i.e.* from butter, cheese or cream, leads to wrong results concerning the mean content of *trans*-octadecenoic acids, since seasonal variations in feeding and lactation will not be taken into account representatively.

2. Goat and Ewe Milk Fats

Other edible sources of *trans*-18:1 acids are ewe and goat milks. Their production may be considered of marginal quantitative importance when compared to that of cow milk at a world level (1 - 2%) [79]. However, ewe and goat milks are not negligible in those countries where climatic and geographic conditions are not well suited to cattle-rearing. For example, 46% of the caprine livestock is concentrated in Greece [79], where the number of cows for milk production is only 1% of cows in the EEC. It is thus necessary to have available quantitative data on *trans*-18:1 acids for those countries where the consumption of such products is comparatively high with regard to cow milk. Greeks, for example, could consume as much as *ca.* 17 g/person/d of fat from ewe and goat milks [79], and only 19 g/person/d of fat from cow milk.

Wolff [79] has studied the content and distribution of *trans*-18:1 isomers in French goat and ewe cheeses. It was assumed in this study that cheese fat was representative of milk fat, because milk fat globules are retained in the cheese matrix during cheese processing. In the first case, a mean *trans*-18:1 acid content of 2.7 ± 0.9% (*n* = 8; minimum, 1.8%; maximum, 4.5%) was determined (Table 1.10). *Trans*-18:1 acids averaged 4.5 ± 1.1% (*n* = 7; minimum, 3.0%; maximum, 6.2%) of total fatty acids in ewe cheese fat [79] (Table 1.10). Though these isomers were previously characterized in earlier studies, only partial quantitative data, unusable for consumption assessments, were reported. In cheeses from both species, the levels of *trans*-18:1 acids may double between the minimum and maximum values. This means that the *trans*-18:1 acid content in goat and ewe milk fats is variable, and it seems that the feed might be of major importance in this respect, as for cows. However, other parameters such as the breed and the

Table 1.10

Content and distribution of *trans*-18:1 isomers in French goat and ewe cheese fat.

Origin	Total *trans*-18:1 acids[a]	n[b]	*Trans*-18:1 isomers[c]						Ref.[d]
			Δ6-Δ9	Δ10 + Δ11	Δ12	Δ13 + Δ14	Δ15	Δ16	
Goat	2.68 ± 0.88	8	12.6 ± 2.9	44.9 ± 4.0	8.7 ± 0.2	18.2 ± 2.5	6.8 ± 1.4	8.8 ± 2.4	[79]
Ewe	4.53 ± 1.11	7	8.3 ± 2.6	56.7 ± 5.3	6.7 ± 1.0	15.6 ± 1.5	5.2 ± 0.5	7.5 ± 1.8	[79]

[a]As weight percent of total fatty acids (mean ± standard deviation).
[b]Number of samples.
[c]As weight percent of total *trans*-18:1 isomers (mean ± standard deviation).
[d]Reference.

stage of lactation may also affect this content. Combining these results with those of cow milk fat, fats can be classified in the following decreasing order with regard to their mean *trans*-18:1 acid content: ewe > cow > goat.

The distribution profiles of *trans*-18:1 isomers in goat and ewe milk fats are summarized in Table 1.10. Identifications were realized by coinjecting the *trans*-18:1 acid fraction, isolated by Ag-TLC, with each individual synthetic *trans*-18:1 isomer with an ethylenic bond between positions Δ5 and Δ15. The *trans*-Δ10 isomer appeared as a small shoulder on the leading edge of the prominent *trans*-Δ11 isomer, which was the main isomer, as in cow milk fat. Apparently, there would be a slightly higher level of vaccenic acid (plus *trans*-Δ10 18:1 acid) in ewe milk fat than in goat milk fat. The distribution of *trans*-18:1 isomers in ewe milk fat is close to that of milk fat from cows in spring (compare Tables 1.9 and 1.10). This would indicate that the profile of *trans*-18:1 acids is related to the feeding conditions of the animals where cows in spring and ewes all throughout the year eat mainly grass. On the other hand, the *trans*-18:1 acid distribution profile in goat milk fat resembles that of cow milk fat in Fall. One can thus hypothesize that the proportion of the *trans*-Δ11 isomer is probably related to the feed, with a trend toward the highest values under grazing conditions. Based on the similarity of the *trans*-18:1 acid profiles in milk fats from the three species, it can be deduced that the biohydrogenation mechanisms that take place in rumen bacteria are similar and that they do not depend on the species.

F. CONCLUSIONS

Except perhaps for vegans, it appears almost impossible to avoid the consumption of *trans*-18:1 acids from natural origin. The distribution profile of these isomers, with the Δ11 being predominant, and as synthesised by the rumen microflora, remains qualitatively unchanged during the metabolic processes that occur between intestinal absorption and the final deposition in tissue lipids. This would indicate that there is no major enzymatic discrimination against any isomer. The analysis of *trans*-18:1 isomers can now be performed routinely by GLC with highly-resolution 100-m (or even 50-m) capillary columns such as the CP-Sil 88 column, with only Ag-TLC as a complementary technique. The distribution of *trans*-18:1 acids in ruminant fats is now well described and their behaviour during capillary GLC is so typical that they can be used as standards for identification purposes, with the Δ11 isomer being used as a landmark.

Despite the recent increased interest in *trans* acids, there has been a lack of adequate methodologies which has led to imprecise data and unproductive discussions. One should not consider *trans* acids as an homogenous category of components; each isomer has its own metabolism and must be considered individually as a specific entity, and this is possible if sufficient care is taken for their analysis. Differences between partially hydrogenated oils and ruminant fats are not in the diversity of individual *trans*-18-1 acids, but only in their distribution profile. So, it might be possible that health effects of partially hydrogenated oils are linked to one or a few given *trans* isomers rather to these components taken as a whole.

REFERENCES

1. Adlof,R.O. and Emken,E.A., *Lipids*, **21**, 543-547 (1986).
2. Akesson,B., Johansson,B.-M., Svensson,M. and Öckerman,P.-A., *Am. J. Clin. Nutr.*, **34**, 2517-2520 (1981).
3. Backderf,R.H. and Brown,J.B., *Arch. Biochem. Biophys.*, **76**, 15-27 (1958).
4. Bayard,C.C. and Wolff,R.L., *J. Am. Oil Chem. Soc.*, **72**, 1485-1489 (1995).
5. Bayard,C.C. and Wolff,R.L., *J. Am. Oil Chem. Soc.*, **73**, 531-533 (1996).
6. Bertram,S.H., *Biochemische Zeitschrift*, **197**, 433-441 (1928).
7. Bickerstaff,R., Noakes,D.E. and Annison,E.F., *Biochem. J.*, **130**, 607-617 (1972).
8. Boatella,J., Rafecas,M. and Codony,R., *Eur. J. Clin. Nutr.*, **47**, S62-S65 (1993).
9. Body,D.R., Shorland,F.B. and Czochanska,Z., *J. Sci. Food Agric.*, **21**, 220-225 (1970).
10. Bolton-Smith,C., Woodward,M., Fenton,S., McKluskey,M.-K. and Brown,C.A., *Brit. J. Nutr.*, **74**, 661-770 (1995).
11. British Nutrition Foundation Task Force, *Trans Fatty Acids*, British Nutrition Foundation (1995).
12. Chen,Z.-Y., Pelletier,G., Hollywood,R. and Ratnayake,W.M.N., *Lipids*, **30**, 15-21 (1995).
13. Christie,W.W. and Moore,J.H., *J. Sci. Food Agric.*, **22**, 120-124 (1971).
14. Conacher,H.B.S. and Iyenbar,J.R., *J. Assoc. Off. Anal. Chem.*, 61, 702-708 (1978).
15. Craig-Schmidt,M.C., in *Fatty Acids in Foods and Their Health Implications*, pp.365-398 (1992) (edited by C.K. Chow, Marcel Dekker, Inc., New-York).
16. DeMan,L. and DeMan,J.M., *J. Am. Oil Chem. Soc.*, **60**, 1095-1098 (1983).
17. De Vries,B. and Jurriens,G., *Fette, Seifen, Anstrichm.*, **65**, 725-727 (1963).
18. Enig,M.G., Atal,S., Keeney,M. and Sampugna,J., *J. Am. College Nutr.*, **9**, 471-486 (1990).
19. Enig,M.G., Pallansch,L.A., Sampugna,J. and Keeney M., *J. Am. Oil Chem. Soc.*, **60**, 1788-1795 (1983).
20. Expert Panel on *Trans* Fatty Acids and Coronary Heart Disease, *Am. J. Clin. Nutr.*, **62** (supplement), 654S-707S (1995).
21. Garton,G.A. and Duncan,W.R.H., *J. Sci. Food Agric.*, **20**, 39-42 (1969).
22. Grossfeld,J. and Simmer,A., *Z. Untersuch. Lebensmittel*, **59**, 237-258 (1930).
23. Gunstone,F.D., Ismail,I.A. and Lie Ken Jie,M.S.F., *Chem. Phys. Lipids*, **1**, 376-385 (1967).
24. Gurr,M.I., *Bull. Int. Dairy Fed.*, **166**, 5-18 (1983).
25. Hartman,L. Shorland,F.B. and McDonald,I.R.C., *Nature*, **174**, 185-186 (1954).
26. Hartman,L. Shorland,F.B. and McDonald,I.R.C., *Biochem. J.*, **61**, 603-607 (1955).
27. Hay,J.D. and Morrison,W.R., *Biochim. Biophys. Acta*, **202**, 237-243 (1970).
28. Hay,J.D. and Morrison,W.R.., *Lipids*, **8**, 94-95 (1973).
29. Heckers,H., Melcher,F.W. and Dittmar,K., *Fette Seifen Anstrichm.*, **81**, 217-226 (1979).
30. Henninger,M., and Ulberth,F., *Milchwissenschaft*, **49**, 555-558 (1994).
31. Homer,D., *Lipidforum Symposium, Göteborg*, 172-179 (1984).
32. Hudgins,L.C., Hirsch,J. and Emken,E.A., *Am. J., Clin. Nutr.*, **53**, 474-482 (1991).
33. Hyvönen,L., Lampi,A.-M., Varo,P. and Koivistoinen,P., *J. Food Comp. Anal.*, **6**, 24-40 (1993).
34. Katan,M.B. and Zock,P.L., *Ann. Rev. Nutr.*, **15**, 473-493 (1995).
35. Katz,I. and Keeney,M., *J. Dairy Sci.*, **49**, 962-966 (1966).
36. Kaufman,H.P. and Mankel,G., *Fette Seifen Anstrichm.*, **66**, 6-13 (1964).
37. Kochbar,S.P. and Matsui,T., *Food Chem.*, **13**, 85-101 (1984).
38. Lanza,E. and Slover,H.T., *Lipids*, **16**, 260-267 (1981).
39. Lanza,E., Zyren,J. and Slover,H.T., *J. Agric. Food Chem.*, **28**, 1182-1186 (1980).
40. Lin,K.C., Marchello,M.J. and Fischer,A.G., *J. Food Sci.*, 49,1521-1524 (1984).
41. Lluch,M.C., Pascual,J., Parcerisa,J., Guardiola,F., Codony,R., Rafecas,M. and Boatella,J., *Grasas y Aceites*, **44**, 97-100 (1993).
42. Lluch,M.C., Roca de Vinyals,M., Parcerisa,J., Guardiola,F., Codony,R., Rafecas,M. and Boatella,J., *Grasas y Aceites*, **44**, 195-200 (1993).
43. Lund,P. and Jensen,F., *Milchwissenschaft*, **37**, 645-647 (1982).
44. Lund,P. and Jensen,F., *Milchwissenschaft*, **38**, 193-196 (1983).
45. Mansour,M.P. and Sinclair,A.J., *Asia Pacific J. Clin. Nutr.*, **3**, 155-163 (1993).

46. Molkentin,J. and Precht,D., *Z. Ernährungswiss.*, **34**, 314-317 (1995).
47. Molkentin,J. and Precht,D., *Chromatographia*, **41**, 267-272 (1995).
48. Moore,C.W., *J. Soc. Chem. Ind.*, **38**, 320-325 (1919).
49. Morris,L.J., *J. Lipid Res.*, **7**, 717-732 (1966).
50. Ohlrogge,J.B., Emken,E.A. and Gulley,R.M., *J. Lipid Res.*, **22**, 955-960 (1981).
51. Ottenstein,D.M., Bartley,D.A. and Supina,W.R., *J. Chromatogr.*, **119**, 401-407 (1976).
52. Ottenstein,D.M., Witting,L.A., Silvis,P.H., Hometchko,D.J. and Pelick,N., *J. Am. Oil Chem. Soc.*, **61**, 390-394 (1984).
53. Parodi, P.W., *J. Dairy Sci.*, **59**, 1870-1873 (1976).
54. Precht,D., *Fat Sci. Technol.*, **93**, 538-544 (1991).
55. Precht,D., *Z. Ernährungswiss.*, **34**, 27-29 (1995).
56. Precht,D., Frede,E., Hagemeister,H. and Timmen,H., *Fette Seifen, Anstrichm.*, **87**, 117-125 (1985).
57. Precht,D. and Molkentin,J., *Kieler Milchwirtsch. Forschungsber.*, **46**, 249-261 (1994).
58. Precht,D. and Molkentin,J., *Schriftenreihe des Bundesministeriums für Ernährung, Landwirtschaft und Forsten, Reihe A: Angewandte Wissenschaft*, Landwirtschaftsverlag GmbH Münster, **445**, 26-37 (1995).
59. Precht,D. and Molkentin,J., *Food-Nahrung*, **39**, 343-374 (1995).
60. Precht,D. and Molkentin,J., *Milchwissenschaft*, **52**, 564-568 (1997).
61. Precht,D. and Molkentin,J., *Int. Dairy J.*, **61**, 791-809 (1996).
62. Rao,P.C. and Daubert,B.F., *J. Am. Chem. Soc.*, **70**, 1102-1104 (1948).
63. Sampugna,J., Pallansch,L.A., Enig,M. and Keeney,M., *J. Chromatogr.*, **249**, 245-255 (1982).
64. Shorland,F.B., Body,D.R. and Gass,J.P., *Biochim. Biophys. Acta*, **125**, 217-225 (1966).
65. Slover,H.T., Lanza,E., Thompson,R.H., Davis,C.S. and Merola,G.V., *J. Food Comp. Anal.*, **1**, 26-37 (1987).
66. Smith,L.M., Dunkley,W.L., Franke,A. and Dairiki,T., *J. Am. Oil Chem. Soc.*, **55**, 257-261 (1978).
67. Sommerfeld,M., *Prog. Lipid Res.*, **22**, 221-233 (1983).
68. Steinhart,H. and Pfalzgraf,A., *Z. Ernährungwiss.*, **31**, 196-204 (1992).
69. Stender,S., Dyerberg,J., Holmer,G., Ovesen,L. and Sandström,B., *Clinical Science*, **88**, 375-392 (1995).
70. Strocchi,A. and Holman,R.T., *Riv. Ital. Sostanze Grasse*, **48**, 617-622 (1971).
71. Strocchi,A., Lercker,G. and Losi,J., *Rev. Fr. Corps Gras*, **20**, 625-630 (1973).
72. Swern,D., Knight,H.B. and Eddy,C.R., *J. Am. Oil Chem. Soc.*, **29**, 44-46 (1952).
73. Twitchell,E., *J. Ind. Eng. Chem.*, **13**, 806-807 (1921).
74. Ulberth,F. and Haider,H.-J., *J. Food Science*, **57**, 1444-1447 (1992).
75. Ulberth,F. and Henninger,M., *J. Dairy Res.*, **61**, 517-527 (1994).
76. Van de Voort,F.R., Ismail,A.A. and Sedman,J., *J. Am. Oil Chem. Soc.*, **72**, 873-880 (1995).
77. Willett,W.C., Stampfer,M.J., Manson,J.E., Colditz,G.A., Speizer,F.E., Rosner,B.A., Sampson,L.A. and Hennekens,C.H., *Lancet*, **341**, 581-585 (1993).
78. Wolff,R.L., *J. Am. Oil Chem. Soc.*, **71**, 277-283, (1994).
79. Wolff,R.L., *J. Am. Oil Chem. Soc.*, **72**, 259-272 (1995).
80. Wolff,R.L. and Bayard,C.C., *J. Am. Oil Chem. Soc.*, **72**, 1197-1201 (1995).
81. Wolff,R.L., Bayard,C.C. and Fabien,R.J., *J. Am. Oil Chem. Soc.*, **72**, 1471-1483 (1995).
82. Wood,R., in *Dietary Fats and Health*, pp. 341- 358 (1983) (edited by E.G. Perkins and W.T. Wisek, AOCS Press, Champaign, IL).
83. Wood,R., in *Fatty Acids in Foods and Their Health Implications*, pp. 663-688 (1992) (edited by C.K. Chow, Marcel Dekker, Inc., New-York).

CHAPTER 2

TRANS *FATTY ACIDS AND THE POTENTIAL FOR LESS IN TECHNICAL PRODUCTS*

R.G. Ackman[1] and T.K. Mag[2]

[1] *Canadian Institute of Fisheries Technology, DalTech, Dalhousie University, P.O. Box 1000, Halifax, Nova Scotia, B3J 2X4, and* [2] *T. Mag/Associates Consulting Inc., 35 Old Church Road, King City, Ontario, L7B 1K4*

A. TO REFINE OR NOT TO REFINE, IS THAT THE QUESTION?

1. Introduction

Are there "natural" *trans* fatty acids? This rhetorical question should be asked by those who are concerned about the impact of *trans* fatty acids on their well-being or desiring a greater life-span. Often they are persuaded that by avoiding natural fats altered by man through hydrogenation they will eliminate *trans* fatty acids from their diets. They should also then avoid dairy products and red meats from beef and sheep. These fats naturally contain up to 5% *trans* acids. At this point there is not much left to eat except olive and tropical oils, both low in polyunsaturated fatty acids. The tropical oils are condemned by some as posing the same health risks as *trans* acids and olive oil is rather a regional commodity. For much of western society that leaves only the retail vegetable salad oils from seeds grown in temperate regions. The latter are refined and, as will be explained, one part of this process can introduce low levels of *trans* acids in the two common all-*cis* polyunsaturated fatty acids, linoleic (18:2n-6) and α-linolenic (18:3n-3). These *trans* ethylenic bonds are "artifacts" of the deodorization process. A "cold pressed" oil is often thought of by unsuspecting consumers as "natural", but that merely means that it is not solvent-extracted. It may well also be refined, or semi-refined, to present a clear and light-coloured oil to the consumer. Whether or not such products contain *trans* acid artifacts of polyunsaturated fatty acids from the process of deodorization will be left unsaid on the label. As long ago as 1979, Sebedio and Ackman [9] showed that oil carefully recovered from canola seed in the laboratory contained certain geometric isomers of α-linolenic acid with *trans* bonds identical to those produced by deodorization, and other seed oils now appear to contain the same materials as well as *trans* forms of oleic (*cis*-9-octadecenoic) acid. Are they "natural"? Leafy vegetables all contain *trans*-3-hexadecenoic acid in very low proportions of total mass but we are unaware of any studies on the potential health risk of this *trans* fatty acid when present in salad constituents. The reader will have to decide on what is a "natural" *trans* acid.

2. Improved Methods Always Create New Problems

Active analytical investigations of the fatty acids, sterols, tocopherols etc. of refined but non-hydrogenated vegetable oils were common starting about 1950. The low sensitivity of infra-red equipment [1] for measuring isolated or methylene-interrupted *trans* acids precluded low levels of *trans* acids from being recognized. Fortunately Fourier-*trans*form infrared spectroscopy has eliminated this problem [2,3]. In contrast, the conjugated dienoic acids of raw or refined edible oils were readily determined in the ultraviolet spectrum with strong absorption at 234 mm. It was not often mentioned that one of the two ethylenic bonds in such fatty acids would very likely be *trans* in geometry, a consequence of the shift of a *cis* bond from the former Δ9 or Δ12 positions to the adjacent Δ10 or Δ11 positions. There the matter of natural or process-derived *trans* acids in edible oils rested for the edible oils industry and for nutritionists until the development of low-erucic acid rapeseed oils just prior to 1970 [4,5].

In 1973, author Ackman was asked by the Government of Canada to examine the newly developed rapeseed oils with low erucic acid contents for unsuspected but minor fatty acids that might create health hazards. The first step applied was the relatively novel technique of open-tubular (capillary) gas-liquid chromatography (GLC) [6]. This study of methyl esters was fortuitously carried out initially on a retail quality "canbra" salad oil and two peaks (A and C) bracketing the peak for the methyl ester of alpha-linolenic acid (18:3n-3) attracted attention (Figure 2.1). The original unrefined oils were available for comparison, and did not show these peaks. Fortunately for the soon-to-be canola industry retail soybean oil purchased in the USA showed the same two peaks associated with α-linolenic acid and crude soybean oil did not. It was possible to apply the then relatively rare technologies of NMR and Raman spectroscopy to isolates of 18:3 fatty acids from thin-layer chromatography on silicic acid impregnated with silver nitrate to show that these two artifacts of refining had two *cis* bonds and one *trans* bond [6]. Refining of edible oils in Canada formerly included deodorization at 252°C for 2.5-3 hours, but conditions are now somewhat milder [7]. We easily reproduced the formation of components A and C by heating oils containing α-linolenic acid (*cis*-9,*cis*-12,*cis*-15-18:3) at 230°C and detected (Figure 2.2) two other 18:3 artifacts, X and Y(not shown), with two *trans* bonds and one *cis* bond [6]. We used both BDS (butanediol succinate polyester) and SILAR-5CP, (an early mixed phenyl-alkyl cyanosilicone) as liquid phases.

Publication of this discovery was soon followed by a paper from a leading food industry company to indicate that these *trans*-acid artifacts in refined salad and cooking oils were of little consequence. In part this belief rested on the existence of metabolic pathways for catabolizing all likely positional ethylenic bonds, either *cis* or *trans*, for energy [8]. In another project we were later [9] able to show by capillary GLC that the *cis*-9,*cis*-12,*trans*-15-18:3 (the original artifact component A) could be detected in oils extracted in the laboratory directly from seeds of *Brassica napus* or B. *campestris* (Figure 2.3). Our conclusion was that since component A was 5 or 6 times as plentiful as component C in these isolates, the Δ15 bond of 18:3n-3 was the position most exposed to the unknown seed factors producing trace amounts of such *trans* fatty acids *in situ*. Possibly a lipoxygenase and a dehydroperoxidase should be suspected as agents in these cases. This GLC work also showed both *cis* and *trans* 15:1n-10 to be common in these oils at the 0.1-0.2% level. Unfortunately, other edible seed oils do not seem to have been investigated as thoroughly for such minor *trans* components.

3. The Current Situation on Refined but Unhydrogenated Edible Oils

The concurrent formation of *trans* isomers of bonds in linoleic acid (18:2n-6), along with those of 18:3n-3, was noted in our original publication [6], but the relative proportions of artifacts were lower compared to the presumably more stable all-*cis* 18:2n-6. A number of conjugated 18:2 isomers can also be produced by heating sunflower seed oil at 275°C under nitrogen [10]. In a thorough investigation of fatty acid isomerization, Pudel and Denecke [11] have shown that temperatures of 300°C or more are needed to form the *trans*,*trans*-dienoic isomers of 18:2n-6. Their technology of admitting water to heated oils to mimic older industrial deodorization conditions [12] resembled

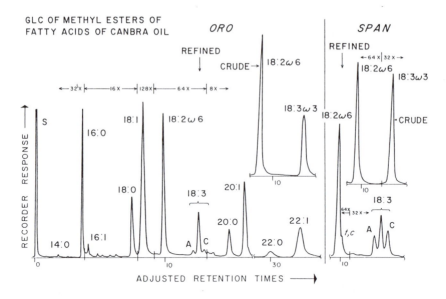

Fig. 2.1. Comparison of partial GLC records of analyses on butanediolsuccinate polyester of cultivars of two crude and refined low erucic acid oils grown circa 1973. A and C are the mono *trans* geometrical artifacts from deodorization, bracketing the original all-*cis* 18:3n-3. Reproduced with permission from the *J. Am. Oil Chem. Soc* [6].

our original work in Halifax, although the water is not necessary to effect geometrical isomerization of either linoleic or α-linolenic acids. Pudel and Denecke were able to show that iron, copper, and phospholipids had no effect on isomerization of α-linolenic acid "in contrast to all expectations". Time-temperature relations explored by these authors with rapeseed oil (low erucic) showed that 200-220°C was the critical temperature range to produce geometrical isomerization for both C_{18} polyunsaturated fatty acids. Somewhat the same ranges were found by Billek [13].

   R.L. Wolff has published extensively on the GLC conditions for determining these isomeric fatty acids [e.g. 14-16]. The use of CP™Sil 88 columns (100% cyanopropyl polysiloxane) as the liquid phase, combined with both methyl and isopropyl esters of the different possible isomers of 18:3n-3, solves most problems [17]. For example the FAME (fatty acid methyl esters] have a coincidence of 9t,12c,15t- with 9c,12t,15t-18:3. With the isopropyl esters these are separately resolved but the 9c,12c,15t-isomer, which gave a good FAME peak is now moved to partly overlap with 9t, 12c, 15t. This ester behaviour difference is also illustrated elsewhere [18,19]. The elution order of the FAME of 18:2 isomers was 9t, 12t; 9c, 12t; 9t,12c; 9c, 12c on columns of this type [10]. Unfortunately, if *cis*-11-20:1 is present (found for example in soybean and canola oils) there may be a coincidence with the normally present and natural all-*cis*-18:3n-3 [20]. Use of the polar liquid phase BPX 70 has been described also by Chardigny *et al.* [18].

   The benefits of α-linolenic acid in human nutrition for subjects of all ages

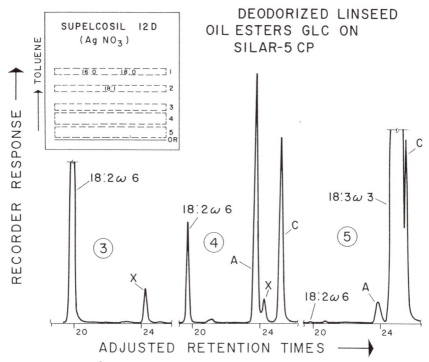

Fig. 2.2. Three partial gas-liquid chromatograms of fatty acid methyl esters prepared from linseed oil deodorized in the laboratory. A *trans* bond confers extra mobility in thin-layer chromatography on silver nitrate impregnated silicic acid, explaining the position of 18:3n-3 geometrical isomers A and C, and X (with two *trans* bonds). Reproduced with permission from the *J. Am. Oil Chem. Soc.* [6].

are now undeniable [21-24]. The amount of α–linolenic acid in the diets of North Americans where soybean and canola oils dominate the market is "sufficient" and may even be beneficial [25]. In fact infant formulas have received special attention in this respect and of course the vegetable oil ingredients are refined thoroughly. *Trans* isomers of 18:2n-6 and 18:3n-3 were found in such U.S. formulas, the latter showing 2.6 to 23.5% of total 18:3n-3 [26]. The same pattern was obvious in retail salad oils in the USA [27]. An interesting application of the GLC technology would be an examination for 18:3n-3 *trans* artifacts to detect refined soybean oil as an adulterant in virgin olive oil and an inter-laboratory exercise on this has been reported [28]. An EU regulation (2568/91) stipulates the *trans* isomer content of "native" olive oil, limited to 0.05% of the oil for each of 18:1 and 18:2n-6 + 18:3n-3 (See Figure 2.3 whence the area % as applied by this regulation). Geometrical isomers are easily prepared by heating common edible oils or can be prepared with nitrous acid as described by Wolff and Sebedio [19]. In the latter paper, the authors dealt with the difficult case of γ-linolenic acid (18:3n-6). The difficulty stems from the very little difference in retention time between $\Delta 6$ and $\Delta 9$ FAME, in either *cis* or *trans* form. The elution orders are discussed

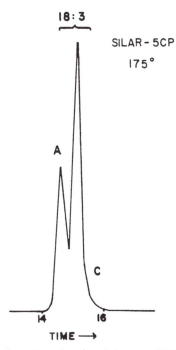

Fig. 2.3. Partial gas-liquid chromatography of methyl esters of fatty acids from a silver-nitrate impregnated silicic acid thin-layer chromatography plate. The area was between the all *cis* diene and triene bonds and shows both A; *cis*-9,*cis*-12,*trans*-15 and C; *trans*-9,*cis*-12,*cis*-15 isomers of α-linolenic acid. The oil was recovered from canola seeds extracted in the laboratory. Reproduced with permission from the *J. Am. Oil Chem. Soc.* [9].

for the highly polar cyanosilicone columns but are given also for the less polar polyglycol-based DB-WAX where they suggest 6c,9c,12c; 6t,9c,12c; 6c,9c,12t; and 6c,9t,12c plus 6t,9c,12t. Although of interest to medical biochemists, health and nutrition interests, and analysts, most of the 18:3n-6 of vegetable origin in our diets is found in less common foods [29], and is consumed in very small amounts compared to the large (if deliberate) amounts of other polyunsaturated C_{18} fatty acids that disappear annually.

Popular food-frying operations undoubtedly are major dietary sources of 18:3n-3 *trans* isomers, depending on the type of oil consumed [30]. Some will be produced during the oil refining, and in possible "light" hydrogenation to reduce total 18:3 to about 3% for improved stability. Others are artifacts of the cooking process itself. Hopefully, badly abused frying oils are always discarded [31].

In the calculations of total *trans* acid consumption in the USA the salad oils with their trace of natural *trans* acids or the artifact *trans* acids from refining were ignored and attention was focused on hydrogenated oils [32,33]. Unfortunately the exaggerated public alarm over all dietary *trans* acids continues, as evidenced in a column in the Toronto Globe and Mail on July 9, 1997 that read, in part:

"Like other oils found on supermarket shelves, canola has been
refined, bleached, deodorized and heated to extremely high
temperatures to squeeze out every tiny modicum of oil from the
seed. Unfortunately, the latter process radically changes the
chemical composition of mass-produced vegetable oils, producing
harmful (to the heart) *trans*-fatty acids and other deleterious oil
chemicals."

Not one of the several "experts" contributing such erroneous views to the
column mentioned the recent study conducted by de Lorgeril *et al.* and
published in *Lancet* [34], where refined and even hydrogenated canola oils in
the diet were shown to improve the heart health of a large number of human
subjects at risk. Such "experts" almost invariably are selling a "natural" edible
oil, or a blend of oils, usually allegedly cold (expeller) pressed and unrefined.
Author Ackman has lately found "semi-refined" appearing on vegetable oil
products in health food stores in Canada, a rather weak admission that some
clean-up of natural oils may be required.

Fortunately a timely paper by Brühl [35] has determined the *trans* acids in
rapeseed (sic) oils processed under different conditions as well as in a variety
of actual oils recovered from dried seeds. In all of the oils from a variety of
seed sources, *trans* acids were present in both the monoethylenic (18:1) fatty
acids and the polyunsaturated fatty acids (18:2 + 18:3). In some cases, such as
soy and rapeseed (Figure 2.4), the 18:1 *trans* proportion exceeded that from
the polyunsaturated acids. It is believed that these two seeds may have been
strongly heated during drying. This suggests that the "natural" *trans* fatty acids
in the C_{18} group could be the result of lipoxygenase [36] or
hydroperoxide/dehydroperoxidase enzyme systems in the living seed. Whether
or not these could break down on strong heating, producing *trans* acids, is
possibly a question of temperature.

In another study by Wolff [37], walnut oil, a popular source of α-linolenic
acid in Germany, was examined. Heating cold-pressed oil under vacuum
produced a mixture of isomers similar to the four artifacts found in common
vegetable oils (Figure 2.5). In the high temperature industrial deodorization
process, steam is usually the carrier used to sweep away objectionable
materials. Some experiments, including our own original work [6], supplied
water to oil heated to a suitable temperature, usually at 200°C or higher. Brühl
[35] mentions "steam washing" allowed in oil processing in Germany, but at
100°C for 3 hours it only slightly increased *trans* acids and did not produce
the marked figures for *trans* artifacts found with more common deodorization
procedures conducted at higher temperatures [13,38]. Thus temperature is
clearly the most critical factor. Our other work on fish oil fatty acids showed
geometrical isomerization in the more highly unsaturated acid all-*cis*-
5,8,11,14,17-eicosapentaenoic [39,40]. The n-3 bond is especially labile, as in
α-linolenic acid, and isomerization is rapid above 200°C in vacuum or under
nitrogen. Figure 2.2 shows the relative positions in silver nitrate-thin-layer

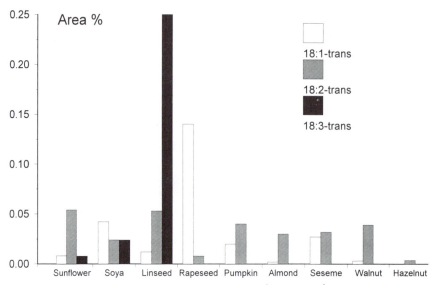

Fig. 2.4. "Natural" *trans* fatty acids (area % of GLC analysis) in fatty acids of oils recovered in the laboratory from retail seeds compared to those (linseed, rapeseed) stated to have been heated during the drying process. Redrawn from ref. 35.

chromatography (TLC) of the *trans* ethylenic 18:3n-3 isomers recoverable from esters of a deodorized vegetable oil. Some may prefer high-performance liquid chromatography (HPLC) with a silver ion column for analyses and this technology has also been described [41].

4. A Matter of Biochemistry

We live in a world where enzymes are continuously at work in all plant and animal life. These remarkable organic catalysts are also present and active in bacteria and the *trans* fatty acids which they produce may be assimilated into higher organisms. The latter may also have adapted some of the functions of the bacterial enzymes for their own purposes, so it appears that occurrence of *trans* acids in plant seed oils is "natural" and the low levels of *trans* acids produced cannot be removed or avoided. Other microbial or plant seed oil *trans* fatty acids are also "natural", for example the conjugated (dienoic) linoleic acids (CLA) [36] recently rehabilitated from hazard to health protection status [42]. The balance between our intakes of natural and unnatural materials may be the key to defining possible hazards [43,44]. Refining edible oils merely <u>increases</u> the levels of certain *trans* artifact fatty acids already naturally present in seed oils but is part of the contemporary standard of living now accepted by most consumers. As long as there is an adequate supply of all-*cis* polyethylenic fatty acids available there does not seem to be much health risk from the natural or artifact polyunsaturated fatty acids containing *trans* ethylenic bonds [44].

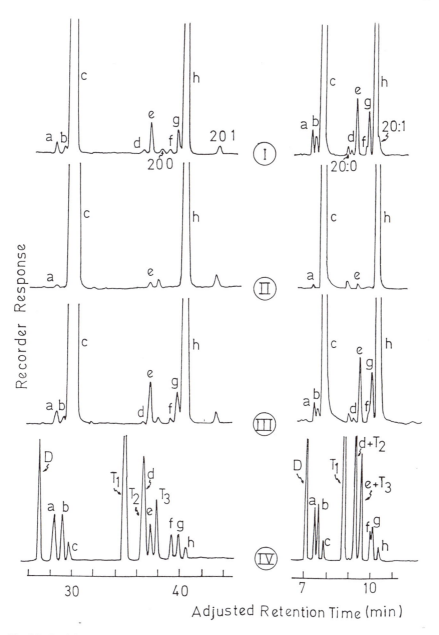

Fig. 2.5. Partial analyses at 150°C (left) or 175°C (right) on a CP™Sil 88 column of methyl esters of fatty acids from I, refined walnut oil; II, cold-pressed walnut oil; III, oil from II heated to 240°C; IV elaidinized linseed oil. Letter identifications are as follows: (a) *cis*-9,*trans*-12 18:2; (b) *trans*-9,*cis*-12 18:2; (c) *cis*-9,*cis*-12 18:2; (d) *trans*-9,*cis*-12,*trans*-15 18:3; (e) *cis*-9,*cis*-12,*trans*-15 18:3; (f) *cis*-9,*trans*-12,*cis*-15 18:3; (g) *trans*-9,*cis*-12,*cis*-15 18:3; (h) *cis*-9,*cis*-12,*cis*-15 18:3; (D) *trans*-9,*trans*-12 18:2; (T1) *trans*-9,*trans*-12,*trans*-15 18:3; (T2) *cis*-9,*trans*-12,*trans*-15 18:3; (T3) *trans*-9,*trans*-12,*cis*-15 18:3; Peaks D, T1, T2 and T3 are not present in walnut seed oil samples G, F3 and F3. Reproduced with permission from *Sciences des Aliments* [37].

5. What About the Consumer?

The natural *trans* acids described above make a modest contribution to the western diet compared to those from both visible (dairy products) and invisible (beef, mutton etc.) animal fats, even though the latter are mostly monoethylenic. A very much larger contribution has been provided for decades by the products of hydrogenation of vegetable oils. Nature provides *trans* ethylenic bonds in the fatty acids of a bewildering variety of fats and lipids (see Appendix). Bacteria and plants achieve this with an apparently minimal expenditure of energy. Enzymes easily perform functions that in our factories absorb incredible amounts of energy. Sometimes this energy can produce unexpected results, as discussed for the deodorization conditions for vegetable oils that can convert certain *cis* ethylenic bonds to *trans* forms. For the consumer, hydrogenation technology also can do this deliberately, or it can eliminate some ethylenic bonds of edible oils completely in a large and complex world-wide industrial system. Consumption of hydrogenated products has been lucidly summarized for the U.S. population in a recent position paper [45]. A year earlier a supplement to the same journal led to considerable discussion in print [46-49]. The European positions are clearly established [43,50,51]. In the face of the numerous biomedical uncertainties and differing points of view it is appropriate to briefly review the technology involved in producing the artificial *trans* acids which are now made available to the consumer in both visible (margarine, spreads, refined salad and cooking oils) and invisible (shortening in baked goods) fats. Most consumers and many scientists do not understand that the oils and fats industry has not been idle in respect to *trans* acids but has for more than two decades sought ways to control their proportion in technical food products.

B. COMMERCIAL PRODUCTION OF *TRANS* FATTY ACIDS

1. Formation of Trans Isomers in Commercial Hydrogenation of Edible Oils

Among the sources of *trans* fatty acid isomers in the supply of edible fat products, partially hydrogenated fats are by far the most important. In current hydrogenation practice, the hydrogenated fats produced have *trans* isomer concentrations in the range of 10-60%, with concentrations of 25-45% being most common. In contrast, deodorization of edible oils commonly produces up to 3% of *trans* isomers. Animal fats from ruminants, such as beef tallow and butter, contain from 5-8% of *trans* isomers.

The introduction of hydrogenation of edible oils early in the 20th century was based on the Normann patents in Germany and the UK [52]. The catalyst used was nickel metal. A feature of nickel-based catalysts in the hydrogenation of unsaturated carbon chains of fatty acids is the conversion of many *cis* ethylenic bonds to *trans* isomers (and also to positional isomers), along with the addition of hydrogen at some ethylenic bonds to decrease overall unsaturation. As is well known, *trans* acids almost invariably have higher melting points than their *cis* analogues. A popular misconception used to be that the objective of hydrogenation was to produce saturated fatty acids. In fact a proportion of *trans* fatty acids with similar high melting points may produce a fat with superior physical properties. Generally, partial elimination of the *cis,cis*-methylene-interrupted structures of polyunsaturated fatty acids (usually linoleic or α-linolenic) increases the stability of the products towards autoxidation. This is especially important in frying oils.

The use of nickel-based, solid (heterogeneous) catalysts has remained essentially unchanged in the edible oils industry up to the present. It is only the chemical form in which nickel is used as the starting material to produce the catalyst, the method of reduction (dry or wet), and the use of inert carriers, such as silicas and aluminas, which have undergone changes over time. These changes have resulted in improved filtration and other handling properties of the catalysts, but have not significantly modified the mode of action of nickel. Hence, geometrical and positional isomerization in the course of nickel-catalysed hydrogenation has remained essentially unchanged as well.

2. Theoretical Basis of Trans Isomer Formation in Edible Oil Hydrogenation

Allen and Kiess in 1955 postulated that the formation of *trans* isomers and positional isomers with the customary nickel catalysts is due to a hydrogenation-dehydrogenation mechanism at the surface of the nickel catalyst [53]. Subsequent work by Allen and others has served to confirm this postulate.

Nickel-catalysed hydrogenation of *cis* monoethylenic bonds is believed to occur via adsorption of atomic hydrogen and adjacent adsorption of the whole double bond system of a fatty acid at the catalyst surface. This may be followed by addition of hydrogen to both ends of the ethylenic bond (saturation) followed by desorption, or only at one end, depending on hydrogen availability. There may then be addition of a second hydrogen, or removal of the first hydrogen atom, or even of another, from or near the partially saturated bond, with possible migration and *trans* isomerization of the ethylenic bond. Eventually these processes are followed by desorption of the modified fatty acid from the catalyst.

With the methylene-interrupted ethylenic bond systems of dienes and polyenes, there may first be an ethylenic bond migration by the above mechanism to form a conjugated diethylenic bond system. This structure is very susceptible to hydrogenation to a monoethylenic bond. This may take place before additional hydrogenation and isomerization can occur, resulting in, for example, two isolated ethylenic bonds, one *trans* and one *cis*, in what was formerly a molecule with three ethylenic bonds (e.g. α-linolenic acid). Commercial hydrogenation practice, including process design, is based on these mechanisms.

In edible oil hydrogenation practice, it is common to use combinations of

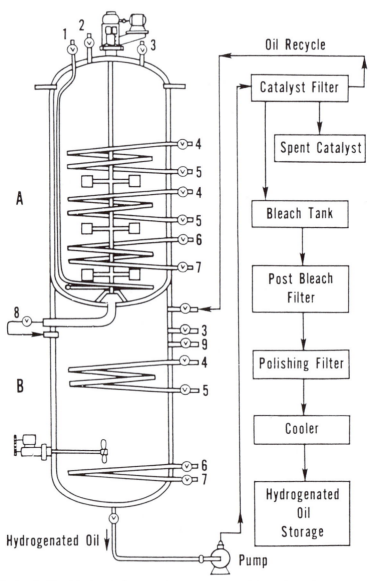

Fig. 2.6. A typical batch hydrogenation apparatus. Reproduced with permission from the *J. Am. Oil Chem. Soc.* [55].

low hydrogen pressures, moderate agitation intensity, and high temperatures to achieve conditions of relative hydrogen scarcity. This results in high selectivity towards achieving partial hydrogenation of multiple ethylenic bond systems and produces high *trans* isomer formation. The term used for this set of hydrogenation parameters is "selective conditions". A fourth factor, namely, high catalyst concentration to supply a large amount of catalytic surface area to compensate for hydrogen scarcity in the oil/hydrogen/catalyst system is usually not used for reasons of catalyst cost.

The melting behaviour of selectively hydrogenated fats high in *trans* isomers and in positional isomers makes them suitable for many edible fat product applications. The reason for this is that isomerization of fatty acid moieties raises their melting point, but not as much as complete double bond saturation. Scholfield published melting points of *cis* and *trans* isomers of C_{18} fatty acids [54], and the melting points for *trans* C_{18} isomers are 20-30°C higher than for the corresponding *cis* isomer. The melting points of C_{18} fatty acid isomers are the main interest in connection with commercial hydrogenated fat products in North America since this chain length is the most common among fatty acids in canola and soybean oils, the two commercially most important hydrogenated edible oils.

The practical effect of this change in melting behaviour in partial hydrogenation of fats is that fats can be produced that have desired firmness below human body temperature, but good melt-away at body temperature. This makes for good functional properties in many fat applications such as for margarine and for frying.

The reverse of two of the above three hydrogenation process operating parameters, that is, high hydrogen pressures and low temperatures, is also used in commercial practice. The term used in the industry to describe this set of hydrogenation process parameters is "non-selective conditions". *Trans* isomer formation is reduced, but it is still very substantial under these conditions and the requirement for hydrogenated fats produced in this way is very small. These fats contain more saturated, high-melting fatty acids than do selectively hydrogenated fats. High intensity agitation, which also reduces *trans* isomer formation, is rarely used for reasons of higher costs for catalyst, energy consumption and hydrogenation equipment.

3. Commercial Hydrogenation Equipment, Process Conditions and Catalysts

i. Hydrogenation equipment: Hydrogenation is most often performed in the batch mode. Many parameters formerly in wide use are covered in an excellent review by the late T.L. Mounts [55]. The need is for a great variety of products with different melting properties from very firm to still fluid and pumpable. The batch reactor is necessary to satisfy these different functional requirements and is the main component of any commercial installation. Ancillary equipment is required to remove the catalyst from the oil as shown in Figure 2.6 for a typical batch hydrogenation reactor [56]. There are a great many different designs in use, but the essential features are as follows: the reactor is a vertical pressure vessel with a height several times its diameter, equipped with heating and cooling coils and an agitator. There is of course piping to introduce hydrogen gas and the catalyst, either dry or already slurried in oil. In addition, a vacuum generating system is required to evacuate air from the hydrogenation reactor before introducing hydrogen gas at the beginning of a hydrogenation run, and to evacuate hydrogen gas from the reactor at the end of a hydrogenation run. The usual instrumentation to indicate and control pressure and temperature is required. The vessel is generally constructed of mild steel for a working pressure of about 700 kPa above ambient pressure.

A key aspect of reactor design is the agitator. Chemical engineering principles for mixing of a reaction system consisting of a liquid (the oil), a solid (the nickel catalyst), and a gas (hydrogen) suggest the use of a flat plate turbine design. This produces high shear and vortexing for efficient contact of the three phases and was usually the chosen design. In the last few decades, a great many other agitator designs, mostly propeller type designs giving high circulation rates, have come into successful use. The aim of these has been to achieve lower usage of agitator drive energy, but they must maintain reactor performance in respect to reaction rate and selectivity and, therefore, also, *trans* isomer formation at levels already established for that production plant. This is important for achieving expected production rates and fat melting behaviour. The latter can be affected by even slight changes in the degree of isomerization.

ii. Hydrogenation process conditions and catalysts. Hydrogenation conditions in commercial practice refer to temperature, pressure, catalyst type and concentration and agitation intensity. The usual range of values is summarized in Table 2.1. Temperature and pressure are the two parameters, which are usually manipulated to achieve desired changes in selectivity and *trans* isomer formation within the range of variation that is achievable with nickel catalysts. In respect to catalyst type, nickel on various solid supports is used (or reused). The amounts of nickel used in commercial operations are in the order of 0.005-0.01% as nickel in oil, usually as a 20-25% concentration on a silica carrier.

Investigations over the years were aimed at improving nickel catalysts with respect to activity, selectivity, filterability and ease of handling. Past investigations into variations in catalyst production procedures and in the choice of catalyst supports have led to the belief that there is not much scope for large changes in selectivity, which has long been a prime concern of the processing industry, or in *trans* isomer formation, which is a more recent concern.

Agitation of the oil/catalyst/hydrogen gas system, as is apparent from the discussion on hydrogenation equipment above, is fixed by reactor and agitator design. Practical limitations (power consumption, mechanical stability of large agitator assemblies) result in reactors being under-agitated; that is, the rate of migration of hydrogen and ethylenic bonds to the catalyst surface controls the

Table 2.1
Typical commercially used hydrogenation process conditions.

Parameter		Selective	Non-selective
Temperature	°C	200	165
Pressure	kPa gauge	300	30
Catalyst conc.	% as Ni	0.005	0.008
Agitation, 6-blade turbine 2-4 sets, rev./min		80	80

reaction rate. This results in greater selectivity with respect to hydrogenation of multiple ethylenic bond moieties and greater *trans* isomerization. Selectivity and *trans* formation are reduced if agitation to the point of elimination of mass transfer of hydrogen and ethylenic bonds to the catalyst surface as the rate controlling step is eliminated. Usual agitator speeds in commercial practice are in the order of 80-100 revolutions/min. It has been shown in extensive plant scale hydrogenation experience that mixing intensities achievable with rotary agitators in plant scale batch reactors are not high enough to increase or decrease the selectivity (and *trans* isomer formation) of nickel-catalysed hydrogenations significantly. Variable speed agitators have therefore not found wide application in the industry.

4. Concentrations of Trans Isomer Fatty Acids in Commercially Hydrogenated Canola and Soybean Oils

Table 2.2 shows data by Bansal and deMan [57] that are typical of *trans* isomer concentrations in canola oils commercially hydrogenated down to different iodine values (IV) under "selective" and "non-selective" hydrogenation conditions.

It can be seen that with selective conditions (temperature 200-215°C, pressure 100-200kPa), *trans* isomer formation is promoted compared to non-selective conditions (temperature 165-180°C, pressure 300 kPa). However, even under the non-selective conditions, the *trans* isomer concentrations are relatively high, leading to high *trans* acid concentrations in edible oil products made from these oils. Compared to selective conditions the change in *trans* isomer concentration due to the change in temperature and pressure and to a slight increase in catalyst concentration results in a lowering of *trans* isomer content of the order of 24-33% at similar iodine values.

Table 2.3 gives typical *trans* isomer concentrations found in partially hydrogenated soybean oils produced commercially. Essentially selective conditions were used. The data show that *trans* isomer formation in soybean oil is somewhat lower than in hydrogenated canola oils. The higher levels of *trans* in hydrogenated canola oil are thought to be due to the presence of 2-5 mg/kg of sulphur compounds in the refined and bleached oil being hydrogenated. These compounds act as catalyst poisons with the effect of increasing *trans* isomerization during hydrogenation. Non-selective conditions

Table 2.2

Trans isomer content of selectively and non-selectively partially hydrogenated canola oils of certain iodine value (IV).

Hydrogenation conditions	Iodine value	Selective *trans* isomers%	Non-selective reduction in % *trans*
Selective	102.7	29.5	–
Selective	88.8	46.5	–
Selective	81.6	55.6	–
Non-selective	104.8	19.7	33
Non-selective	84.9	35.3	24
Non-selective	71.7	42.0	–

Adapted from Ref. [57].

Table 2.3

Trans isomer content of commercially hydrogenated soybean oils processed to certain IV under different temperature (T) and pressure (P) conditions.

Hydrogenation conditions (T, °C/P, kPa)	Iodine value	*Trans* isomers, %
215/35	109	16.4
215/270	96.7	27.7
215/230	86.0	29.2
215/270	76.1	37.7
215/270	65.2	44.5

with soybean oil result in lower *trans* levels than selective conditions (data not shown), similar to the experience with canola oil.

In Northern Europe, South America, Mexico and Japan, fish oils are used in edible oil products. Hydrogenation is even more essential to improve oxidative stability with the highly unsaturated fish oils than with the common vegetable oils, and also to harden the oil for proper functionality in margarines and baking fats. Table 2.4 lists the *trans* isomer contents of three hardened oils of marine origin from Japan [58]. Usually, the iodine value of these oils must be reduced considerably to achieve adequate stability. This reduction is in effect mostly of the *cis*-methylene interrupted polyunsaturated fatty acids, primarily 20:5n-3 and 22:6n-3. The three fish oils of Table 2.4 will also contain substantial proportions of 20:1 and 22:1 fatty acids [59,60]. These are, in most instances, simply converted to *trans* forms. In consequence, the concentration of *trans* isomers in these oils tends to be relatively high, 40-50%, that is, at the same level as highly hydrogenated canola and soybean oils intended for stick margarines. Menhaden and sardine oils initially have relatively little 20:1 and 22:1 and the progress of this process has been described for menhaden oil in considerable detail by Sebedio and Ackman [61]. Such products have recently been granted GRAS (generally regarded as safe) status in the USA.

Table 2.4

Trans isomer content of fish oils hydrogenated under similar or different
temperature (T) and pressure (P) conditions

Oil	Hydrogenation conditions (T, °C/P, kPa)	Iodine value	Trans isomers, %
Cod liver			
	180/200	65.2	44.1
	215/200	67.2	47.6
Mackerel			
	180/200	75.6	48.8
Capelin			
	180/200	58.2	41.6
Adapted from Ref. [58].			

5. Possibilities of Suppressing Trans Isomer Formation in Hydrogenation

Two approaches with potential commercial significance have been investigated over the years, i.e. i. Catalysts other than nickel, and ii. Changes in nickel catalyst performance coupled with changes in hydrogenation conditions.

i. Catalysts other than nickel. With the concern about *trans* isomer fatty acids since the 1970's, deliberate attempts were made to find catalyst systems that would result in negligible, or, at least, greatly reduced *trans* isomerization. The investigations were concerned with

(a) the use of metals other than nickel and the use of a variety of solid supports to produce heterogeneous catalysts, and

(b) homogeneous catalysts and, also mixtures of homogeneous and heterogeneous catalysts.

The literature on catalysts other than nickel for edible oil hydrogenation is quite extensive. A representative selection of examples is reviewed briefly for an appreciation of the results and problems associated with these catalyst systems.

Among heterogeneous catalysts, a copper-ammonia complex on alumina as the support in the hydrogenation of linolenate was described by Koritala in 1970 [62]. This work was aimed at improved selectivity for polyunsaturates, rather than suppressing *trans* isomer formation. Experience on a laboratory scale showed that selectivity with respect to polyunsaturates was excellent. No saturated fatty acids were formed and *trans* isomerization was very low when linolenate in soybean oil was reduced from 7-9% before hydrogenation to 1-3%. Low *trans* isomer formation and no increase in saturates resulted in an oil with very little solid fat at 0°C, which was very desirable.

Other copper-based catalysts were developed and tested by Koritala [62], Koritala *et al.* [63], Kirschner and Lowrey [64] and a number of additional investigators. In the 70's and 80's, the North American edible oil industry did use a proprietary copper chromite catalyst to produce a more flavour-stable soybean oil (α-linolenic acid reduced to < 3%) containing only small amounts of high melting components. The same temperatures, but higher pressures than those used in selective hydrogenation with nickel were applied with this type of catalyst. Eventually, the low activity of the copper catalysts and, especially, the danger of leaving traces of copper in the oil, decreasing stability, eliminated the use of copper catalyst. For these reasons, copper catalysts are unlikely to find favour as low *trans*-producing catalysts

Chromium on silica and precipitated chromium, unsupported, as hydrogenation catalysts were described by Bernstein *et al.* [65]. Some success was achieved in avoiding *trans* isomerization, but the catalysts on silica were very difficult to prepare. All had impractically low activity even at hydrogen pressures of 1700 kPa.

Palladium black and palladium supported on carbon, alumina and on barium sulphate was used in the hydrogenation of canola and soybean oil by Hsu *et al.* The products were analysed to determine isomerization characteristics [66]. When high pressures (5000 kPa) and low temperatures (70°C) were used, *trans* isomerization was reduced by 50% with the palladium-on-alumina catalyst compared to commercial nickel catalyst. The high pressures would require re-equipping commercial plants with new and much more expensive reactors. Catalyst losses would have to be very carefully controlled to avoid exorbitant catalyst costs. No commercial application in edible oils is known. The reduction in *trans* isomerization that can be achieved would probably not be considered commensurate with the extra cost.

Platinum, palladium, rhodium, ruthenium, iridium and osmium and mixed metal catalysts involving platinum and the above metals were investigated by Rylander [67]. The most noteworthy findings with respect to isomerization were that some mixed metal compositions reduced *trans* isomer formation by the order of 50 %. This is about the same as was found by Hsu *et al.* [66] with palladium. Hydrogen pressures required to be effective were found to be high. Van der Plank *et al.* of Unilever [68-70] impregnated a resin with palladium chloride. This resulted in a very selective catalyst which produced little positional but substantial geometric isomerization. Gold as a catalyst supported on alumina or silica in the hydrogenation of canola oil was investigated by Caceres *et al.* [71]. Somewhat reduced *trans* isomerization was obtained, but only at pressures in the order of 3500 kPa.

Homogeneous catalysts have so far not found use in the edible oil industry for hydrogenation. They are very difficult to remove compared to heterogeneous catalysts. Nevertheless, in the search for catalysts that do not form *trans* isomers, a number of researchers have investigated oil-soluble metal complexes, and also, mixed heterogeneous/homogeneous catalysts.

Bello *et al.* [72] investigated a ruthenium complex (dichlorodicarbonylbis(triphenylphosphine) ruthenium (II)) to determine isomerization in the hydrogenation of canola oil. Compared to commercial nickel catalyst, the ruthenium-complex showed minimal *trans* isomer formation and had very good activity and acceptable selectivity. The main disadvantages were the poor ability to remove the catalyst by filtration and the need to use relatively high pressures (5000 kPa).

Diosady *et al.* [73] and Koseoglu [74] investigated hydrogenation of canola oil with a methylbenzoate-chromium-carbonyl complex. *Trans* isomer contents were in the range of 2.0-6.5% when linolenate was reduced from 9-11% before hydrogenation to 0-4.3% after. This compares with 20-35% *trans* isomers formed with commercial nickel catalysts. There was excellent selectivity. Pressures required were in the range of 2,500-5,400 kPa. The catalyst was difficult to remove from the oil. Similar results were obtained by these two authors with a benzene-chromium-carbonyl and a toluene-chromium-carbonyl complex.

Heterogeneous commercial nickel catalyst and homogeneous methyl benzoate-chromium-carbonyl complexes were used together in an attempt to combine the advantages of the high activity of the nickel catalysts with the superior selectivity and the low *trans* isomer formation of the methyl benzoate-chromium-carbonyl complexes [74]. It was possible to obtain a hydrogenated canola oil of 3% linolenate with about 13% *trans* content instead of the 20-30% obtained with commercial nickel. The chromium-carbonyl complex had to constitute at least 10% of the mixed catalyst, which would present a catalyst-removal problem.

It is apparent from this discussion that changes from heterogeneous, nickel-based catalysts to other catalytic metals, and to homogeneous catalysts or mixtures of these two types of catalysts are not promising, or are not feasible at present for technical or economic reasons.

ii. Changes in nickel catalysts and in hydrogenation conditions. Recently, the catalyst industry has renewed earlier attempts at modifying heterogeneous nickel catalysts and begun to investigate the performance of such catalysts in connection with greater changes in temperature, gas pressure and catalyst concentration than have commonly been used, but that can still be considered as being within the range of economic feasibility.

The modified nickel catalysts are termed "flexible". They have the property of resisting poisoning very well, which is typical of the non-selective nickel catalysts used in the hydrogenation of fatty acids for non-food uses, from which they are derived. These "flexible" catalysts display good selectivity with respect to multiple ethylenic bonds even under non-selective conditions, without a high formation of *trans* isomers. This is a new feature. A high degree of selectivity is very important in commercial hydrogenation.

Hasman published data in 1995 from the laboratory-scale hydrogenation of canola oil with such a "flexible" catalyst when used at various temperatures, pressures, and concentrations [75]. These data are summarized in Table 2.5 together with data obtained with a commonly used and selective catalyst under normally used selective hydrogenation conditions and catalyst concentration. The data in Table 2.5 show that the "flexible" catalyst under these normally

Table 2.5

Fatty acid composition of *trans*-suppressed canola oils at 90 and 60 IV, nickel catalyst.

Hydrogenation conditions	catalyst	% Reduction in *trans*	Fatty Acids (w/w%)				
			trans	C18:0	C18:1	C18:2	C18:3
90 IV							
normal[a]	selective	control	30.2	2.5	81.1	10.4	0.7
normal[b]	flexible	30	21.2	3.0	78.5	11.9	0.7
1[c]	flexible	50	15.2	3.3	79.0	11.6	0.9
2[d]	flexible	60	12.2	7.4	68.2	13.4	3.0
3[e]	flexible	70	9.0	10.3	64.1	14.7	3.6
60 IV							
normal	selective	control	52.5	26.9	69.0	0	0
normal	flexible	23	40.4	17.0	84.1(?)	0	0
1	flexible	33	35.0	25.6	68.0	0.4	0
2	flexible	54	24.0	26.0	63.2	3.2	0.1
3	flexible	66	17.5	28.7	58.4	5.2	0.2

[a] normal condition, selective catalyst: 204°C, 103 kPa, 0.05 % Ni as G-95D catalyst
[b] normal condition, flexible catalyst: 204°C, 182 kPa, 0.0568 % Ni as G-135A catalyst
[c] condition 1: 160°C, 310 kPa, 0.0909% Ni as G-135A catalyst
[d] condition 2: 104°C, 620 kPa, 0.1818% Ni as G-135A catalyst
[e] condition 3: 77 °C, 1717 kPa, 0.500% Ni as G-135A catalyst

Adapted from Ref. [75]; G-95D and G-135A are trade names of United Catalyst Inc. of Louisville Kentucky, USA.

used selective hydrogenation conditions produced a hydrogenated canola oil at 90 IV with a 30% lower *trans* isomer content, without a loss in selectivity.

When the temperature was lowered from 204°C to 160°C and the pressure raised from 103 kPa to 310 kPa, and flexible catalyst concentration increased from 0.06% nickel to 0.09 % (condition 1 in the table), the *trans* isomer content of the resulting 90 IV oil was reduced by 50%, still without a loss in selectivity. These are the process conditions used in industry for "non-selective" hydrogenation, except for the use of the "flexible" catalyst and a somewhat higher catalyst concentration.

Further lowering of temperatures to 104°C and 77°C and further increases in pressure to 620 kPa and 1717 kPa and nickel concentrations to 0.18% and 0.50% (conditions 2 and 3 in the table) produced 60% and 70% lower *trans* isomer concentrations at 90 IV. Unfortunately, at these conditions a noticeable reduction in selectivity also occurred, as shown by the increased formation of stearic acid, which rose from about 3% to 7.4 and 10.3% respectively.

When hydrogenation was carried out to 60 IV, there was still a reduction of 23% in *trans* isomer concentration due to the "flexible" catalyst alone and reductions from 33-66% when using reduced temperatures, increased pressures and increased catalyst concentrations. A reduction of 33% in *trans* isomers using a temperature, pressure and catalyst concentration combination that would normally be applied in conventional, non-selective hydrogenation

Table 2.6

Fatty acid composition of *trans*-suppressed soybean oils at 100 and 70 IV, nickel catalyst.

Hydrogenation conditions	catalyst	% Reduction in *trans*	Fatty Acids (w/w%)				
			trans	C18:0	C18:1	C18:2	C18:3
100 IV							
normal[a]	selective	control	29	4	61	25	4
normal[b]	flexible	negligible	27	4	63	25	0
1[c]	flexible	negligible	28	4	61	26	0
2[d]	flexible	45	16	9	44	32	4
3[e]	flexible	59	12	10	40	34	4
70 IV							
normal[a]	selective	control	44	14	76	0	0
normal[b]	flexible	negligible	46	14	76	2	0
1[c]	flexible	36	28	15	69	5	0
2[d]	flexible	39	27	16	62	10	0
3[e]	flexible	45	20	20	54	13	1

[a] normal condition, selective catalyst: 204°C, 103 kPa, 0.005% Ni as G-95D catalyst
[b] normal condition, flexible catalyst: 204°C, 103 kPa, 0.005% Ni as G-135A catalyst
[c] condition 1: 204°C, 310 kPa, 0.005% Ni as G-95A catalyst
[d] condition 2: 104°C, 620 kPa, 0.04% Ni as G-135A catalyst
[e] condition 3: 77°C, 1717 kPa, 0.11% Ni as G-135 A catalyst

Adapted from Ref. [75]; G-95D and G-135A are trade names of United Catalysts Inc. of Louisville, Kentucky, USA.

(condition 1) was not significantly different from that obtained with conventional nickel-catalysts. Using temperatures in the range of 77-104°C and higher pressures in the range of 620-1717 kPa gave reductions in *trans* isomers of 54-66%, while good selectivity was maintained. Hydrogenation reactors that accommodate pressures in the order of 600-2000 kPa are still relatively economical to construct. In Table 2.6, results are listed from similar tests with soybean oil by Hasman [75].

Essentially no reductions in *trans* isomer content were obtained from the use of the flexible catalyst alone. Lowering temperatures and raising pressures and catalyst concentration to the values usually used for non-selective conditions had no effect at IV 100. Further changes produced reductions in *trans* isomer concentrations of 45-59% at IV 100. At 70 IV, lower temperatures and higher pressures resulted in *trans* reductions of 36-45%. It appears that the *trans* isomer reductions achievable with soybean oil are smaller than with canola oil under otherwise similar conditions.

Indications from this work are that it will be possible for industry to achieve significant reductions in *trans* isomer content of some hydrogenated oil products with modified nickel catalysts and using alternative hydrogenation conditions that can be accommodated in existing equipment. This approach can probably be further developed with time. The possible reductions in *trans* isomer content indicated in Table 2.5 and Table 2.6 should eventually result in a significant lowering of the *trans* isomer content of dietary fat products available in North America.

C. CONCLUSIONS

The human animal has evolved mechanisms to digest, absorb and catabolize a wide variety of foods. Radical changes in diet took place with the introduction of agriculture thousands of years ago, more recently with the movement to cities in the early stages of the industrial revolution, again with the introduction of modern rail and ship *trans*portation systems, and finally with freezer technology and manufactured foods. A moderate level of *trans* acids can be reached by modified hydrogenation processes. Some natural sources of *trans* acids will remain, even if unsuspected by the consumer in search of allegedly natural foods [35], but can be combined with a good supply of all-*cis* essential fatty acids as part of the comfortable life-style we have achieved.

ABBREVIATIONS

FAME, fatty acid methyl esters; GLC, gas liquid chromatography; TLC, thin-layer chromatography; IV, iodine value.

REFERENCES

1.   Fritsche,J. and Steinhart,H., in *New Techniques and Applications in Lipid Analysis*, pp. 234-255 (1997) (edited by R.E. McDonald and M.M. Mossoba, AOCS Press, Champaign, IL).
2.   Sleeter,R.T. and Matlock,M.G., *J. Am. Oil Chem. Soc.*, **66**, 121-127 (1989).
3.   Ulberth,F. and Haider,H-J., *J. Food Sci.*, **57**, 1444-1447 (1994).
4.   Kramer,J.K.G., Sauer,F.D. and Pigden,W.J. (eds) *High and Low Erucic Acid Rapeseed Oils*, 582 pp. (1983) (Academic Press, Toronto).
5.   Shahidi,F. (ed.), *Canola and Rapeseed: Production, Chemistry, Nutrition and Processing Technology*, 355 pp. (1990) (Van Nostrand Reinhold Company Inc., New York).
6.   Ackman,R.G., Hooper,S.N. and Hooper,D.L., *J. Am. Oil Chem. Soc.*, **51**, 42-49 (1974).
7.   Mag,T.K., in *Canola and Rapeseed: Production, Chemistry, Nutrition and Processing Technology*, pp. 251-276 (1990) (edited by F. Shahidi, Van Nostrand Reinhold Company Inc., New York).
8.   Emken,E.A., *J. Am. Oil Chem. Soc.* **60**, 995-1004 (1983).
9.   Sebedio,J.-L. and Ackman,R.G., *J. Am. Oil Chem. Soc.*, **56**, 15-21 (1979).
10. Sebedio,J.-L., Grandgirard,A. and Prevost,J., *J. Am. Oil. Chem. Soc.*, **65**, 362-366 (1988).
11. Pudel,F. and Denecke,P., *Oléag. Corps Gras Lipides*, **4**, 58-61 (1997).
12. Teasdale,B.F. and Mag,T.K., in *High and Low Erucic Acid Rapeseed Oils*, pp. 197-229 (1983) (edited by J.K.G. Kramer, F.D. Sauer, and W.J. Pigden, Academic Press, Toronto)
13. Billek,G., *Fat Sci. Technol.*, **94**, 161-172 (1992).
14. Wolff,R.L., *J. Chromatogr. Sci.*, **30**, 17-22 (1992).
15. Wolff,R.L., *J. Am. Oil Chem. Soc.*, **69**, 106-110 (1992).
16. Wolff,R.L., *Oléag. Corps Gras Lipides*, **2**, 391-400 (1995).
17. Sebedio,J.L., *Chromatography and Analysis*, October 1991 pp. 9-11.
18. Chardigny,J.-M., Wolff,R.L., Mager,E., Bayard,C.C., Sébédio,J.-L., Martine,L. and Ratnayake,W.M.N., *J. Am. Oil Chem. Soc.*, **73**, 1595-1601 (1996).
19. Wolff,R.L. and Sébédio,J.-L., *J. Am. Oil Chem. Soc.*, **71**, 117-126 (1994).
20. Wolff,R.L., *J. Am. Oil Chem. Soc.*, **71**, 907-909 (1994).
21. Bourre,J.-M., Dumont,O. and Clément,M., *Oléag. Corps Gras Lipides*, **2**, 254-263 (1995).

22. Renaud,S., *Oléag. Corps Gras Lipides*, **3**, 169-172 (1996).
23. Mendy,F., *Oléag. Corps Gras Lipides*, **2**, 36-45 (1995).
24. McLennan,P.L., *Am. J. Clin. Nutr.*, **57**, 207-212 (1993).
25. Hunter,J.E., *Am. J. Clin. Nutr.*, **51**, 809-814 (1990).
26. O'Keefe,S.F., Wiley V. and Gaskins,S., *Food Res. Internat.*, **27**, 7-13 (1994).
27. O'Keefe,S., Gaskins-Wright,S., Wiley,V. and Chen,I.-C., *J. Food Lipids*, **1**, 165-176 (1994).
28. Amelio,M., Amelotti,G., Cozzoli,O., Faraone,A., Mariani,C., Mattei,A., Marzo,S., Morchio,G., Serani,A., Spinetti,M. and Tiscornia,E., *Riv. Ital. Sostanze Grasse*, **70**, 561-566 (1993).
29. Ackman,R.G., in *Fish, Fish Oil and Human Health*, pp. 14-24 (1992) (edited by J.C. Frölich and C. von Schacky, Klinische Pharmakologie, Clinical Pharmacology, Vol. 5, W. Zuckschwerdt Verlag).
30. Sebedio,J.L., Catte,M., Boudier, M.A., Prevost,J. and Grandgirard,A., *Food Res. Internat.*, **29**, 109-116 (1996).
31. Hunter,J.E. and Applewhite,T.H., *J. Am. Oil Chem. Soc.*, **70**, 613-617 (1993).
32. Enig,M.G., Atal,S., Keeney,M. and Sampugna,J., *J. Am. Coll. Nutr.*, **9**, 471-486 (1990).
33. Hunter,J.E. and Applewhite,T.H., *Am. J. Clin. Nutr.*, **54**, 363-369 (1991).
34. de Lorgeril,M., Renaud,S., Mamelle,N., Salen,P., Martin,J.-L., Monjaud,I. Guidollet,J., Touboul,P. and Delaye,J., *Lancet*, **343**, 1454-1459 (1994).
35. Brühl,L., *Fett/Lipid*, **98**, 380-383 (1996).
36. Martini,D. and Iacazio,G., *Oléag. Corps Gras Lipides*, **2**, 374-385 (1995).
37. Wolff,R.L., *Sciences des Aliments* **13**, 155-163 (1993).
38. Tirtiaux,A. and Gibon,V., *Oléag. Corps Gras Lipides*, **4**, 45-51 (1997).
39. Wijesundera,R.C. and Ackman,R.G., *J. Chromatogr. Sci.*, **27**, 399-404 (1989).
40. Wijesundera,R.C., Ratnayake,W.M.N. and Ackman,R.G., *J. Am. Oil Chem. Soc.*, **66**, 1822-1830 (1989).
41. Juanéda,P., Sebedio,J.L. and Christie,W.W., *J. High Resolut. Chromatogr.*, **17**, 321-324 (1994).
42. Belury,M.A., *Nutr. Rev.* **53**, 83-89 (1995).
43. Entressangles,B., *Oléag. Corps Gras Lipides*, **2**, 162-169 (1995).
44. Chardigny,J.-M., Blond,J-P., Bretillon,L., Mager,E., Poullain,D., Martine,L., Vatèle,J.-M., Noël,J.-P. and Sébédio, J.-L., *Lipids*, **32**, 731-735 (1997).
45. ASCN/AIN Task Force on *Trans* Fatty Acids. *Am. J. Clin. Nutr.* **63**, 663-670 (1996).
46. Katan,M.B., *Am. J. Clin. Nutr.*, **62**, 518-519 (1995).
47. Leveille,G.A., *Am. J. Clin. Nutr.*, **62**, 520-521 (1995).
48. Newtel,P.J., *Am. J. Clin. Nutr.*, **62**, 522-523 (1995).
49. Willett,W.C. and Ascherio,A., *Am. J. Clin. Nutr.*, **62**, 524-526 (1995).
50. Wolff,R.L., *J. Am. Oil Chem. Soc.*, **71**, 277-283 (1994).
51. Wolff,R.L., *Oléag. Corps Gras Lipides*, **1**, 209-218 (1994).
52. Normann,W., Brit. Pat. 1,515 (1903).
53. Allen,R.R. and Kiess,A.A., *J. Am. Oil Chem. Soc.*, **32**, 400-405 (1955).
54. Scholfield,E.R., in *Geometrical and Positional Fatty Acid Isomers, Analysis and Physical Properties of Isomeric Fatty Acids*, p. 41 (1979) (edited by E.M. Emken and H.J. Dutton, American Oil Chemists' Society, Champaign, IL).
55. Mounts,T.L., in *Handbook of Soy Oil Processing and Utilization*, pp. 131-144 (1980) (edited by D.R. Erickson, E.H. Pryde, O.L. Brekke, T.L. Mounts and R.A. Falb, American Soybean Association, St. Louis and American Oil Chemists' Society, Champaign).
56. Allen,R.R., *J. Am. Oil Chem. Soc.*, **55**, 792-795 (1978).
57. Bansal,J.D. and deMan,J.M., *J. Food Science*, **47**, 2004-2007, 2014 (1982).
58. Suzuki,K., and Murase,Y., *Yukaguku*, **44**, 503-508 (1985).
59. Ackman,R.G., *Chem. Ind. (London)*, 139-145 (1988).
60. Sebedio,J.-L., Langman,M.F., Eaton,C.A. and Ackman,R.G., *J. Am. Oil Chem. Soc.*, **58**, 41-48 (1981).
61. Sebedio,J.-L. and Ackman,R.G., *J. Am. Oil Chem. Soc.* **60**, 1986-1991 (1983).
62. Koritala,S., *J. Am. Oil Chem. Soc.* **47**, 106-107, (1970).
63. Koritala,S.J., Butterfield,R.O. and Dutton,H.J., *J. Am. Oil Chem. Soc.* **47**, 266-268 (1970).
64. Kirschner,E. and Lowrey,E.R., *J. Am. Oil Chem. Soc.* **47**, 467-469 (1970).
65. Bernstein,P.A., Graydon,W.F. and Rubin,L.J., *J. Am. Oil Chem. Soc.*, **66**, 680-684 (1989).
66. Hsu,N., Diosady,L.L. and Rubin,L.J., *J. Am. Oil Chem. Soc.*, **65**, 349-356 (1988).
67. Rylander,P.N., *J. Am. Oil Chem. Soc.* **47**, 482-486 (1970).

68. van der Plank,P., van Oosten,H.J. and van Dijk,L., *J. Am. Oil Chem. Soc.*, **56**, 45-49, (1979).
69. van der Plank,P. and van Oosten,H.J., *J. Am. Oil Chem. Soc.*, **56**, 50-53, (1979).
70. van der Plank,P. and van Oosten,H.J., *J. Am. Oil Chem. Soc.*, **56**, 54-57, (1979).
71. Caceres,L, Diosady,L.L. Graydon,W.F. and Rubin,L.J., *J. Am. Oil Chem. Soc.*, **62**, 906-910 (1985).
72. Bello,C., Graydon,W.F. and Rubin, L.J., *J. Am. Oil Chem. Soc.*, **62**, 1587-1592 (1985).
73. Diosady,L.L., Graydon,W.F., Koseoglu,S.S. and Rubin,L.J., *J. Can. Inst. Food Sci. Technol.*, **17**, 218-223 (1984).
74. Koseoglu,S.S., Ph. D. Thesis, Univ. of Toronto, Dept. of Chem. Eng., Food Eng. Group (1984).
75. Hasman,J.M., *INFORM*, **6**, 1206-1213, (1995).

CHAPTER 3

WORLDWIDE CONSUMPTION OF TRANS-*FATTY ACIDS*

Margaret C. Craig-Schmidt
Department of Nutrition and Food Science, Auburn University, Auburn, AL 36849, USA

A. Introduction
B. Methods Used to Estimate *Trans*-Fatty Acid Consumption
    1. Estimates based on "food disappearance" or market share data
    2. Analysis of dietary consumption data of a representative population
    3. Laboratory analysis of duplicate portion or composite diets
    4. Estimates based on the *trans*-fatty acid content of biological tissues
C. Estimates of *Trans*-Fatty Acids in the Diet of Various Populations
    1. North America
    2. United Kingdom
    3. "Continental" Europe
    4. Eastern European/Former Soviet Countries: Croatia, Serbia, Russia and Poland
    5. Nordic Countries
    6. Israel
    7. India
    8. Australia and New Zealand
    9. Asian Pacific
D. Future Trends

A. INTRODUCTION

Trans-fatty acids have two primary sources in the diet: biohydrogenation in the rumen of animals and commercial hydrogenation of vegetable and marine oils [30]. As a result of biohydrogenation, meat and dairy products from ruminant animals contain 1 to 8% of the fat as *trans*-fatty acids. Commercial hydrogenation results in products such as margarine, shortening and frying fats that contain variable amounts of *trans*-fatty acids up to 40 to 50% of total fatty acids. Bakery goods, snack foods and other commercial products made with industrially hydrogenated fat, as well as deep-fat fried fast foods, are all major sources of *trans*-fatty acids in the diet. Thus, populations consuming relatively large amounts of commercially hydrogenated fat tend to have a greater content of *trans*-fatty acids in the diet than do populations whose dietary fat is primarily butter and other ruminant-derived fats or olive oil and other non-hydrogenated vegetable oils.

The purpose of this review is to summarize data on the world-wide consumption of *trans*-fatty acids. Earlier reviews or government reports which summarize estimates of *trans*-fatty acids in the diet have focused for the most part on data for individual countries or on regions of the world. These reviews include a report by the Life Sciences Research Office for the U.S. Food and Drug Administration compiled by Senti [103], the International Life Sciences Institute Expert Panel on *Trans*-Fatty Acids and Coronary Heart Disease [61], the American Society for Clinical Nutrition/American Institute of Nutrition Task Force on *Trans*-Fatty Acids [4], the International Life Sciences Institute Expert Panel on *Trans*-Fatty Acids and Early Development [62], two reports by the British Nutrition Foundation [18,19], the Danish Nutrition Council Report by Stender *et al.* [107,108], the FAO/ WHO Report of the Joint Expert Consultation on Fats and Human Nutrition [41] and reviews by Emken [33], Enig [34,35], Gurr [46,47,48], Sanders [100], Wahle and James [124], Schaafsma [101], Mensink and Katan [84], Precht and Molkentin [93], and Becker [12].

B. METHODS USED TO ESTIMATE *TRANS*-FATTY ACID CONSUMPTION

Several approaches have been taken in estimating *trans*-fatty acids in the diet: (1) estimates based on "food disappearance" or market share data, (2) analysis of dietary consumption data of a representative population, (3) laboratory analysis of duplicate portion or composite diets, and (4) estimates based on the *trans*-fatty acid content of biological tissues. Each method has inherent advantages and disadvantages, but taken together, they can give reasonable, comparative estimates of *trans*-fatty acid consumption throughout the world.

1. Estimates Based on "Food Disappearance" or Market Share Data

Government statistics or in some cases, industry records indicate the amounts of commodities that are available for human consumption in a given year. Using these data, plus the population of the country, one can calculate the amount of margarine "consumed" in grams per day per person. Estimates based on these food disappearance data or market share data have the advantage in that they account for the availability of fat for the total population. However, the estimates are dependant on pairing the market share data for a given commodity (*e.g.* margarine) with the "average" or "typical" *trans*-fatty acid content for that commodity. Because values for the *trans*-fatty acid content of partially hydrogenated oils vary widely among brand names as well as with changes in processing methods over time, the values for the *trans*-fatty acid content of each commodity is at best a rough estimate. Estimates based on market share data must also be corrected for wastage [59], and factors used for wastage are also approximations.

Senti [103], in the report prepared for the United States Food and Drug Administration, estimated the average consumption of *trans*-fatty acids in the United States in 1983 to be 8.3 grams per person daily (Table 3.1). This estimate was based on United States Department of Agriculture production and sales data for both animal and vegetable fats and the laboratory analyses of Enig *et al.* [39] and Slover *et al.* [105] for average values of *trans*-fatty acids in the various commodity groups. Assumptions or auxiliary data were necessary to factor in values for the proportion of stick and tub margarine available in 1983, the wastage value for fat from retail cuts, the wastage value for discard or frying fats and the proportion of salad oils hydrogenated at the time. The value of 8.3 grams per person daily was similar to the value of 7.65 grams per person daily determined by Hunter and Applewhite [57] who used a similar approach. The value determined by Senti was, however, much less than the 13.3 g/person/day value reported by Enig *et al.* [36], who assumed a greater value for the *trans*-fatty acid content of shortening and commercial frying fats and made different assumptions with respect to fat wastage. Thus, estimates of *trans*-fatty acid consumption based on food disappearance or market share data can vary widely depending on the factors used to account for wastage and the values assumed for the *trans*-fatty acid content of various fats. Because these estimates represent availability in the food supply rather than individual consumption, they tend to overestimate *trans*-fatty acid consumption.

2. Analysis of Dietary Consumption Data of a Representative Population

Methods to assess the diets of individuals in a population include diet recalls, diet records, and food frequency questionnaires [78]. These data can then be combined with data from food composition tables to give the average daily consumption of a nutrient or other component in the diet. One advantage of these methods is that *consumption* by the individual rather than *availability* to the individual is assessed. Diet recalls and diet records document what is eaten usually during one to seven days, with three-day diet recalls or records being commonly used. Food frequency methods assess the usual diets of individuals over several months. All of these methods are subject to errors in the ability of the individual to recall or record accurately the amounts and kinds of foods which have been consumed. This dietary methodology is also dependant on accurate values for the content of the component in each food item consumed. For *trans*-fatty acid values of food, the data bases are limited and subject to inherent inaccuracy because of changes in food processing, the wide variability in *trans*-fatty acid content of similar types of foods, and

TABLE 3.1

Estimated per capita daily consumption of *trans*-fatty acids in the United States in 1983 using government commodity production and sales data[a].

| | | *Trans* fatty acids | |
Commodity	Fat intake[b] (g/day)	Concentration (%)	Intake (g/day)
Butter	5.1	3.4 [c]	0.17
Milk products	18.3	3.4 [c]	0.62
Meat + edible fat (beef)	18.1	5.8 [d]	1.42
Total animal and dairy fats			2.21
Margarine — hard	6.37	23.9 [e]	1.52
— soft	2.59	16.2 [e]	0.42
Shortening and oils	18.3	16.3 [f]	2.98
Salad oil	30.8	10.0 [f]	3.08
Total vegetable fats			8.00
Total dietary *trans*-fatty acids			**10.21**[g]
			(8.3)[h]

[a] Adapted from Senti, 1985 [103].
[b] Based on data from Human Nutrition Information Service, U. S. Department of Agriculture (USDA) sent to Senti as a personal communication.
[c] Based on data of Enig et al., 1983 [39].
[d] Based on data of Slover et al., 1985 [105].
[e] Based on data of Slover et al., 1985 [105] and Enig et al., 1983 [39]. Quantities of stick and tub margarine available in 1983 were based on information provided by the National Association of Margarine Manufacturers.
[f] Based on data of Enig et al., 1983 [39] as well as the assumption that only the soybean oil component was hydrogenated and that soybean oil constituted 83% of all salad and cooking oils.
[g] Value not corrected for wastage.
[h] Value which takes into account a wastage factor for frying oils (11 g per capita daily) and discard of fat from beef retail cuts (assumed to be 75% of the separable fat).

differences in the analytical methods used to determine the *trans*-fatty acid content of each food item. In general, estimates of dietary *trans*-fatty acids tend to be underestimated by this method.

The semi-quantitative food frequency questionnaire has been used by Willett and colleagues to assess the *trans*-fatty acid content of groups within the United States population [6,60,80,116,126]. In a cross-sectional study by Troisi *et al.* [116], the "Willett" questionnaire, consisting of 116 food items with serving sizes, was administered to 748 United States men aged 43 to 85 years. Fatty acid intake was computed by multiplying the frequency of consumption of each food item by the fatty acid composition for the portion size specified and summing across all foods. Calculations took into account the type of margarine (stick, tub or diet) and the types of fats or oils used for frying, cooking and baking by each individual. The *trans*-fatty acid content of food items used in this study was based primarily on United States Department of Agriculture tables [28], and values used for various types of foods are shown in Table 3.2. Mean energy-adjusted intake of *trans*-fatty acids

TABLE 3.2

Trans-fatty acid values for selected foods used in combination with a
semi-quantitative food frequency questionnaire to calculate *trans*-fatty acid
consumption in U.S. men [116]

Food	*Trans*-fatty acids
	g/serving
Margarine (5 g)[1]	
Stick	0.8
Diet	0.6
Tub	0.3
Beef (124-186 g)	1.3
Fried potatoes (124 g)	2.69
Oil and vinegar dressing (15mL)	0.8
Ready-made cookie (one)	0.6
Ready-made sweet roll, coffee	
cake, pastry (one)	1.6
Hamburger patty (124 g)	0.8
Mayonnaise (15 mL)	0.5
Cheese (31 g)	0.3
Beef sandwich (93 g)	0.55

[1] United States Department of Agriculture value for margarines made
with soybean oil [28]

for the entire sample (n = 748) was 3.4 ± 1.2 g/day (± sd). *Trans*-fatty acid
intake constituted 5.5% of total fat and 1.6% of total energy. Use of the food
frequency questionnaire method has the advantage of allowing the diet of a
large number of subjects to be assessed but generally suffers from incomplete
or inaccurate *trans*-fatty acid values in the food composition database used.
The values for *trans*-fatty acid consumption obtained by Willett and
colleagues using the food frequency questionnaire have been less than those
obtained from availability data for the United States [36,38,57,58,103].

The assessment of *trans*-fatty acid intake on the basis of individual
consumption data has the advantage of allowing one to determine the range
consumed by a single individual or among individuals in a group. Even in a
selected, highly homogeneous population, the individual variation in *trans*-
fatty acid consumption is great. For example, van den Reek *et al.* [118] used
7-day diet records to assess the diets of eight, apparently healthy, U.S.
adolescent Caucasian girls, aged 12 to 15 years. The mean 18:1t content of
the diet was 5.3% of total fatty acids, with a range of 3.0 to 9.1%. Also, the
variation was wide for a single subject. For example, over the 7-day period,
the amount of total *trans*-fatty acids in the daily diet of one subject ranged
from 2 to 12% of total fatty acids [119]. In the United Kingdom, *trans*-fatty
acid consumption among subjects ranged from less than 1 g/day to over 20
g/day [19]. However, the distribution was skewed toward the lower end of the
spectrum with most of the daily intake values ranging from 2 to 7 g of *trans*-
fatty acids.

3. Laboratory Analysis of Duplicate Portion or Composite Diets

In the duplicate portion technique, diets identical to those consumed are collected and analysed in the laboratory. The advantage of this method is that the *trans*-fatty acid content of the diets can be determined very accurately, especially if analytical methodology, such as that recently employed by Ratnayake and colleagues is used [24,25,95]. On the other hand, laboratory analysis is very expensive, and therefore, only a limited number of diets can be analysed. Commonly, the duplicate portion method is applied to a small, highly select sample that is not representative of the entire population. It is therefore important to interpret estimates obtained by the duplicate portion technique within the context of the population from which the diets were collected.

An example of *trans*-fatty acid consumption values obtained by the duplicate portion technique is the study of Åkesson *et al.* [3]. These investigators determined intake of *trans*-octadecenoic acid (18:1t) by analysis of duplicate portions of diets collected from various groups in Sweden (Table 3.3). Using this method, it is possible to distinguish between groups with different dietary habits, and it is clear that vegans consume much less *trans*-fatty acids than do omnivores (0.9 g/day vs. 2.8-4.9 g/day). Thus, the *trans*-fatty acid content of the diets of a small segment of a population can be determined with a high degree of accuracy by the duplicate portion technique. However, only as the group is reflective of the population as a whole can these values be deemed to represent the diet of that country.

Alternately, "typical" composite diets are formulated based on other diet assessment methods and the *trans*-fatty acid content of the diets determined by analytical methods. For example, the *trans*-fatty acid content of the diets of a number of countries has been determined using diet composites in the Seven Countries Study [74]. Food tables with detailed information on the fatty acid composition of foods in the seven countries being compared were lacking; therefore, in 1987, the investigators decided to construct retrospectively equivalent food composites representing the average food intake of the 1958 to 1964 baseline survey of the 16 cohorts in the Seven Countries Study. These composite diets were then analysed by a single laboratory for fatty acid composition including *trans*-fatty acids (Figure 3.1). The amount of 18:1t in the daily diet ranged from 5.5 g in the Netherlands to 0.2 g in Japan. Thus, comparative analyses of the *trans*-fatty acid content of the diet in widely variant populations can be done using this method. The major drawback in the Seven Countries Study was the long interval (25 years) between the original dietary data collection and the chemical analyses of the composite diet; however, relative comparisons between the countries appear to reflect traditional dietary patterns.

4. Estimates Based on the Trans-Fatty Acid Content of Biological Tissues

Estimates of the *trans*-fatty acid content of the diet can be extrapolated also

TABLE 3.3

Intake of *trans*-octadecenoic acid in Swedish diets determined by analysis of duplicate portions of mixed diets[1]

	No. of subjects	No. of daily portions	18:1t	18:1t
			% of total fatty acid	*g/day*
1968, men	10	70	5.0	4.9
1968, women	10	70	5.0	2.8
1975, women	6	36	5.1	3.0
1978, vegans	6	24	1.8	0.9
1980. lacto-vegetarians	6	24	3.9	3.0

[1] Data from Åkesson *et al.* [3]

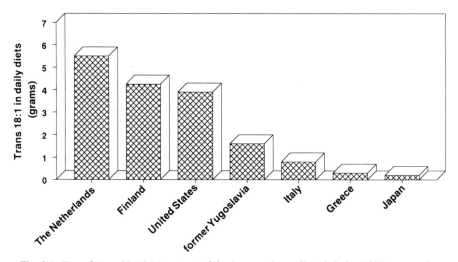

Fig. 3.1. *Trans*-fatty acid (18:1t) content of food composites collected during 1987 representing the average food intakes of the 16 cohorts of the Seven Countries Study. Data represent one cohort each for The Netherlands and the United States; two each for Finland, Greece and Japan; and three for Italy. Cohorts in the former Yugoslavia included two in Croatia and three in Serbia. (Adapted from Kromhout *et al.* 1995[74]).

from the concentration of *trans*-fatty acids found in biological tissues. Two "tissues," human milk and adipose tissue, have been used. The *trans*-fatty acid content of human milk reflects the *trans*-fatty acid content of the mother's diet on the previous day [2,29]; adipose tissue, on the other hand, is believed to reflect long-term consumption patterns [13,53].

   i. Human milk. The *trans*-fatty acid content of human milk varies across populations (Figure 3.2) [2,15,22-25,27,29,40,43,56,64,72,73,77,92,129]. From these comparative data, one can get a rough estimate of the relative amounts of *trans*-fatty acids being consumed in various countries at the time the milk samples were obtained. Moreover, the distribution of positional *trans*-isomers of the fatty acids in breast milk appears to reflect the predominant source of

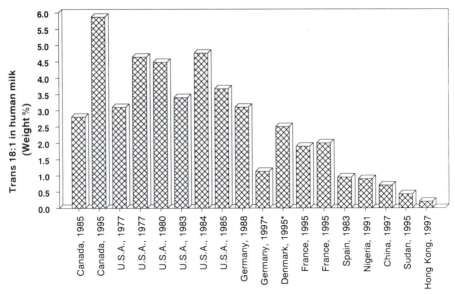

Fig. 3.2. *Trans*-fatty acid (18:1t) content of human milk samples from various countries. * Values for Germany, 1997, and Denmark, 1995, reported as total *trans*-fatty acids rather than 18:1t. Dates refer to date of publication. (Sources for data are as follows: Canada, 1985 from Chappell *et al.* [22]; Canada, 1995 from Chen *et al.* [25]; U.S.A., 1977 from Picciano [92] and Aitchison *et al.* [2]; U.S.A., 1980 from Clark *et al.* [27]; U.S.A., 1983 from Hundrieser, *et al.* [56]; U.S.A., 1984 from Craig-Schmidt *et al.* [29]; U.S.A., 1985 from Finley *et al.* [40]; Germany, 1988 from Koletzko *et al.* [72]; Germany, 1997 from Genzel-Boroviczeny *et al.* [43]; Denmark, 1995 from Jorgensen [64]; France, 1995 from Chardigny *et al.* [23] and Wolff [129]; Spain, 1983 from Boatella *et al.* [15]; Nigeria, 1991 from Koletzko *et al.* [73] ; China, 1997 from Chen *et al.* [24]; Sudan, 1995 from Laryea [77]; Hong Kong, 1997 from Chen *et al.* [24]).

trans-fatty acids in the maternal diet. The presence of relatively large amounts of vaccenic acid (*trans*-11 18:1) and the relative amount of *trans*-16 18:1 indicate that milk and other dairy products are the major sources of *trans*-fatty acids in the maternal diet. This was the case for lactating women in France [23,129]. On the other hand, a more equal distribution of *trans*-11 18:1 and *trans*-10 18:1 indicates that the *trans*-fatty acids in the maternal diet are derived primarily from commercially hydrogenated fats. This was the case for lactating women in Canada [25].

In making inter-country as well as intra-country comparisons, it is important to keep in mind the methodology used in determining the *trans*-fatty acid content of the milk samples. For example, Ratnayake and colleagues [25] have used a combination of capillary gas chromatography and silver nitrate thin-layer chromatography, a method in which excellent separation of geometric and positional isomers of fatty acids is achieved. Thus, the relatively high value in Figure 3.2 for the 18:1t content of Canadian human milk samples [25] is due in part to sophisticated analytical methodology compared to direct gas chromatographic techniques which were used in determining most of the other values illustrated in Figure 3.2. Exceptions to this would be the values obtained by Wolff [129] and Chardigny *et al.* [23], who also used techniques similar to that of Ratnayake in determining the *trans*-fatty acid content of French human milk samples. Wolff [129] has suggested that values of 18:1t obtained by direct gas chromatography should be multiplied by a factor of 1.25 to make them comparable to those obtained by the method of Ratnayake and colleagues. Thus, in comparing *trans*-fatty acid values for human milk, it is important to use data obtained by similar analytical methods whenever possible or to make corrections such as those suggested by Wolff [129] for differences in methodology.

Although it is clear that the *trans*-fatty acid content of human milk reflects the composition of the maternal diet [2,29], few attempts have been made to formalize this relationship. In 1984, Craig-Schmidt *et al.* [29] quantified the relationship between the amount of 18:1t in the maternal diet and the amount of 18:1t in human milk samples obtained the following day. This linear relationship can be expressed by the equation:

$$y = 1.49 + 0.42x$$

where y = 18:1t in human milk, expressed as percent of total fatty acids and x = 18:1t in the maternal diet of the previous day, expressed as % of total fatty acids. The 18:1t in the milk was highly correlated ($r = 0.909$) with 18:1t in the diet. The equation was based on analysis of 80 milk samples from eight United States lactating women who were 2 months postpartum. The women consumed diets of known *trans*-fatty acid content during two five-day experimental periods in which milk samples were collected daily. Use of this equation in combination with analysis of baseline milk samples enabled Craig-Schmidt *et al.* [29] to estimate the 18:1t content of the women's self-chosen diets to be 7.8% of the total fatty acids, a value consistent with other estimates of the *trans*-fatty acid content of the United States diet at the time [38,75].

In a similar manner, Wolff [129] has plotted 18:1t values for human milk from four countries against the estimated daily per capita 18:1t intake for each of these countries to obtain the following relationship:

$$y = 0.20 + 0.76x$$

where y = 18:1t in human milk, expressed as percent of total fatty acids and x = daily per capita consumption of 18:1t, expressed as g/person/day. In deriving this equation, Wolff used *trans*-fatty acid values for human milk based on his own data for France [129], the data of Boatella [15] for Spain, the data of Koletzko [71] for Germany and the data of Craig-Schmidt [29] for the United States, standardized to account for differences in methodology. One of the limitations of this equation is that the human milk data and the dietary data were not obtained from the same subjects, as was the case in the study of Craig-Schmidt *et al.* [29].

The regression equation derived by Craig-Schmidt *et al.* [29] has been used by the International Life Sciences Institute Expert Panel on *Trans*-Fatty Acids and Early Development [62] to calculate the *trans*-fatty acids in the diet of lactating women in the United States diet to be 4.2 g/day or 5.3% of total dietary fat. The human milk data used in this estimation was that of Finley *et al.* [40] for the *trans*-fatty acid content (18:1t) of 57 human milk samples.

Similarly, Chen *et al.* [25] used the equation of Craig-Schmidt *et al.* [29] in combination with analysis of 198 human milk samples collected in 1992 from nine provinces in Canada to estimate the *trans*-fatty acid content of the Canadian diet. These investigators calculated total *trans*-fatty acids in the Canadian diet to be 10.6 g/person/day or 4.0% of energy. It should be noted that this estimate was based on the total *trans*-fatty acids in the milk and not on the 18:1t content of the milk on which the original equation was based. Moreover, the equation of Craig-Schmidt [29] was based on experimental maternal diets formulated to be adequate in all nutrients, including energy, and may not be applicable to milk samples obtained from women who are losing weight [22].

ii. Human adipose tissue. The *trans*-fatty acid composition of adipose tissue from subjects in a number of countries has been determined and is believed to reflect the amount of *trans*-fatty acids in the diet [13,53]. The isomeric *trans*-fatty acid distribution in adipose tissue may also reflect the type of fat in the diet. For example, Chen *et al.* [26] found that the *trans*-isomeric distribution of 18:1 in adipose tissue from Canadian subjects differed from that of butter but was similar to that in partially hydrogenated vegetable oils, indicating that these oils were the predominant source of *trans*-fatty acids in the Canadian diet.

Shown in Figure 3.3 are concentrations of 18:1t in adipose tissue obtained from subjects in several countries [26,37,60,67,79,97,121,125]. Comparison of these data can indicate trends in habitual dietary intake, e.g. that the *trans*-fatty acid intake in the United Kingdom has declined in recent years or that the intake of *trans*-fatty acids in the early 1990's was less in Germany than in the United States. However, caution must be exercised in interpreting these data. In making comparisons among these studies, one needs to be cognizant always of the time frame in which the samples were collected, as well as the analytical methodology used. For example, the value shown for Canada in Figure 3.3 is the greatest of any in the figure in part because the method of fatty acid analysis used by Ratanayake and colleagues [26] separates *trans*-isomers more completely than the direct gas chromatography methodology used by most investigators. Studies in which adipose tissue samples were collected in the same way, within approximately the same period of time and analysed by the same method are therefore particularly valuable in making comparisons among various countries. One such study is the EURAMIC study [5], in which 18:1t in adipose tissue sampled in 1991 and 1992 from men from eight European countries and Israel was analysed by a common laboratory (Figure 3.4). A gradient was observed, with values of 18:1t ranging from 2.3% in the Netherlands to 0.43% in Spain, reflecting the relative amounts of commercially hydrogenated fats consumed in these countries. Similarly, in another investigation recently conducted as part of the EURAMIC study, standardized methodology was used in collecting and analysing adipose tissue samples from postmenopausal women in five European countries [69]. Results from this study indicated that women in Ireland and the Netherlands were

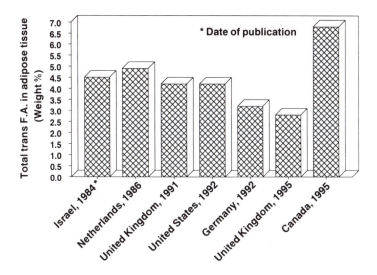

Fig. 3.3. Total *trans*-fatty acids in human adipose tissue samples from various countries. Dates refer to date of publication. (Sources for data are as follows: Israel, 1984 from Enig, *et al.* [37]; Netherlands, 1986 from van Staveren *et al.* [121] and Katan *et al.* [67]; United Kingdom, 1991 from Wahle *et al.* [125]; United States, 1992 from Hunter *et al.* [60]; Germany, 1992 from Leichsenring *et al.* [79]; United Kingdom, 1995 from Roberts *et al.* [97] and Canada, 1995 from Chen *et al.* [26]).

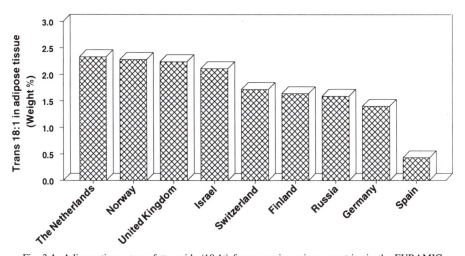

Fig. 3.4. Adipose tissue *trans*-fatty acids (18:1t) from men in various countries in the EURAMIC Study (1991-1992). Data are weighted means of both "cases" and control subjects from each country. (Adapted from Aro *et al.*, 1995 [5]).

consuming more *trans*-fatty acids than women in Switzerland and Germany and confirmed that the Spanish diet contained very small amounts of *trans*-fatty acids compared to the diet in northern European countries (Figure 3.5). Thus, adipose tissue can serve as a valuable biomarker of habitual *trans*-fatty

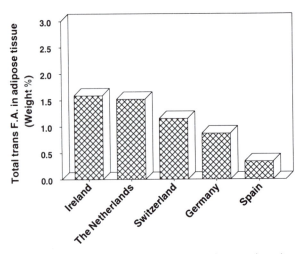

3.5. Adipose tissue total *trans*-fatty acids from postmenopausal women in various countries in the EURAMIC Study (1990-1992). Data are means of control subjects from each country. (Adapted from Kohlmeier *et al.*, 1995 [69]).

acid intake, particularly if comparisons are made using standardized collection procedures and analytical methodology.

Several attempts have been made to quantify the relationship between *trans*-fatty acids in adipose tissue and the diet. Some authors, *e.g.* Hudgins *et al.* [54], have assumed a 1:1 correlation between *trans*-fatty acids in adipose tissue and diet, while other investigators [61] have assumed a 1:2 relationship between the adipose tissue *trans*-fatty acid content and that in the diet. Still other investigators have derived equations describing this relationship in more detail. On the basis of studies in animals, Enig *et al.* [36] derived the following equation:

$$y = 0.97 + 0.44x$$

where y = % adipose tissue 18:1t and x = % dietary 18:1t. Others [62] have based the equation for the relationship between dietary *trans*-fatty acid intake and concentrations in adipose tissue on data from human studies:

$$y = 1.52 + 2.40x$$

where y = *trans*-fatty acids in the diet in g/day and x = % 18:1t in adipose tissue. This relationship was based on studies of *trans*-fatty acid consumption and the content of adipose tissue samples taken from subjects in eight countries: the Netherlands, Norway, United States, United Kingdom, Israel, Finland, Germany, and Spain [5,54,58,59,63,67,79,80,85,97,112,120,121,125], but studies with a small number of subjects were not included [1,26,37,51,89].

Estimation of *trans*-fatty acid intake on the basis of the *trans*-fatty acid content of adipose tissue is subject to some of the same limitations as the estimates based on human milk concentrations. First, use of older analytical methods leads to underestimation of the *trans*-fatty acid content of adipose tissue and thus of diet. However, investigations such as those in the EURAMIC study [5] in which all samples were analysed in the same laboratory can yield valuable comparative information even if older, less precise methods of fatty acid analysis were used. Second, the proportion of *trans*-fatty acids stored in the adipose tissue versus those metabolized may be dependent on dietary factors such as caloric intake, and this relationship needs further investigation. Third, the equations used to quantify the relationship between tissue concentrations and dietary intake need to be verified. In addition, the *trans*-fatty acid content may be different for different sampling sites, *e.g.* abdominal versus lateral thigh [26], thus making quantitative comparisons between some studies difficult.

C. ESTIMATES OF *TRANS*-FATTY ACIDS IN THE DIET OF VARIOUS POPULATIONS

1. North America

Both the United States and Canada have comparatively large amounts of *trans*-fatty acids in the food supply, with the majority derived from commercial hydrogenation of vegetable oils. It is estimated that as much as 90% of the *trans*-fatty acids in the North American food supply is derived from hydrogenated vegetable oils as opposed to ruminant products [30]. For example, in 1981 Brisson [17] estimated that the proportion of total *trans*-fatty acids derived from commercial hydrogenation in Canada was 94%. Other estimates ranging from 78 to 91% of total dietary *trans*-fatty acids coming from vegetable sources have been made for the United States [29,36,58,103]. Per capita consumption of *trans*-fatty acids from primary food sources in the United States is illustrated in Figure 3.6.

i. United States. Recently, three expert panels [4,61,62] have reviewed the available data on *trans*-fatty acid consumption in the United States. The International Life Sciences Institute Expert Panel on *Trans*-Fatty Acids and Coronary Heart Disease [61] concluded that *trans*-fatty acids contribute about 4 to 12% of total dietary fat intake (or 2 to 4% of total energy intake) to the food supply in the United States. Similarly, the American Society for Clinical Nutrition/American Institute of Nutrition Task Force [4] reported the estimated per capita consumption of dietary *trans*-fatty acids from both vegetable and animal sources to be 8.1 to 12.8 g/day, representing 2 to 4% of energy intake. The International Life Sciences Institute Expert Panel on *Trans*-Fatty Acids and Early Development [62] concluded that median values of about 8% of total fatty acid intake and about 6.4 g/person/day appear to be "reasonable best-guess" estimates for the *trans*-fatty acid content of the diets of mothers in the United States. Thus, the "consensus" values of all three panels are in rough agreement with each other, but indicate a range of intake.

All of the approaches discussed in Section B above have been used in estimating the *trans*-fatty acid content of the United States diet (Table 3.4) [2,6,29,32,36,38,54,57,58,60,74,75,80,103,116,118,119,126,]. One of the first estimates was that of Kummerow [75] who used 1971 United States Department of Agriculture Household Consumption Data [117] and "typical" values for visible fats to estimate the *trans*-fatty acid content of the United States diet as 8% of total fat. This value is in agreement with the 1978 value of Enig *et al.* [38] who also estimated 8% of total fat or 12.1 g/day based on fat availability in the United States food supply. Values based on disappearance or availability data range from 6.8 to 13.3 g/day per capita [32,36,38,57,58,103] and tend to be higher than values estimated by methods which are based on individual consumption.

Estimates of the *trans*-fatty acid content of the diet in the United States also have been made based on laboratory analysis of diet collections [2,29,74]. Aitchison *et al.* [2] based their estimate on laboratory analysis of three "duplicate portions" diets collected from each of eleven lactating U.S. women. Even within this limited population, a wide range of values was found, *i.e.* 0 to 4.4 g/day as the minimum value and 2.3 to 9.9 g/day as the maximum value for the three days of diet collection for each subject. This corresponded to an average of 5.0% of total fat with a range of 1.3 to 8.3%. Craig-Schmidt *et al.* [29] defined extremes for the *trans*-fatty acid content of home-cooked daily meals in the United States. In this study, diets were formulated using fats purchased from grocery stores. Five daily diets made with sources of commercially hydrogenated fats, *e.g.* margarine, contained an average of 9.7 g/day of 18:1t, whereas the same five daily diets made with sources of "naturally" hydrogenated fats, *e.g.* butter, contained only 0.9 g/day. This corresponded to a range of 1.0 to 11.8% of total fatty acids as 18:1t in the diet. Similarly, the *trans*-fatty acid content of seven diets collected from each of eight Caucasian adolescent females in the Southern United States ranged from 0.5 to 8.0 g/day or 2.0 to 17% of total fatty acids [119]. Thus, analysis of individual daily diets has the advantage of defining possible ranges of *trans*-fatty acid intake, but the disadvantage of assessing the diet of only a small number of subjects who are not necessarily representative of the population as a whole.

An alternate approach is to analyse a food composite that is considered typical of the diet of a given population. This approach was taken by Kromhout *et al.* [74] in the Seven Countries Study (Figure 3.1). For the United States Railroad cohort in this study, a value of 3.9 g/day of dietary 18:1t was found, with cohorts from only Finland and the Netherlands having greater values.

Self-reported dietary data, in combination with food composition tables listing the *trans*-fatty acid content of various food items, have been used also in determining the quantities of *trans*-fatty acids being consumed in the United States. Using the 18:1t content of 220 selected food items grouped into 35 categories [39], van den Reek *et al.* [118] analysed seven-day diet records to estimate the *trans*-fatty acid content of the diets of eight adolescent girls.

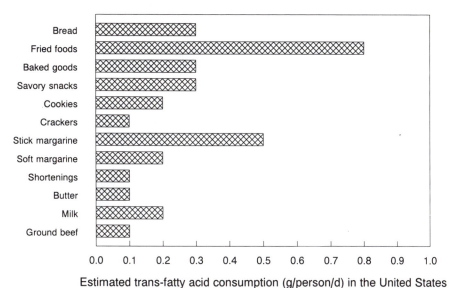

Estimated trans-fatty acid consumption (g/person/d) in the United States

Fig. 3.6. Sources of *trans*-fatty acids in the United States diet. Estimates were based on United States Department of Agriculture 3-day food intake surveys (1989-1990 and 1990-1991) and tabulated *trans*-fatty acid composition data. Fried foods include home and food service combined; margarine (stick and soft) does not include use as ingredients in foods listed elsewhere in figure. (Adapted from the report of the International Life Sciences Institute Expert Panel on *Trans*-Fatty Acids and Coronary Heart Disease, 1995 [61]).

Values of 5.3 ± 0.44% of total fatty acid or 2.8 ± 0.26 g/day were obtained. Average caloric intake of these girls was 1489 ± 64 with 35% of calories from fat or about 60 g of dietary fat. Obviously, the total amount of fat in the diet should be considered when comparing estimates of *trans*-fatty acid consumption expressed as grams per day. Some of the estimates of Enig *et al.* [36] for the *trans*-fatty acid content of the American diet represent extreme values in part because of very high fat consumption by individual subjects. For example, Enig *et al.* [36] reported that *trans*-fatty acid consumption could be as high as 14.3 g/day for women and 20.6 g/day for men. The reported fat consumption values for these subjects were 179 g for women and 258 g for men, values which would represent extremes in total fat consumption for individuals in the United States. Thus, the discrepancy in *trans*-fatty acid consumption of approximately 3 g/day by van den Reek *et al.* [118] and 20.6 g/day reported by Enig *et al.* [36] reflect differences in total fat intake perhaps more than differences in the *trans*-fatty acid content of the food supply.

The Willett semi-quantitative food frequency questionnaire has been used by a number of investigators in estimating the *trans*-fatty acid content of habitual diets of various United States populations [6,60,80,116,126]. In these studies, average values for *trans*-fatty acids in the diets of adult men and women ranged from 3 g/day to 4 g/day. (See Table 3.4.) Thus, the estimates derived from food frequency questionnaires tend to be much less than values derived from disappearance data.

TABLE 3.4

Estimates of *trans*-fatty acid consumption in North America.

Country	Diet / Dietary data	Population	Trans-fatty acid intake			Method used	Reference
			g/person/day	energy %	% total F.A.		
USA	Fat disappearance data (1971)	General U.S. population			8.0	Household Consumption Data (USDA, 1971) [117] and typical *trans* content of visible fats; assumed 25-35% *trans* content of stick margarines,15-25% for tub margarines, 20-30% for shortenings and 0-15%for salad oils.	Kummerow, 1975 [75]
USA	Duplicate portion diets	Lactating U.S. women, (n=11)	min.= 0.0-4.4 max.= 2.3-9.9 per person		5.0 (range= 1.3-8.3)	Laboratory analysis of three daily diet collections from each subject	Aitchison *et al.*, 1977 [2]
USA	Amount available in U.S. food supply	General U.S. population	12.1		8.0	Estimates based on government fats and oils availability data and known *trans*-fatty acid content of foods	Enig *et al.*, 1978 [38]
USA	Fat disappearance data	General U.S. population	6.8 (18:1t)			Estimates based on consumption of hydrogenated soybean oil containing 20% average *trans*-fatty acid	Emken, 1981 [32]
USA	Self-selected diets	Lactating U.S. women (n=8)			7.8	Estimates based on equation indicating relationship of 18:1t in diet and human milk of the following day	Craig-Schmidt *et al.*, 1984 [29]

Country	Diet / Dietary data	Population	*Trans*-fatty acid intake			Method used	Reference
			g/person/day	energy %	% total F.A.		
USA	Diets formulated using normal foods	Diets adequate for lactating women • With hydrogenated fat (n=5) • Without hydrogenated fat (n=5)	9.7 0.9		11.8 1.0	Laboratory analysis (18:1t) of experimental diets made with and without commercial sources of hydrogenated fats	Craig-Schmidt *et al.*, 1984 [29]
USA	Food disappearance data (1983)	General U.S. population • With correction for discarded fat ... • Without correction of discarded fat ...	8.3 10.2		5.9 6.1	Estimates based on availability of fat in the U.S. food supply as indicated by U.S. production and sales data	Senti, 1985 [103]
USA	Duplicate diet analysis	Caucasian adolescent females 12-15 years of age (n=8)	3.14±0.26 (18:1t)		6.5±0.42 (18:1t)	Laboratory analysis of duplicate portion diets (7 daily diets per subject)	van den Reek *et al.*, 1986 [119]
USA	Diet records	Caucasian adolescent females, 12-15 years of age (n=8)	2.8±0.26		5.3±0.44	Calculations from 7-day weighed diet records and 18:1t content of foods from Enig *et al.* (1983) [39]	van den Reek *et al.*, 1986 [118]
USA	Fat availability in U.S. food supply (1980, 1984)	General U.S. population	7.6			Estimates based on retail sales for household use, amount of fat used by food service industry and amount of fats used by bakeries and other food processors and typical values for *trans* fatty acid content of foods.	Hunter & Applewhite, 1986 [57]

Country	Diet / Dietary data	Population	*Trans*-fatty acid intake			Method used	Reference
			g/person/day	energy %	% total F.A.		
USA	Amount available for total population	General US population	13.3 (range= 12.5-15.2)		range=5-15	Estimates based on government (USDA-ERS) fats and oils data and weighted averages for *trans*-content of categories of fats and oils	Enig *et al.*, 1990 [36]
USA	Amount available for selected age groups	Adults, age 20-65 years; Lipid Research Clinics men (with fat intakes of 40-258 g fat/day).............. women (with fat intakes of 31-179 g fat/day).............	2.4-20.6 1.9-14.3			Estimates based on reported fat intake values and 8% of total fat as *trans*-fatty acids	Enig *et al.*, 1990 [36]
USA	Food availability in U.S. food supply (1989)	General U.S. population	8.1			Estimates based on industry market size and share data combined with ISEO (Institute of Shortening and Edible Oils) data for typical amounts of *trans* fatty acids in products	Hunter & Applewhite 1991[58]
USA	Habitual diet (1981)	Adult Caucasian males, aged 23-78 (n=76)	range=1.3-7.6			Values based on one-to-one relationship between *trans*-content of adipose tissue and diet assuming a 3000 calorie diet with 38% of calories as fat	Hudgins *et al.*, 1991 [54]
USA	Habitual diet (1986-1987)	Boston-area post-menopausal U.S. women free from cancer, aged 36-81 years (n=115)	3.44±1.84 (total) 2.83±1.63 (18:1t)		5.83±1.68 (total) 4.77±1.56 (18:1t)	Willett semiquantitative food-frequency questionnaire containing 116 food items combined with USDA nutrient composition data and other food composition data	London *et al.*, 1991 [80]

Country	Diet / Dietary data	Population	*Trans*-fatty acid intake g/person/day	energy %	% total F.A.	Method used	Reference
USA	Habitual diet (1986-1987)	Boston area men, aged 40-75 years (n=118) initial questionnaire.......... final questionnaire..........	3.0±1.6 3.4±1.6		4.7±1.5 5.4±1.5	Willett semiquantitative food-frequency questionnaire containing 131 food items (administered twice) combined with USDA nutrient composition data and other food composition data	Hunter *et al.*, 1992 [60]
USA	Habitual diet (1986-1987)	Boston area men, aged 40-75 years (n=118)			4.2±0.9	Estimates based on fat aspirates	Hunter *et al.*, 1992 [60]
USA	Habitual diet (1987-1990)	U.S. men, aged 43-85 (n=748)	3.4±1.2	1.6 (range= 0.3-3.8)	5.5	Willett semiquantitative food-frequency questionnaire containing 116 food items combined with USDA nutrient composition data and other food composition data	Troisi *et al.*, 1992 [116]
USA	Habitual diet (1980)	Women from Nurses' Health Study (n=85,095)	4.0±1.9 (range= 2.4-5.7)		5.8±1.8	Willett semiquantitative food-frequency questionnaire containing 61 food items combined with USDA nutrient composition data and other food composition data	Willett *et al.*, 1993 [126]

Country	Diet / Dietary data	Population	*Trans*-fatty acid intake			Method used	Reference
			g/person/day	energy %	% total F.A.		
USA	Habitual diet (1982-83)	Boston area white men and women,<76 years of age				Willett semiquantitative food-frequency questionnaire containing 116 food items combined with USDA nutrient composition data and other food composition data	Ascherio, et al., 1994 [6]
		men..........	4.36±2.33	1.5	4.3		
		women..........	3.61±2.24	1.7	4.8		
		energy adjusted mean					
		lowest quintile..........	3.0-3.1				
		highest quintile..........	6.7-6.8				
USA	Food composites (1987) of typical intake as assessed in the Seven Countries Study (1958-1964)	U.S. Railroad cohort (n=30)	3.9 (18:11)			Laboratory analysis of food composites	Kromhout et al., 1995 [74]
Canada	Typical Canadian diet (1977)	General Canadian population	9.1 (avg.) 17.5 (max.)		9.5	Calculated from average *trans*-fatty acid content of selected foods and Canadian government consumption data, 1977	Brisson, 1981 [17]
		Hypothetical Man, aged 20-39	11.1 (avg.) 17.1 (max.)				
Canada	Usual diet (1992)	Lactating Women (n=198)	10.6	4.0		Calculated from human breast milk analysis and equation of Craig-Schmidt, et al. [29] for the relationship of milk *trans*-18:1 to dietary *trans*-18:1	Chen et al., 1995 [25]
		Low *trans*-milk (n=18)	3.0	1.1	3.7		
		Medium *trans*-milk (n=155)	10.1	3.9	13.2		
		High *trans*-milk (n= 25)	20.3	7.7	26.0		

Country	Diet / Dietary data	Population	*Trans*-fatty acid intake			Method used	Reference
			g/person/day	energy %	% total F.A.		
Canada	Diet based on government survey	Male, age 18-34	12.5	3.7		Estimates based on fat and calorie intakes reported in the Nova Scotia Dietary Survey and assuming 10.4% of the total fat as *trans*-18:1 (range 0.6-32.3%), which was calculated using the Craig-Schmidt, *et al.* [29] equation and *trans*-18:1 levels found in Canadian human milk	Ratnayake & Chen, 1995 [94]
		age 35-49	9.6	3.7			
		age 50-64	9.1	3.7			
		age 65-74	7.8	3.5			
		Female, age 18-34	7.0	3.7			
		age 35-49	6.4	3.7			
		age 50-64	5.6	3.4			
		age 65-74	5.2	3.4			
		Average	8.4 (range = 0.5-26.1)	3.7 (range = 0.2-11.3)			

The *trans*-fatty acid content of human milk has been used also as a basis for estimating *trans*-fatty acid consumption in the United States. In 1984, Craig-Schmidt *et al.* [29] estimated that the usual diet of lactating women participating in their study contained 7.8% of total fatty acids as 18:1t. This estimate was based on an equation describing the relationship of 18:1t in the maternal diet to 18:1t in milk of the following day. Using this regression equation and the data of Finley *et al.* [40] for the *trans*-fatty acid content of 57 human milk samples, the International Life Sciences Institute Expert Panel on *Trans*-Fatty Acids and Early Development [62] calculated *trans*-fatty acids in the diet of lactating women in the United States diet to be 4.2 g/day or 5.3% of total dietary fat. Wolff [129] has pointed out that on the basis of human milk composition, *trans*-fatty acid consumption in the United States appears to greater than that in Germany, France or Spain.

The *trans*-fatty acid content of adipose tissue taken from subjects in the United States has been determined by a number of investigators [1,54,60,80,81,89]. In all of these studies, the *trans*-fatty acid content of adipose tissue taken from adult men and women was approximately 4.0% of total fatty acids. Using the quantitative relationships between *trans*-fatty acids in adipose tissue and diet (Section B.4.*ii*), one can calculate *trans*-fatty acid consumption in the United States to be approximately 7 to 8% of dietary fat or 5 to 11 g/person /day, depending upon the assumption one makes as to average fat intake in the United States.

ii. Canada. The average intake of *trans*-fatty acids in Canada was estimated by Brisson in 1981 [17] to be 9.1 g/person daily with possible maximum values of 17.53 g/person daily (Table 3.4). Of this amount, 94% of the dietary *trans*-fatty acids was derived from hydrogenated vegetable oils with an average of only 0.56 g/person/day of *trans*-fatty acids coming from ruminant sources.

More recent estimates of Ratnayake and colleagues are in general agreement with Brisson's earlier estimates. Chen *et al.* [25], estimated *trans*-fatty acid intake from concentrations of *trans*-fatty acids in 198 human milk samples collected in 1992 from nine provinces of Canada. Using the equation of Craig-Schmidt *et al.* [29], describing the linear relationship between *trans* 18:1 in the milk and the maternal diet, these investigators estimated the intake of total *trans*-fatty acids from various dietary sources by Canadian lactating women to be 10.6 ± 3.7 g/person/day. Individual variation was wide with minimal values of 3.0 g/person/day and maximal values of 20.3 g/person/day estimated for the diets of lactating women. The profile of the *trans*-isomers in Canadian human milk suggested that partially hydrogenated vegetable oils were the major dietary sources of these *trans*-fatty acids. Using government survey data in combination with their results from the study in lactating women, Ratnayake and Chen [94] estimated the *trans*-fatty acid intake in the general Canadian population to range from 0.5 to 26.1 g/person/day with a mean of 8.4 g/person/day (Table 3.4). Thus, it appears that the average Canadian consumes 8 to 10 g of *trans*-fatty acids per day derived primarily from partially hydrogenated soybean and canola oils.

2. United Kingdom

Several authors, including Gurr [46-48] and Wahle and James [124] have reviewed the data on *trans*-fatty acid consumption in the United Kingdom. Two comprehensive reports have been published by the British Nutrition Foundation Task Force on *Trans*-Fatty Acids, the first in 1987 [18] and the second in 1995 [19]. In the 1987 report, the *trans*-fatty acid content of the diet in the United Kingdom was estimated to be 7 g/person/day or about 6% of total dietary fat (Table 3.5). The 1995 estimate was only slightly less, with 4 to 6 g/person/day reported. This represented 2% of total dietary energy. The percentage of total dietary fat as *trans*-fatty acids was the same as in the earlier report, *i.e.* 6%.

The main sources of *trans*-fatty acids in the diet of the United Kingdom are partially hydrogenated oils and fats. It is estimated that margarines, spreads, frying oils and some bakery products contribute at least 45% of the total *trans*-fatty acids in the diet [19]. Ruminant meat fats, milk fats and dairy products contribute approximately 35%. Estimated intake of *trans*-fatty acids from various commodities is shown in Figure 3.7. For the Scottish diet, Bolton-Smith *et al.* [16] determined that industrially hydrogenated *trans*-fatty acids make up nearly 58% of the total *trans*-fatty acid intake for men and 61% for women. Of the commercially hydrogenated fat in the diet, approximately 60% comes from cakes, biscuits and sweets, while about 20% comes from hard margarines. Distribution of dietary sources within the naturally derived *trans*-fatty acids was 27% for red meat, 20% for milk, 18-19% for butter and 13-16% for cheese. Thus, compared to the diet of people in the United States, a greater proportion of *trans*-fatty acids in the diet of people in the United Kingdom comes from ruminant sources.

In contrast to margarines in the United States, some margarines and shortenings in the United Kingdom have been made traditionally from partially hydrogenated fish oils. The presence of the highly unsaturated fatty acids, eicosapentaenoic acid (20:5n-3) and docosahexaenoic acid (22:6n-3), in marine oils results in the formation of a complex mixture of geometric and positional fatty acid isomers when the oils are partially hydrogenated. Typically, partially hydrogenated fish oils contain 25 to 45% *trans*-fatty acids [55], with large amounts of *trans*-fatty acids having a chain length of 20 or 22 carbons [102]. The proportion of hydrogenated fish oil in fats and oils incorporated into foods in the United Kingdom was approximately 26% from 1980 to 1985 but dropped to 22% in 1986 [18] and to 19% in 1992 [18]. Because of economic reasons and consumer health concerns, hydrogenated fish oils continue to be less commonly incorporated into the food supply of the United Kingdom. Vegetable oils which are used as raw materials for commercial hydrogenation in the United Kingdom include soybean, rapeseed and sunflower seed oils [19].

Estimates of *trans*-fatty acids in the diet of the United Kingdom have been made by a number of investigators [16,18,19,46,47] (Table 3.5). The early estimate of Gurr [46] as 12 g/person/day is greater than later estimates which generally average 4 to 7 g/person/day [16,18,19,47]. Most of the data on which these estimates were made is from food composition data from the Laboratory of the Government Chemist for the Ministry of Agriculture, Fisheries and Food and government dietary intake data from surveys such as the National Food Survey, the Total Diet Study and the Dietary and Nutritional Survey of British Adults [45]. Results from these studies are detailed in the British Nutrition Foundation report of 1995 [19] and summarized in Table 3.5. In general, there appears to be a slight downward trend in the consumption of *trans*-fatty acids in the United Kingdom from foods eaten in the home from 1980 to 1992. Men in the United Kingdom have a greater intake of *trans*-fatty acids expressed as g/day than women, but on an energy basis or as a percentage of total fatty acids, there is little difference due to gender. Similarly, there was only slight variation by region of the country or socioeconomic group. The estimate of *trans*-fatty acid intake based on laboratory analysis of items purchased in 1991 to represent the typical diet was slightly less (3.57 g/person/day) than that determined in other ways. Thus, the overall estimate of the average intake of *trans*-fatty acids in the United Kingdom as 4 to 6 g/person/day (2% of total dietary energy or 6% of total dietary fat) appears to be representative of the population as a whole [19].

Estimates of dietary *trans*-fatty acids based on a food frequency questionnaire administered to participants in the Scottish Heart Health Study [16] are somewhat greater than the estimates of intake by Scottish adults in the Dietary and Nutritional Survey of British Adults [19,45]. In the Bolton-Smith *et al.* [16] study, the mean *trans*-fatty acid intake was reported as 7.1 g/person/day for men and 6.4 g/person/day for women, compared to 5.6 g/day for men and 4.0 g/day for women using the data of Gregory *et al.* [45]. Both studies were conducted at approximately the same time (1984 to 1986 and 1986 to 1987), so differences in these estimates must be due to differences in dietary assessment methodology and food composition databases.

Like estimates of *trans*-fatty acid intake in North America, extremes in consumption of *trans*-fatty acids have been reported for individuals in the United Kingdom. Based on household food purchasing data from the National Food Survey [21], the 1987 British Nutrition Foundation Task Force [18] estimated that a person conforming to recommendations for prevention of cardiovascular disease would consume only 5 grams of *trans*-fatty acids per day, whereas a person consuming foods in which the fat is predominantly hydrogenated fish oil could consume as much as 27 grams of *trans*-fatty acids per day. At the lower extreme in later estimates of *trans*-fatty acid intake are values in the range of 0.4 to 1.0 g/person/day for women and 0.8 to 1.9 g/person/day for men [16,19]. At the upper extreme are values in the range of 8.8-11.3 g/person/day for men and women in the upper 2.5 percentiles in the Dietary and Nutritional Survey of British Adults [19,45] and values as high as 47.6 g/person/day or 12% of total energy in the Scottish Heart Health Study [16].

Compared with other countries, the United Kingdom has been among one of

TABLE 3.5
Estimates of *trans*-fatty acid consumption in the United Kingdom.

Country	Diet / Dietary data	Population	*Trans*-fatty acid intake			Method used	Reference
			g/person/day	energy %	% total F.A.		
United Kingdom	Typical UK diets	General UK population compiled from published data	12.0		9.3	Calculated values based on average *trans*-fatty acid content of foods and trends in food consumption	Gurr, 1983 [46]
United Kingdom	Household food consumption	General UK Population	7.7		7.5	Based on Ministry of Agriculture, Fisheries and Food, National Food Survey data	Gurr, 1986 [47]
United Kingdom	Household food consumption	General UK Population	7 (estimated range=5–27)		6	Based on Ministry of Agriculture, Fisheries and Food, National Food Survey data	British Nutrition Foundation, 1987 [18]
United Kingdom	Household food consumption	General UK Population 1980............ 1982............ 1984............ 1986............ 1988............ 1990............ 1992............	5.6 5.5 5.2 4.7 5.0 4.6 4.8	2.3 2.3 2.3 2.0 2.3 2.2 2.3		Based on Ministry of Agriculture, Fisheries and Food, National Food Survey data (excludes food eaten outside the home)	British Nutrition Foundation, 1995 [19]
United Kingdom	Model of the average national domestic diet in the UK: collections of representative foods (1991)	General UK Population	3.57	1.7		Based on analysis of foods collected to be representative of the population, Ministry of Agriculture, Fisheries and Food, National Food Survey data, Total Diet Study	British Nutrition Foundation, 1995 [19]

Country	Diet / Dietary data	Population	Trans-fatty acid intake			Method used	Reference
			g/person/day	energy %	% total F.A.		
United Kingdom	Dietary data (1986-1987)	British adults, 16-64 years of age (n=>2,000)	4.9		5.6	Based on Dietary and Nutritional Survey of British Adults, Gregory *et al.*, 1990 [45] (note: range given as lower and upper 2.5 percentiles)	British Nutrition Foundation, 1995 [19]
		men	5.6 (range= 1.9-11.3)				
		women	4.0 (range= 1.0-8.8)				
United Kingdom	Dietary data (1986-1987)	UK men, 16-64 years of age Scotland.................. Northern England........	5.18 5.73	2.02 2.10		Based on Dietary and Nutritional Survey of British Adults, Gregory *et al.*, 1990 [45]	British Nutrition Foundation, 1995 [19]
		Central and South West England, and Wales......	5.65	2.01			
		London and South East England..................	5.56	2.03			
United Kingdom	Dietary data (1986-1987)	UK women, 16-64 years of age Scotland.................. Northern England..........	3.89 3.96	2.06 2.10		Based on Dietary and Nutritional Survey of British Adults, Gregory *et al.*, 1990 [45]	British Nutrition Foundation, 1995 [19]
		Central and South West England, and Wales......	4.10	2.13			
		London and South East England..................	3.93	2.07			
United Kingdom	Household food consumption (1994)	UK men (n=1087) UK women (n=1110)	3-5 (range=1-21) 2-4 (range=1-14)			Based on Ministry of Agriculture, Fisheries and Food, National Food Survey data (includes food eaten outside the home)	British Nutrition Foundation, 1995 [19]

Country	Diet / Dietary data	Population	*Trans*-fatty acid intake			Method used	Reference
			g/person/day	energy %	% total F.A.		
United Kingdom	Habitual diet (1984-1986)	Scottish men and women, 40-59 years of age (n=10,359)				Semi-quantitative food frequency questionnaire used in Scottish Heart Health Study and *trans*-fatty acid food compositonal data from government, industry, and published values	Bolton-Smith, *et al.*, 1995 [16]
		men........................	7.1±3.1 (range= 0.8-33.7)	2.7±0.9 (range= 0.45-11.6)	7.8±2.1 (range= 2.8-21.5)		
		women.......................	6.4±2.9 (range= 0.4-47.6)	3.3±1.1 (range= 0.33-11.6)	8.2±2.3 (range= 1.2-22.4)		

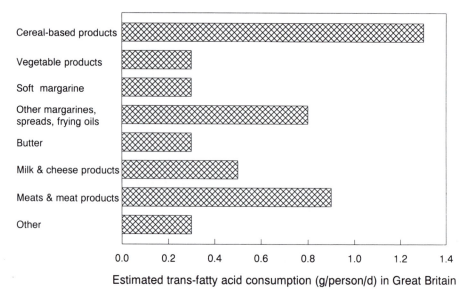

Estimated trans-fatty acid consumption (g/person/d) in Great Britain

Fig. 3.7. Sources of *trans*-fatty acids in the diet of British adults. Estimates were based on government data from 1988-1987. Total *trans*-fatty acid consumption was estimated to be 4.9 g/person/day. (Adapted from the report of the British Nutrition Foundation, 1995 [19]).

the highest in average intake of *trans*-fatty acids. Based on adipose tissue data [5], consumption of *trans*-fatty acids by subjects in Edinburgh, Scotland, was comparable to that in the Netherlands, Norway and Israel, but greater than that in Finland, Russia, Germany and Switzerland and much greater than that of Spain (Figure 3.4). Recent adipose tissue data [97] indicate that dietary *trans*-fatty acids in the United Kingdom may have declined in recent years compared with earlier data [111-114,125] (Figure 3.3).

3. "Continental" Europe

Wide variation exists in the pattern and amount of *trans*-fatty acids consumed in Europe. Estimates of *trans*-fatty acid consumption in countries in continental Europe are summarized in Table 3.6.

The variation in *trans*-fatty acid consumption among the European countries is due to differences in the composition of margarines and other industrially hydrogenated fats as well as to differences in types of dietary fats consumed. Based on analysis of milk fat and other ruminant fats as well as per capita consumption of various fats, Wolff [128,129] has calculated the mean daily per capita consumption of 18:1 *trans*-isomers from milk fat and total ruminant fats by people in the twelve countries of the European Economic Community. Daily intake of *trans*-fatty acids from dairy products and other fats of ruminant origin varies from 0.80 g/person in Portugal to 1.82 g per person in Denmark (Figure 3.8). For the twelve countries in the European Economic Community milk fat contributed an average of 1.16 g/person/day to the total *trans*-fatty

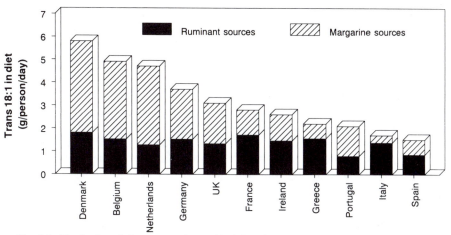

Fig. 3.8. Distribution of dietary *trans*-fatty acids (18:1t) into ruminant and margarine sources for countries in the European Economic Community. (Adapted from Wolff, 1995 [129]).

acid intake [128]. Wolff [129] has also pointed out that a general gradient exists in *trans*-fatty acid consumption from southeastern to northwestern Europe. The proportion of dietary *trans*-fatty acids derived from margarine and other hydrogenated fats increases roughly from the southeast to the northwest in parallel with the increase in total *trans*-fatty acid consumption (Figure 3.8). The fatty acid analysis of adipose tissue from European men [5] (Figure 3.4) and women [69] (Figure 3.5) reinforce this general trend.

    i. The Netherlands. Traditionally, the *trans*-fatty acid content of the diet of the Netherlands has been one of the highest in the world (Table 3.6). In 1986, Brussaard [20] estimated that 17 g/day, equivalent to 5% of dietary energy, was consumed by the average Dutch person. This estimate was based on food disappearance data and published food composition values. Of the 17 g/day in total *trans*-fatty acid consumption, Brussaard [20] estimated that 1.8 g was derived from milk and dairy products, 0.6 g from butter, and 0.4 g from meat. Thus, commercially hydrogenated fat was the major source of *trans*-fatty acid in the Dutch diet. The distribution of *trans*-fatty acids in Dutch foods is shown in Figure 3.9. Hydrogenated fish oil has been used for the manufacture of margarine and shortenings in the Netherlands [84], and its use may account for the high amount of *trans*-fatty acids in the food supply of the Netherlands.

    A somewhat lower value than that estimated by Brussaard [20] was reported by van Dokkum *et al.* [120]. This value of 10 g/day of *trans*-fatty acids in the diet of the Netherlands was found based on laboratory analysis of 24-hour duplicate daily diet samples collected in 1984 and 1985. Expressed as a percentage of dietary energy, the value of 5% was the same as that estimated by Brussaard [20].

    These relatively high values for average *trans*-fatty acid consumption in the Netherlands are confirmed in studies which have compared intakes among several countries. The food composites representing the daily diet of the Dutch cohort in Zutphen, Netherlands, contained the greatest amount of 18:1t (5.5 g) of any of the cohorts studied in the Seven Countries Study [74]. Similarly, the 18:1t content of adipose tissue from Dutch subjects in Zeist, Netherlands, was greater than that from male subjects in any of the other countries in the EURAMIC Study [5] (Figure 3.4). Only Ireland was greater than the Netherlands in adipose tissue concentrations in samples taken from postmenopausal women in the study by Kohlmeier *et al.* [69] (Figure 3.5).

Within the last few years, however, major changes have occurred in the manufacture of margarine in the Netherlands in order to decrease the amount of *trans*-fatty acids available in the food supply. Dutch margarine manufacturers now state the *trans*-fatty acid content of their products on the label [66]. In 1995, Katan [66] predicted that the *trans*-fatty acid content of margarines and spreads would be reduced to less than 5%, and for many products, the content would drop below 1%. He stated that as a result of these changes in the composition of margarine and other edible fats, the average Dutch person would be eating 4 g less of *trans*-fatty acids per day than was typical in the 1980's. There has been some concern, however, that efforts to decrease the *trans*-fatty acid content of margarines would result in greater saturated fat in the product. As shown in Table 3.7, some Dutch margarines have been formulated to contain low *trans*-fatty acids and much less saturated fat than butter [65]. This Dutch margarine contains 85-90% unmodified sunflower oil to provide linoleic acid, plus 10-15% of hard stock high in stearic and palmitic acid to provide firmness and structure. The cholesterol-raising saturated fats, lauric, myristic and palmitic acid, make up 11% of fatty acids and *trans*-fatty acids make up less than 1% of fatty acids in the product. In comparison, butter has approximately 40% cholesterol-raising saturated fatty acids and 5% *trans*-fatty acids. Thus, general consumption of products such as this would be predicted to lower the *trans*-fatty acid content of the Dutch diet and at the same time have a favorable impact on cardiovascular risk in the population.

ii. Germany. Precht and Molkentin [93] have reviewed sources of *trans*-fatty acids in the German diet with emphasis on margarines and milk fats. The most recent estimates of the intake of *trans*-fatty acids by the German people are in the range of 3 to 4 g/day (Table 3.6). This value appears to be less than that in several other countries including the United States. Precht and Molkentin [93] attribute this to differences in consumption patterns as well as to a lower *trans*-fatty acid content of German margarines. Analysis of adipose tissue in the EURAMIC Study [5] confirmed that habitual intake of 18:1t was less in Germany than in the Netherlands, Norway, United Kingdom, Israel, Switzerland, Finland, and Russia (Figure 3.4). Of the countries studied, only Spain had a lower adipose tissue content of *trans*-fatty acids. Similar results were seen for German women in the study by Kohlmeier [69] (Figure 3.5).

Sources of *trans*-fatty acids in the German diet of the early 1990's have been summarized by Steinhart and Pfalzgraf [109,110]. The consumption of *trans*-fatty acids in various categories of foods is shown in Figure 3.10 for

TABLE 3.6

Estimates of trans-fatty acid consumption for "Continental" Europe and Eastern Europe/former Soviet Union

Country	Diet / Dietary data	Population	Trans-fatty acid intake			Method used	Reference
			g/person/day	energy %	% total F.A.		
Netherlands	Food balance sheets (1981)	General population	17 (range= 5.7-23.8)	5		Calculations based on food disappearance data for the Netherlands and published food composition data	Brussaard, 1986 [20]
Netherlands	Diet samples collected in autumn (1984) (n=56) and in spring, 1985 (n=56)	Dutch Population (n=110)	10.0	5		Laboratory analysis (GC) of 24-hour duplicate diets	van Dokkum et al., 1989 [120]
Netherlands	Food composites (1987) of typical intake as assessed in the Seven Countries Study (1958-1964)	Zutphen cohort (n=45)	5.5			Laboratory analysis of food composites	Kromhout et al., 1995 [74]
Netherlands	Self-selected diets with "Western-type" products	Men (n=38)			4.7±0.27¹ (1.4 - 8.9)	Laboratory analysis of duplicate portions of diets in double-blind crossover experimental design	Mensink & Hornstra, 1995 [83]
	Self-selected diets with palm oil substitutes				2.1±0.16 (1.1 - 4.8)		
Germany	Typical West German diet	General population of West Germany	4.5-6.4	4		Calculations based on 18:1t content of 110 brands of fat purchased in 1973/74 and 1976 [50], on market share of these fats and on average fat consumption including ruminant products	Heckers et al., 1979 [51]

Country	Diet / Dietary data	Population	*Trans*-fatty acid intake			Method used	Reference
			g/person/day	energy %	% total F.A.		
France	Typical French diet (1988-1989)	General French population	4.8			Estimates based on 1988 fat availability data from the Organization for Economic Cooperation and Development, Paris, [88] and food composition data	Somerset, 1994 [106]
France	Typical French diet	General French population	2.8			Estimates based on laboratory analyses of ruminant milk and meat fats, and margarines in combination with per capita consumption values for these commodities.	Wolff [128,129]
France	Typical French diet	Lactating French women (n=10)	3			Estimates based on human milk 18:1t content	Wolff [129]
Spain	Typical Spanish diet (1988)	General Spanish population	2.4			Calculated from laboratory analysis of 378 food samples and official food consumption in Spain	Boatella et al., 1993 [14]
Italy	Food composites (1987) of typical intake as assessed in the Seven Countries Study (1958-1964)	Crevalcore cohort (n=29) Montegiorgio cohort (n=35) Rome cohort (n=49)	1.1 0.5 0.7			Laboratory analysis of food composites	Kromhout et al., 1995 [74]
Greece	Food composites (1987) of typical intake as assessed in the Seven Countries Study (1958-1964)	Crete cohort (n=31) Corfu cohort (n=37)	0.4 0.2			Laboratory analysis of food composites	Kromhout et al., 1995 [74]

Country	Diet / Dietary data	Population	Trans-fatty acid intake g/person/day	energy %	% total F.A.	Method used	Reference
Germany	Diet based on the National Consumption Assay (1991)	German Males............	4.11			Estimates of 18:1t intake based on analysis of 196 German food samples and 7-day diet records in National Consumption Assay	Steinhart & Pfälzgraf, [109,110]
		4-6 years old	2.95				
		7-9 years old	3.44				
		10-12 years old	3.88				
		13-14 years old	4.32				
		15-18 years old	4.38				
		19-35 years old	4.27				
		36-50 years old	4.15				
		51-65 years old	4.30				
		>65 years old	4.22				
		German Females............	3.36				
		4-6 years old	2.72				
		7-9 years old	3.18				
		10-12 years old	3.46				
		13-14 years old	3.51				
		15-18 years old	3.26				
		19-35 years old	3.32				
		36-50 years old	3.24				
		51-65 years old	3.50				
		>65 years old	3.43				
Germany		General German population	3-4			Data from the Diät Verband (Diet Association), 1995	In review article by Gertz, 1996 [44]
Germany	Typical diet plans for young children: Diet with randomly selected ingredients	Hypothetical child, aged 4-7	3.1	1.6		Calculations based on laboratory analysis of 42 food items (spreads and cold cuts) and data of Steinhart & Pfälzgraf (1992) [109] in combination with typical diet plans for young children	Demmelmair *et al.,* 1996 [31]
	Diet selected to minimize *trans*-fatty acid intake	Hypothetical child, aged 4-7	1.5	0.8			

Country	Diet / Dietary data	Population	Trans-fatty acid intake			Method used	Reference
			g/person/day	energy %	% total F.A.		
Croatia	Food composites (1987) of typical intake as assessed in the Seven Countries Study (1958-1964)	Dalmatia cohort (n=24) Slavonia cohort (n=24)	1.1 1.2			Laboratory analysis of food composites	Kromhout et al., 1995 [74]
Serbia	Food composites (1987) of typical intake as assessed in the Seven Countries Study (1958-1964)	Velika Krsna cohort (n=21) Zrenjanin cohort (n=40) Belgrade cohort (n=41)	2.2 1.3 2.0			Laboratory analysis of food composites	Kromhout et al., 1995 [74]

German males. Approximately 44% of the *trans*-fatty acids are derived from meat, milk, cheese and other ruminant fats. Thus, the consumption of *trans*-fatty acids in Germany appears to be equally divided between fats of ruminant origin and commercially hydrogenated oils.

In 1979, Heckers *et al.* [51] estimated that per capita consumption of *trans*-fatty acids in Germany was 4.5 to 6.4 g/day. This estimate was based on laboratory analysis of margarines and other fats and on market share data. The estimate of Heckers *et al.* [51] is greater than later estimates of 3.4 g/day for women and 4.12 g/day for men reported by Steinhart and Pfalzgraf [109,110] which were based on 7-day diet records. This is consistent with the observation that, in general, estimates based on market share data tend to be greater than those based on diet records.

Pfalzgraf and Steinhart [91] documented that the *trans*-fatty acid content of German margarines is decreasing. Most of the 24 samples which these investigators analysed had a *trans*-fatty acid content of less than 10% of total fatty acids. Only the margarines made from sunflower oil contained greater amounts of *trans*-fatty acids. Thus, consumption of *trans*-fatty acids in Germany may be expected to be even less than the estimates currently available.

Demmelmair *et al.* [31] recently analysed 42 food items typically consumed by young German children and found that with a diet selected to minimize *trans*-fatty acids, consumption could be as low as 1.5 g/day. It is therefore possible at the present time for German children to consume a diet containing less *trans*-fatty acids than the average value of 2.7 to 3.0 g/day estimated by Steinhart and Pfalzgraf [109] for children 2 to 4 years old. Moreover, recent analysis of German human milk [43] indicates a lower consumption of *trans*-fatty acids in the maternal diet than previously observed [72] (Figure 3.2).

iii. Switzerland. Switzerland was one of the countries included in the EURAMIC study [5]. Adipose tissue samples taken from men in Zurich, Switzerland, exhibited 18:1t values which were slightly greater than those in samples from men in Berlin, Germany. Comparable levels of habitual *trans*-fatty acid consumption to that in Switzerland were found in Finland and Russia (Figure 3.4). Similarly, adipose tissue taken from postmenopausal women in Switzerland exhibited less *trans*-fatty acids than Ireland or the Netherlands, but more than Germany and Spain. [69] (Figure 3.5).

iv. France. Somerset [106] estimated the *trans*-fatty acid consumption in France to be 4.8 g/person/day. This estimate was based on apparent consumption data from the Organization for Economic Cooperation and Development [88] and the *trans*-fatty acid content assumptions of Hunter and Applewhite [58]. Contributions according to commodity grouping to total per capita *trans*-fatty acid consumption in France were as follows: ruminant and dairy products, 1.84 g/day; margarine and processed fats, 1.68 g/day; vegetable oils and fats, 1.33 g/day; other animal fats, 0.11 g/day.

For France, Wolff [128] estimated that the intake of 18:1 *trans*-isomers from dairy fat is 1.5 g/person/day, contributing more to total *trans*-fatty acid

TABLE 3.7

Comparison of a Dutch zero-*trans*, low-saturated fat margarine with butter[1]

Fatty acid[2]	Zero-*trans* margarine	Butter
Lauric, myristic, and palmitic acids	11	41
Stearic acid	10	12
Trans-fatty acids	0	6
Cis-oleic acid	16	22
Cis, cis-linoleic acid	60	1

[1] Data from Katan [65]
[2] g/100 g fatty acids

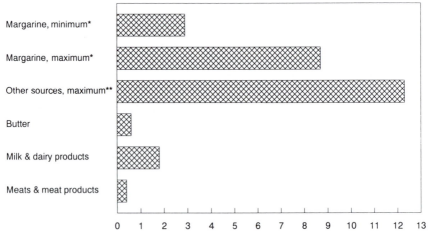

Fig. 3.9. Sources of *trans*-fatty acids in the diet of the Netherlands. Government food availability data of 1981 were used to estimate *trans*-fatty acid consumption. *Minimum assumes tub margarine was used; maximum assumes stick margarine was used. **Range was zero to 12.3 depending on type of fat used. Total *trans*-fatty acid consumption was estimated to be 5.7 to 23.8 g/person/day. (Adapted from Brussaard, 1986 [20]).

consumption than margarine, which is estimated to contribute 1.0 to 1.1 g/person/day. If the contribution from meat and meat products is added, the total 18:1t consumption in France is estimated to be 2.8 g/person/day [129]. Thus, ruminant fats account for about 60% of the *trans*-fatty acids in the French diet, whereas margarines contribute about 40% (Figure 3.8). Wolff also estimated *trans*-fatty acid consumption in France to be about 3 g/person/day based on the analysis of human milk samples from ten lactating French women. The fact that vaccenic acid was the major 18:1 *trans*-isomer in these human milk samples was consistent with dairy fats being the major contributor to *trans*-fatty acids in the French diet.

The estimates of Somerset [106] and Wolff [129] are in agreement with each

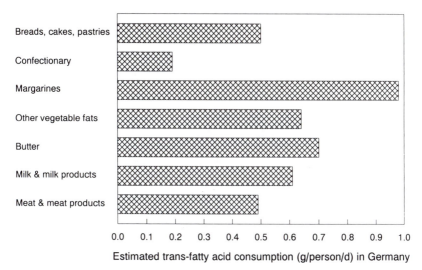

Fig. 3.10. Sources of *trans*-fatty acids in the diet of German males. Estimates were based on the German National Consumption Assay in 1991 and the *trans*-fatty acid content of 196 German food samples. Total *trans*-fatty acid consumption was estimated to be 4.1 g/person/day. (Adapted from Steinhart and Pfalzgraf, 1992 and 1994 [109,110]).

other for the amount of *trans*-fatty acids in the diet contributed by dairy and ruminant products, *i.e.* 1.7 to 1.8 g/day. However, the proportion of *trans*-fatty acids contributed by industrially hydrogenated products to the French diet differs between the two estimates.

v. Spain. In contrast to the fats and oils consumed in Northern European countries, olive oil accounts for the greatest proportion of dietary fat intake in Spain. Boatella *et al.* [14] found that refined olive oil contains only 0.5 g per 100 grams of total fatty acids as *trans*-isomers (16:1t + 18:1t + 18:2t) and therefore contributes only very small amounts of *trans*-fatty acids to the Spanish diet. On the basis of 1991 government food consumption data in combination with the laboratory analysis of 378 Spanish food products grouped into 11 categories, Boatella *et al.* [14] estimated that the mean intake of *trans*-fatty acids in Spain was 2.4 g/person/day (Table 3.6). This value is less than that in countries where the major fats consumed are industrially hydrogenated oils. In Spain, meat and meat products, primarily beef, are the major sources of *trans*-fatty acids in the diet but still comprise less than 1 g/person/day (Figure 3.11). The low intake of *trans*-fatty acids in Spain compared to other countries is confirmed by the relatively low 18:1t content of adipose tissue [5,69] (Figure 3.4 and Figure 3.5) and human milk [15] samples obtained from Spanish subjects (Figure 3.2).

vi. Italy and Greece. In Mediterranean countries, olive oil is used extensively as a dietary fat. It is not surprising then that composite diets for Italy and Greece contained less 18:1t than any of the other countries in the Seven Countries Study except Japan [74] (Figure 3.1). In contrast to other European countries, ewe and goat milk in Greece contribute 0.63 g or about

45% to the total daily consumption of 18:1t from dairy products (1.34 g /person/day) [129].

4. Eastern European/Former Soviet Countries: Croatia, Serbia, Russia and Poland

As part of the Seven Countries Study, two cohorts from Croatia and three from Serbia were studied [74]. Although there was variation among cohorts, consumption of 18:1t in these countries was found to be intermediate between countries, such as the United States, Finland and the Netherlands, in which industrially hydrogenated fats have traditionally been a major source of dietary fat and the Mediterranean countries where olive oil is consumed in large quantities (Figure 3.1). This finding is consistent with the EURAMIC study [5] in which adipose tissue samples from Russian men contained less 18:1t than those from the Netherlands (Figure 3.4). Adipose tissue from men in Finland, however, appeared to have approximately the same *trans*-fatty acid content as those from Russian men.

Wadolowska *et al.* [123] have assessed the fat consumption patterns of Polish women during the perinatal period. They found that greater than 70% of the subjects used margarine rather than butter as a spread. Additionally, margarine was used more often than butter in preparing baked goods. Thus, the authors concluded that Polish infants may be exposed to relatively high levels of *trans*-fatty acids.

5. Nordic Countries

The intake of *trans*-fatty acids in the Nordic countries has been reviewed recently by Becker [12] and by Vessby *et al.* [122]. Consumption of *trans*-fatty acids for a person in Denmark, Finland, Iceland, Norway and Sweden was estimated to vary from 2 to 10 g/day, corresponding to 1 to 3% of dietary energy [12]. In Denmark, Finland, Iceland and Sweden, dairy products and ruminant fats contribute about 50 to 60% of total *trans*-fatty acids in the diet, while margarines and shortenings are the major sources (about 80%) in the Norwegian diet [12]. In most of the Scandinavian countries, vegetable oils have been used as the major raw material in industrial hydrogenation; however, in Norway, marine oils were used extensively in margarine manufacture until the 1960's [12]. Hydrogenated marine oil is still used in margarines and shortenings to a limited extent, particularly in Norway [122].

In 1994-1995, the food industry in the Nordic countries responded to concern about the possible adverse effects of *trans*-fatty acids voiced by the scientific community [107,122], as well as by the mass media, and in this period, a sharp decrease in *trans*-fatty acid availability in the food supply occurred. In comparing the recent data on consumption of *trans*-fatty acids in the Nordic countries, Vessby *et al.* [122] stated that consumption in Finland, Sweden and Denmark corresponded to 1% of dietary energy or less, while

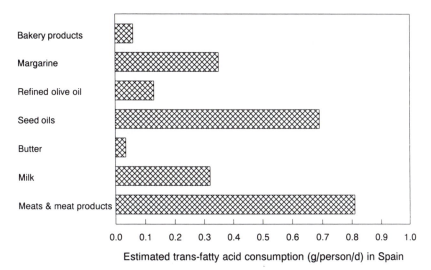

Fig. 3.11. Sources of *trans*-fatty acids in the Spanish diet. Estimates were based on government food consumption data. Total *trans*-fatty acid consumption was estimated to be 2.4 g/person/day. (Adapted from Boatella *et al.*, 1993 [14]).

intake in Norway and Iceland was probably somewhat higher. In Sweden, the mean daily intake of *trans*-fatty acids was estimated to be 2-3 grams/person/day; in Finland, average intake was estimated to be below 2 g/person/day. The Nordic Nutrition Recommendations, ratified in 1996 by the Nordic Council of Ministers, stated that the dietary intake of *trans*-fatty acids together with saturated fatty acids should not exceed 10% of energy. Historical as well as current estimates of *trans*-fatty acid intake in the Nordic countries are summarized in Table 3.8.

i. Finland. One of the earliest estimates of *trans*-fatty acid consumption in Finland was that of Toponen [115] who estimated intake to be 5.6 g/day. The estimate of Koivistoinen and Hyvonen [70] was less at approximately 3 g/day. In the Seven Countries Study, Kromhout *et al.* [74] estimated that the average consumption of 18:1t in Finland was 3.6 to 4.9 g/person/day in agreement with the earlier estimates [70,115]. The estimate in the Seven Countries study was based on analysis of composite diets made from food collected in 1987, but formulated based on 1959 intake data. Becker [12] points out that the original intake data [98] indicate that almost all of the *trans*-fatty acids must have originated from butter and other dairy products. This estimate of 18:1t consumption in Finland is greater than later estimates probably due to a pronounced decrease in the per capita availability of butter from 18 kg/person/year in 1965 to less than 8 kg/person/year in 1990 [11,12]. Among the countries in The Seven Countries study [74], Finland's consumption of *trans*-fatty acids was intermediate between that of the Netherlands and the United States (Figure 3.1).

In the EURAMIC study [5], the 18:1t level in adipose tissue sampled in

1991 and 1992 was used as a basis for comparative consumption of *trans*-fatty acids in nine countries. The 18:1t levels in samples taken from men in Helsinki, Finland, were less than in those from Norwegian men (1.64% compared to 2.28%), in general agreement with estimates of *trans*-fatty acid intake determined by other methods (Figure 3.4).

The most recent estimates [52,68] indicate that *trans*-fatty acid consumption in Finland is less than 2 g/day. Laboratory analysis of diets composited in 1989 on the basis of 1987 Finnish food balance records by Heinonen *et al.* [52] indicated that the average daily consumption of *trans*-fatty acids in Finland was 1.7 g/person, an amount equivalent to 1.9% of total fat. Approximately four-fifths of the dietary *trans*-fatty acids were isomers of 18:1 and one-fifth was composed of 18:2 *trans*-isomers. Similarly, Kleemola *et al.* [68] estimated *trans*-fatty acid consumption to be 1.4 to 1.9 g/day (0.7% of energy) for the average person in Finland on the basis of 3-day diet records obtained in 1992. Thus, among the Nordic countries, Finland appears to have one of the lowest levels of *trans*-fatty acids in the diet.

ii. Denmark. The Task Force on *Trans*-Fatty Acids appointed in 1993 by the Danish Nutrition Council [107,108] was charged with estimating the Danish consumption of *trans*-fatty acids. Based on per capita consumption data of 1991, this group estimated that *trans*-fatty acids from margarines and other hydrogenated fats was 2.5 to 2.7 g/day, accounting for approximately half of the total intake of 5.0 to 5.7 as *trans*-fatty acids for the average Danish person (Figure 3.12).

Additionally, the Danish Nutrition Council [107] estimated that 150,000 people, or 5% of the adult population in Denmark, were consuming more than 5 g/day of *trans*-fatty acids from margarine, and the committee recommended reduction of *trans*-fatty acids such that even this segment of the Danish population who were consuming large amounts of margarine would have a low *trans*-fatty acid intake. In the 1995 report [107] it was recommended therefore that all Danish margarine products be reduced to 5% or less *trans*-fatty acids. The report also encouraged the marketing of *trans*-free products. Such a reduction would limit average consumption to 2 g/day of *trans*-fatty acids derived from industrially hydrogenated products.

The task force [107] documented that per capita consumption of *trans*-fatty acids in Denmark had decreased over the previous 15 years, from 8.5 g/day in 1976 to 6.0 g/day in 1985 and further, to 5.0 g/day in 1991. This decrease in *trans*-fatty acid consumption appears to reflect both a decreased consumption of table and frying margarines as well as decreased *trans*-fatty acids in hydrogenated products [107]. For example, Danish table margarines with a high *trans*-fatty acid content comprised a larger part of the total consumption of margarine in 1985 (70%) compared with 1990 (58%) [107].

More recently, Ovesen *et al.* [90] confirmed that even further changes had been made in the composition of Danish margarines between 1992 and 1995. The content of 18:1t in semi-soft margarine brands covering the entire Danish market had decreased from 9.8 ± 6.1% in 1992 to 1.2 ± 2.2% in 1995. Also, an increase in "*trans*-free" margarines was found in 1995 compared with 1992. These investigators calculated from sales figures that the supply of 18:1t plus saturated fatty acids had decreased over this three year period by 1.4 g/day and had been replaced by *cis*- monounsaturated and polyunsaturated fatty acids.

Partially hydrogenated marine oils, however, were still being used in the manufacture of some Danish margarines and shortenings in 1995. Out of the 77 brands analysed by Ovesen *et al.* [90], 13 brands contained more than 5% long-chain fatty acids with greater than 20 carbons. These brands had an average of 15% total *trans*-monoenoic fatty acids, of which close to 50% were 20:1t and 22:1t.

iii. Sweden. Based on the content of *trans*-fatty acids in edible fats and oils as analysed by Fuchs and Kuivinen [42], the *trans*-fatty acid intake in Sweden was estimated in 1980 to be 9 g/day or 3% of dietary energy. Laboratory analysis of duplicate portion diets collected in 1968 and 1975 indicated a somewhat lower intake, *i.e.* 4.9 g/day for Swedish men and 2.8 to 3 g/day for Swedish women [3]. Åkesson *et al.* [3] also analysed diets collected from vegans and lactovegetarians. Vegans appeared to have small amounts of 18:1t in the diet, *i.e.* 0.9 g/day, whereas samples from lactovegetarians contained 3 g/day of 18:1t.

Becker [12] has documented changes in the intake of *trans*-fatty acids in Sweden. In 1984, per capita availability of *trans*-fatty acids in the Swedish food supply was estimated to be 7 g/day [10], compared to 3 g/day in 1994-1995 [12]. Thus, a substantial decrease in dietary *trans*-fatty acids in Sweden has occurred since the 1980's. The availability of margarine in the food supply has not decreased substantially in this period, however [11]; therefore, the decrease in *trans*-fatty acid availability must be due to changes in margarine composition, with substitution of non-hydrogenated fats for hydrogenated fats [12].

iv. Norway. In contrast to the other Nordic countries, *trans*-fatty acids in the Norwegian diet are derived primarily from hydrogenated vegetable and marine oils used in margarines and bakery products [63]. Dairy and meat products contribute only 16 to 17% of the total dietary *trans*-fatty acids in Norway [63], compared to 50-60% in the other Nordic countries [12,122]. The intake of *trans*-fatty acids from various categories of foods as assessed by household consumption survey in 1989 to 1991 is illustrated in Figure 3.13. Out of 7.8 g/day of *trans*-fatty acids consumed by the average person in Norway, only 1.2 g was derived from dairy products and meat products.

Johansson *et al.* [63] has calculated the content of *trans*-fatty acids in the Norwegian diet based on information from the food industry, on food consumption as assessed by food balance sheets, household consumption surveys and on a national dietary survey of 18-year old students (Table 3.8). The estimates based on household consumption data indicate that per capita *trans*-fatty acid consumption has decreased in Norway from 12.6 g/day in 1958 to 9.7 g/day in 1977-79 and further to 7.8 g/day in 1989-91. Students in the national dietary survey of 1993 also were consuming on the average 7.8 g/day or 2.2% of dietary energy. Thus, even though a reduced consumption of

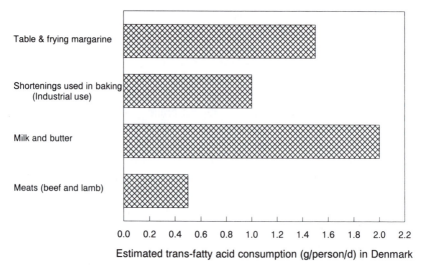

Fig. 3.12. Sources of *trans*-fatty acids in the Danish diet. Estimates were based on per capita statistics of 1991. Total *trans*-fatty acid consumption was estimated to be 5.0 g/person/day. (Adapted from the report of the Danish Nutrition Council, Stender *et al.*, 1995 [107]).

margarine has resulted in a decreased content of *trans*-fatty acids in the Norwegian diet in the last 30 years [63], *trans*-fatty acid consumption in Norway appears to have been the highest of any of the Nordic countries. Greater concentrations of 18:1t in adipose tissue samples obtained in the EURAMIC study [5] for Norway as opposed to Finland are consistent with this. Compared to the other countries included in the EURAMIC study, only the 18:1t adipose tissue concentrations in samples from the Netherlands were greater than those from Norway (Figure 3.4).

   v. Iceland. Published data on *trans*-fatty acid consumption in Iceland is very limited. Becker [12] reports that per capita availability of *trans*-fatty acids in Iceland was 6.5 g/day in 1990 and 5.9 g/day in 1994 (personal communication, 1995, from Holmfridur Thorgeirsdóttir, Icelandic Nutrition Council, to Becker [12]). Thus, the consumption of *trans*-fatty acids in Iceland appears to be one of the greatest among the Nordic countries, with only Norway traditionally having a greater intake.

6. Israel

   In 1984, Enig *et al.* [37] estimated *trans*-fatty acid consumption in Israel to be 6.5 g/person/day on the basis of government food balance sheets for 1980 and laboratory analysis of Israeli margarines and other fats (Table 3.9). The mean *trans*-fatty acid content of subcutaneous fat samples taken from eight people in Israel was 4.5% with a range of 1.9-6.6% [37]. This value was roughly comparable to the mean value of 4.0% reported by Ohlrogge *et al.* [89] for adipose tissue samples taken in the United States. Adipose tissue sampled from 108 Israeli men in 1991-1992 as part of the EURAMIC study contained 2.1% 18:1t [5]. The lower value than that reported earlier may be due in part to the fact that Enig *et al.* [37] included *trans*-isomers of 16:1 and 18:2, as well as 18:1, in the analysis, to differences in sampling, or to the time difference in the two studies. In the EURAMIC study [5], *trans*-fatty acid consumption as reflected by concentrations in the adipose tissue was among the higher levels observed, with men from Israel, the United Kingdom, Norway, and the Netherlands all having 2% or more *trans*-fatty acids in the adipose tissue (Figure 3.4). Thus, consumption of *trans*-fatty acids in the Israeli population appears to be in line with other populations who consume relatively large amounts of margarine and other industrially hydrogenated fats.

7. *India*

As reported by the FAO/WHO Joint Expert Consultation on Fats and Oils in Human Nutrition [41], the intake of hydrogenated fat having 55% *trans*-fatty acids was 2.04 g/day for the average person in India. This estimate was based on data from the National Council for Applied Economic Research, 1991 [86]. It was also estimated that an adult living in the Indian states with the highest levels of consumption could have an intake of approximately 11 grams or 4% of energy per day of *trans*-fatty acids. Consumption of *trans*-fatty acids in urban populations in India is greater than that in rural populations due to greater consumption of ghee (clarified butter) and vegetable ghee (hydrogenated vegetable fat) in urban populations [104].

8. *Australia and New Zealand*

Only a few estimates of *trans*-fatty acid intake are available for Australia (Table 3.9). These estimates have been reviewed briefly by Somerset [106], Samman [99], and Roberts [96].

In Australia, the major commercial fats are palm oil (0% *trans*-fatty acids) and tallow (3% *trans*-fatty acids) [87]. Mansour and Sinclair [82], however, point out that their analysis of some brands of margarines in 1992, compared to the composition reported in 1982 [127], indicates a shift away from animal blends and saturated fats such as coconut oils and palm oils toward the increased use of linoleic acid-rich partially hydrogenated vegetable oils.

Typical sources of *trans*-fatty acids in the Australian diet [87] are shown in Figure 3.14. Approximately 40 to 60% of the *trans*-fatty acid intake in the Australian diet is derived from domestic margarines with the remainder coming from dairy and beef fats [82,87]. This is equivalent to about 1 to 1.5% energy.

Mansour and Sinclair [82] have estimated the total *trans*-fatty acid intake in the Australian diet to be between 2.7 and 4.8 g/day for the average person (Table 3.9). These estimates were based on their analysis of 13 margarines, 5 butter/dairy blends and 2 animal fats collected in 1992 in combination with the 1991/92 Australian Bureau of Statistics' apparent consumption data for butter,

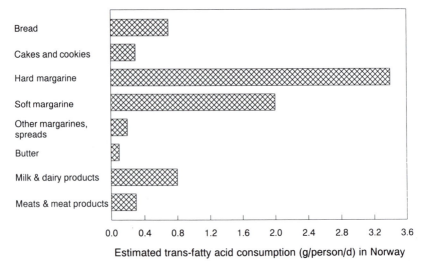

Estimated trans-fatty acid consumption (g/person/d) in Norway

Fig. 3.13. Sources of *trans*-fatty acids in the Norwegian diet (1989-1991). Estimates were based on household consumption surveys. Total *trans*-fatty acid consumption was estimated to be 4.5 g/person/day. (Adapted from Johansson *et al.*, 1994 [63]).

table margarine, other margarines, whole milk, cheese and meat [7]. Minimal and maximal values were calculated based on the range of *trans*-fatty acids in the margarines which they analysed.

Using two approaches different from that of Mansour and Sinclair [82], Noakes and Nestle [87] estimated *trans*-fatty acid consumption in Australia to be 4.0 to 6.4 g/day for males and 3.2 to 6.4 g/day for females (Table 3.9). In the first approach, simulated diets based on Australian food and nutrient intake data [8] were used to estimate *trans*-fatty acids in the typical Australian diet; in the second approach, food frequency data [9] were used. In both approaches, intakes were calculated assuming two scenarios for dietary fat in Australia: a) margarine (15% *trans*) and tallow (3% *trans*) and b) margarine (12% *trans*) and palm oil (0% *trans*). These investigators also predicted that if industrially hydrogenated fats (30% *trans*) were to replace tallow and palm oil as the major source of commercial fats in Australia, then the *trans*-fatty acid content of the typical Australian diet would increase to values between 7.1 to 13.6 g/day. These values would be in line with some of the estimates for the United States and the United Kingdom where hydrogenated fats are used extensively as commercial fats.

Thus, in general, the estimates of *trans*-fatty acids in the Australian diet are considered to be less than those in the diet of the United States. This is attributed to the use of palm oil and tallow instead of hydrogenated soybean oil in commercial baking and frying fats, as well as to a lower content of *trans*-fatty acids in margarines [87]. Somerset [106], however, cautions against drawing this conclusion based on estimates of *trans*-fatty acid intake that were determined using different methods. In a comparison of Australia, the United

TABLE 3.8

Estimates of *trans*-fatty acid consumption in the Nordic countries.

Country	Diet / Dietary data	Population	Trans-fatty acid intake			Method used	Reference
			g/person/day	energy %	% total F.A.		
Finland			5.6				Toponen, 1983 [115]
Finland			3.0				Koivistoinen & Hyvonen, 1991 [70]
Finland	Composite diets (1987-89)	General Finnish population	1.7		1.9	Laboratory analysis of composite diets (1989) based on average food consumption data from national food balance sheets for 1987	Heinonen et al., 1992 [52]
Finland	3-day diet records (1992)		1.4-1.9	0.7			Kleemola et al., 1994 [68]
Finland	Food composites (1987) of typical intake as assessed in the Seven Countries Study (1958-1964)	East Finland cohort (n=30) West Finland cohort (n=30)	4.9 3.6			Laboratory analysis of food composites	Kromhout et al., 1995 [74]
Denmark	Food balance sheets (1976, 1985, & 1991)	General Danish Population 1976 1985 1991	 8.5 6.0 5.0			Calculated values based on average consumption of food items (per capita statistics, 1976, 1985, & 1991) and average of published values for *trans*-fatty acid content of food types	Stender et al., 1995 [107]
Sweden		General Swedish population	9		3		Fuchs & Kuivinen, 1980 [42]

Country	Diet / Dietary data	Population	Trans-fatty acid intake			Method used	Reference
			g/person/day	energy %	% total F.A.		
Sweden	Normal Diet 1968 1968 1975 Vegans (1978) Lactovegetarians(1980)	Swedish adults Men (n=10) Women (n=10) Women (n=6) Mixed (n=6) Mixed (n=6)	4.9 2.8 3.0 0.9 3.0		5.0 5.0 5.1 1.8 3.9	Laboratory analysis of duplicate portions	Åkesson et al., 1981[3]
Sweden			7				Becker, 1988 [10]
Sweden	Food balance sheets (1994-95)		3.3	1		Calculations based on food supply data and market share data	Becker, 1996 [12]
Norway	Household consumption surveys (1958, 1977-79, 1989-91), food balance sheets (1986), and a nationwide dietary survey (1993)	General estimate for Norwegian population 1958 1977-79 1986 1989-91 1993 (18 year old pupils, n=1564)	2-3 12.6 9.7 10.2 7.8 7.8	7-9 4.3 3.6 2.9 2.9 2.2		Calculations based on information from the food industry, on food consumption according to government data	Johansson et al., 1994 [63]
Iceland	Food balance sheets	General Icelandic population 1990 1994	6 6.5 5.9	1.7			Thorgeirsdóttir 1995, personal communication to Becker, 1996 [12]

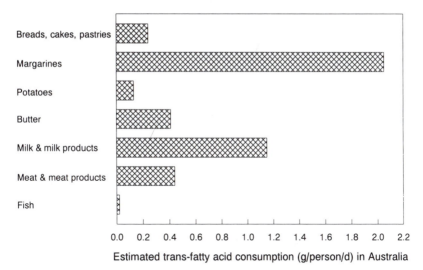

Fig. 3.14. Sources of *trans*-fatty acids in the diet of Australian males. Estimates were based on mean food frequency data and the assumption that Australian dietary fat consisted of margarines (15% *trans*-fatty acids) and commercial fats as tallow (3% *trans*-fatty acids). Total *trans*-fatty acid consumption was estimated to be 4.5 g/person/day. (Adapted from Noakes and Nestel, 1994 [87])

States, and France, Somerset [106] estimated intakes of *trans*-fatty acids for these three countries based on apparent fat consumption data obtained from a single source [88]. Applying assumptions similar to those of Hunter and Applewhite [58] for estimating *trans*-fatty acid consumption from market share data, he found similar values for the availability of *trans*-fatty acids in the diets of Australia and the United States (7.58 and 8.92 g/day, respectively), but less per capita availability in France (4.8 g/day).

Limited information is available on the *trans*-fatty acid content of the New Zealand diet. Recently, however, the *trans*-fatty acid content of margarines and other selected food items in the New Zealand food supply were analysed by Lake *et al.* [76]. On the basis of these analyses, the level of dietary *trans*-fatty acids in New Zealand can be assumed to be similar to that of Australia. Specifically, the 12 New Zealand margarines and table spreads analysed by Lake *et al.* [76] were similar in *trans*-fatty acid content to the Australian margarines analysed by Mansour and Sinclair [82]. Beef tallow, rather than hydrogenated plant-based oils, is commonly used in deep frying in New Zealand as it is in Australia, and the samples of potato chips (French fries) fried in beef drippings were similar in *trans*-fatty acid content to that found in the Australian study [82]. The *trans*-fatty acid content in snack foods, *e.g.* sweet biscuits (cookies), crackers, potato crisps (potato chips), as well as in pastries and cakes were low by comparison with products in the United States and Canada [39,95]. Lake *et al.* [76] conclude that on the basis of the foods analysed in their study, it appears that hydrogenated fats with a high *trans*-fatty acid content are not common in processed New Zealand foods and that the intake of *trans*-fatty acids in New Zealand is likely to be similar to that estimated for Australia [87].

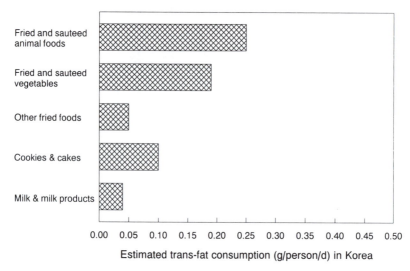

Fig. 3.15. Sources of *trans*-fatty acids in the diet of Korean women. Estimates were based on analysis of foods and snacks by students living in dormitories at a women's university in Seoul, Korea. Total *trans*-fatty acid consumption was estimated to be 0.63 g/person/day. (Adapted from Won and Ahn, 1990 [130]).

9. Asian/Pacific

Estimates of *trans*-fatty acid intake in the Asia /Pacific region (Table 3.9) appear to be relatively low compared to other regions of the world. Human milk samples from Hong Kong and China contain very small amounts of 18:1t compared to human milk samples taken from other parts of the world [24] (Figure 3.2). Consumption of *trans*-fatty acids in Korea and Japan appears to be less than 1 g/person/day [74,130].

i. Korea. Won and Ahn [130] have estimated the intake of *trans*-fatty acids by students living in dormitories of a women's university in Korea to be 0.63 g/person/day. This estimate was made on the basis of three-day diet collections from three dormitories and the laboratory analysis of foods in five categories. The *trans*-fatty acid intake from each category is shown in Figure 3.15.

ii. Japan. Two cohorts from Japan were included in the Seven Countries Study [74] (Figure 3.1). The composite diets for these cohorts contained the least 18:1t of any of the seven countries studied. An 18:1t intake of 0.2 g/day was found for Japan, compared to 5.5 g/day for the Netherlands, the country with the highest *trans*-fatty acid intake in the study.

D. FUTURE TRENDS

The fats and oils industry worldwide is rapidly responding to pressure to give consumers a wide range of choices with respect to household table fats and spreads [49]. There are movements in several countries to mandate the

TABLE 3.9

Estimates of *trans*-fatty acid consumption in Israel, Australia, Korea and Japan

Country	Diet / Dietary data	Population	*Trans*-fatty acid intake			Method used	Reference
			g/person/day	energy %	% total F.A.		
Israel	Typical Israeli diet	General Israeli population	6.5			Calculations based on government apparent consumption data (1980) and analysis of Israeli margarines, oils and cookies	Enig, *et al.*, 1984[37]
Australia	Typical Australian diet (1991-1992)	General Australian population	2.7 (min.) 4.8 (max.)			Calculated minimum and maximal values based on laboratory analysis of Australian margarines, dairy blends, & animal fats and apparent consumption data from the Australian Bureau of Statistics (1991/92)	Mansour & Sinclair, 1993 [82]
Australia	Typical Australian diet (1988)	General Australian population	7.58 (range= 4.73-10.42)			Estimates based on 1988-1989 fat availability data from the Organization for Economic Cooperation and Development, Paris, [88] and food composition data of Wills *et al.*[127]	Somerset,1994 [106]

Country	Diet / Dietary data	Population	Trans-fatty acid intake			Method used	Reference
			g/person/day	energy %	% total F.A.		
Australia	Simulated diets based on Australian (CSIRO) food and nutrient intake data	Males Diet 1	6.4	2.5	6.3	Published food composition data and data on fats provided by industry in combination with theoretical diets	Noakes and Nestel, 1994[87]
		Diet 2	5.4	2.1	5.3		
	Diet 1 assuming margarine=15%*trans* & commercial fats = tallow (3% *trans*)	Females Diet 1	4.4		5.4		
		Diet 2	3.6		4.4		
	Diet 2 assuming margarine=12% *trans* & commercial fats = palm oil (0% *trans*) Mean daily food frequency data (CSIRO); assumptions for Diet 1 & 2 same as above	Males Diet 1	4.5		4.5		
		Diet 2	4.0		4.0		
		Females Diet 1	3.5		4.2		
		Diet 2	3.2		3.9		
Korea	Diet collections	Korean university students	0.63			Laboratory analysis of foods collected in Korean dormitories	Won & Ahn [130]
Japan	Food composites (1987) of typical intake as assessed in the Seven Countries Study (1958-1964)	Tanushimaru cohort (n=24)	0.1			Laboratory analysis of food composites	Kromhout et al., 1995 [74]
		Ushibuka cohort (n=8)	0.3				

inclusion of the *trans*-fatty acid content of margarines and other products on the food label [65,66]. In Denmark, legislation limits the level of *trans*-fatty acids in processed edible oil products. In Europe and Canada, "*trans*-free" margarines have been marketed for several years, and in the United States, several companies are currently introducing a range of "*trans*-free" products. Some margarines are made "*trans*-free" by using interesterification of palm and palm kernel oils with liquid oils; other spreads on the market have a low *trans*-fatty acid content as a result of blending fully hydrogenated oils with liquid oils, such as high oleic sunflower oils. Canola oil and olive oils, both with a high content of monounsaturated fatty acids, are being incorporated into a variety of table spreads. The end result is that the *trans*-fatty acid content of the diet in most countries eventually could be less than 2 g/person/day, dependent totally on the amount of *trans*-isomers contributed "naturally" by meat, milk ard dairy products.

NOTE ADDED IN PROOF

Recently published results from the TRANSFAIR study [van Poppel,G. *Lancet*, **351**, 1099 (1998)] indicate that *trans*-fatty acid consumption may have decreased in some European countries. *Trans*-fatty acids in the 1995-1996 diet (g/day) were reported as follows: Belgium, 4.1; Denmark, 2.6; Finland, 2.1; France, 2.3; Germany, 2.2; Greece, 1.4; Iceland, 5.4; Italy, 1.6; Netherlands, 4.3; Norway, 4.0; Portugal, 1.6; Spain, 2.1; Sweden, 2.6; and UK, 2.8.

ACKNOWLEDGEMENTS

Appreciation is expressed to the personnel of the Interlibrary Loan Department of the Auburn University Libraries for their help in securing many of the journal articles and government reports needed to write this chapter. The assistance of students, Brenda Holzer, Fang-Yan Du and Kavitha Bejgam, in locating articles is also gratefully acknowledged. Appreciation is also expressed to Jai-Ok Kim, Fennechiena Dane, Kim Davis, and Ursula Heilman for assistance in translation of Korean, Dutch, Norwegian, and German materials into English.

REFERENCES

1. Adlof,R.O. and Emken,E.A., *Lipids*, **21**, 543-547 (1986).
2. Aitchison,J.M., Dunkley,W.L., Canolty,N.L. and Smith,L.M., *Am. J. Clin. Nutr.*, **30**, 2006-2015 (1977).
3. Åkesson,M.D., Johansson,B.-M., Eng,M., Svensson,M., Eng,M. and Öckerman,P.-A., *Am. J. Clin. Nutr.*, **34**, 2517-2520 (1981).
4. American Society for Clinical Nutrition /American Institute for Nutrition Task Force on *Trans* Fatty Acids, *Am. J. Clin. Nutr.*, **63**, 663-670 (1996).
5. Aro,A., Kardinaal,A.F.M., Salminen,I., Kark,J.D., Riemersma,R.A., Delgado-Rodriguez,M., Gomez-Aracena,J., Huttunen,J.K., Kohlmeier,L., Martin,B.C., Martin-Moreno,J.M., Mazaev,V.P., Ringstad,J., Thamm,M., Van't Veer,P. and Kok,F.J., *Lancet*, **345**, 273-278 (1995).

6. Ascherio,A., Hennekens,C.H., Buring,J.E., Master,C., Stampfer,M.J. and Willett,W.C., *Circulation*, **89**, 94-101 (1994).

7. Australian Bureau of Statistics, Apparent consumption of selected foodstuffs, Australia 1991-1992, *Australian government publishing service* (1992) (Canberra).

8. Baghurst,K., Crawford,D. and others, The Victorian nutrition survey II. Nutrient intakes by age, sex, area of residence and occupational status, *CSIRO Division of Human Nutrition* (1987) Adelaide.

9. Baghurst,K., Crawford,D. and others, The Victorian nutrition survey III. The contribution of various foods to nutrient intakes, *CSIRO Division of Human Nutrition* (1987) Adelaide.

10. Becker,W., Vår Näring, **1**, 4-5 (1988).

11. Becker,W. and Enghardt,H., *Scand. J. Nutr./Näringsforskning*, **37**, 118-124 (1993).

12. Becker,W., *Scand. J. Nutr.*, **40**, 16-18 (1996).

13. Beynen,A.C., Hermus,R.J.J. and Hautvast,J.G.A.J., *Am. J. Clin. Nutr.*, **33**, 81-85 (1980).

14. Boatella,J., Rafecas,M. and Codony,R., *Eur. J. Clin. Nutr.*, **47**, S62-S65 (1993).

15. Boatella,J., Rafecas,M., Codony,R., Gibert,A., Rivero,M., Tormo,R., Infante,D. and Sanchez-Valverde,F., *J. Pediatr. Gastroenterol. Nutr.*, **16**, 432-434 (1993).

16. Bolton-Smith,C., Woodward,M., Fenton,S., McCluskey,M.K. and Brown,C.A., *Brit. J. Nutr.*, **74**, 661-670 (1995).

17. Brisson,J., in *Lipids in Human Nutrition: An Appraisal of Some Dietary Concepts*, pp. 41-71 (1981) (edited by Jack K. Burgess, Inc., Eaglewood, New Jersey).

18. British Nutrition Foundation. Trans *fatty acids*, (1987) (London: British Nutrition Foundation).

19. British Nutrition Foundation. Trans *fatty acids*, (1995) (London: British Nutrition Foundation).

20. Brussaard,J.H., *Voeding*, **47**, 108-111 (1986). (in Dutch)

21. Burt,R. and Buss,D.H., *Br. J. Clinical Practice*, **31** (S1), 20-23 (1984).

22. Chappell,J.E., Clandinin,M.T. and Kearney-Volpe,C., *Am. J. Clin. Nutr.*, **42**, 49-56 (1985).

23. Chardigny,J.M., Wolff,R.L., Mager,E., Sebedio,J.L., Martine,L. and Juaneda,P., *Eur. J. Clin. Nutr.*, **49**, 523-531 (1995).

24. Chen,Z.Y., Kwan,K.Y., Tong,K.K., Ratnayake,W.M.N., Li,H.Q. and Leung,S.S.F., *Lipids*, **32**, 1061-1067 (1997).

25. Chen,Z.Y., Pelletier,G., Hollywood,R. and Ratanayke,W.M.N., *Lipids*, **30**, 15-21 (1995).

26. Chen,Z.Y., Ratanayake,W.M.N., Fortier,L., Ross,R. and Cunnane,S.C., *Can. J. Physiol. Pharmacol.*, **73**, 718-723 (1995).

27. Clark,R.M., Ferris,A.M., Fey,N., Hundrieser,K.E. and Jensen,R.G., *Lipids*, **15**, 972-974 (1980).

28. Consumer and Food Economics Institute. Composition of food. United States Department of Agriculture handbook no. 8. Washington, DC: U.S. Government Printing Office, 1976-1989.

29. Craig-Schmidt,M.C., Weete,J.D., Faircloth,S.A., Wickwire,M.A. and Livant,E.J., *Am. J. Clin. Nutr.*, **39**, 778-786 (1984).

30. Craig-Schmidt,M.C. in *Fatty Acids In Foods and Their Health Implications. Fatty acid isomers in foods*, pp. 363-396 (1992) (edited by C. K. Chow, Marcel Dekker, New York).

31. Demmelmair,H., Festl,B., Wolfram,G. and Koletzko,B., *Z. Ernahrungswiss*, **35**, 235-240 (1996).

32. Emken,E.A., *J. Am. Oil Chem. Soc.*, **58**, 278-283 (1981).

33. Emken,E.A., *Ann. Rev. Nutr.*, **4**, 339-376 (1984).

34. Enig,M.G., *Nutr. Quar.*, **17**, 79-95 (1993).

35. Enig,M.G., Trans Fatty Acids in the Food Supply: A Comprehensive Report Covering 60 Years of Research, 2nd Ed. (1995) (Enig Assoc., Silver Spring, MD)

36. Enig,M.G., Atal,S., Keeney,M. and Sampugna,J., *J. Am. Coll. Nutr.*, **9**, 471-486 (1990).

37. Enig,M.G., Budowski,P. and Blondheim,S.H., *Human Nutr.: Clin. Nutr.*, **38C**, 223-230 (1984).

38. Enig,M.G., Munn,R.J. and Keeney,M., *Fed. Proc.*, **37**, 2215-2220 (1978).

39. Enig,M.G., Pallansch,L.A., Sampugna,J. and Keeney,M., *J. Am. Oil Chem. Soc.*, **60**, 1788-1795 (1983).

40. Finley,D.A., Lonnerdal,B., Dewey,K.G. and Grivetti,L.E., *Am. J. Clin. Nutr.*, **41**, 787-800 (1985).

41. Food and Agriculture Organization (FAO), *Fats and oils in human nutrition: Report of a joint expert consultation*, (Volume 57) (1993) (Food and Agriculture Organization of the United Nations and World Health Organization, Rome, Italy).

42. Fuchs,G. and Kuivinen,J., *Vår Föda*, **32** (suppl 1), 31-37 (1980).

43. Genzel-Boroviczeny,O., Wahle,J. and Koletzko,B., *Eur. J. Pediatr.*, **156**, 142-147 (1997).

44. Gertz,C., *Lebensmittelchemie*, **50**, 50-52 (1996).

45. Gregory,J., Foster,K., Tyler,H. and Wiseman,M., *The Dietary and Nutritional Survey of British Adults*, (1990) (HMSO, London, England).
46. Gurr,M.I., *Int'l Dairy Fed.Document*, **166**, 5-18 (1983).
47. Gurr,M.I., *British Nutrition Foundation Nutr. Bul.*, **11**, 105-122 (1986).
48. Gurr,M.I., *Nutr. Res. Rev.*, **9**, 259-279 (1996).
49. Haumann,B.F., *INFORM*, **9**, 6-13 (1998).
50. Heckers,H. and Melcher,F.W., *Am. J. Clin. Nutr.*, **31**, 1041-1049 (1978).
51. Heckers,H., Melcher,F.W. and Dittmar,K., *Fette Seifen Anstrichmittel*, **81**, 217-226 (1979).
52. Heinonen,M., Lampi,A.M., Hyvonen,L. and Homer,D., *J. Food Comp. Anal.*, **5**, 198-208 (1992).
53. Hirsch,J., Farquhar,J.W., Ahrens,E.H.,Jr., Peterson,M.L. and Stoffel,W., *Am. J. Clin. Nutr.*, **8**, 499-511 (1960).
54. Hudgins,L.C., Hirsch,J. and Emken,E.A., *Am. J. Clin. Nutr.*, **53**, 474-482 (1991).
55. Hulshof,P.J.M., van de Bovenkamp,P., Boogerd,L., Bos,J., Germing-Nouwen,C., Kosmeyer-Schuil,T., Hollman,P. and Katan,M.B., *Spijsveten en Oliën*, Dept. of Human Nutrition, Agricultural University Wageningen, (1991).
56. Hundreiser,K.E., Clark,R.M. and Brown,P.B., *J. Pediatr. Gastroenterol. Nutr.*, **2**, 635-639 (1983).
57. Hunter,J.E. and Applewhite,T.H., *Am. J. Clin. Nutr.*, **44**, 707-717 (1986).
58. Hunter,J.E. and Applewhite,T.H., *Am. J. Clin. Nutr.*, **54**, 363-369 (1991).
59. Hunter,J.E. and Applewhite,T.H., *J. Am. Oil Chem. Soc.*, **70**, 613-617 (1993).
60. Hunter,D.J., Rimm,H.B., Sacks,F.M., Stampfer,M.J., Colditz,G.A., Litin,L.B. and Willett,W.C., *Am. J. Epidemiol.*, **135**, 418-427 (1992).
61. International Life Sciences Institute Expert Panel on *Trans* Fatty Acids and Coronary Heart Disease, *Am. J. Clin. Nutr.*, **62** (suppl), 655S-708S (1995).
62. International Life Sciences Institute Expert Panel on *Trans* Fatty Acids and Early Development, *Am. J. Clin. Nutr.*, **66** (suppl), 715S-736S (1997).
63. Johansson,L., Rimestad,A.H. and Andersen,L.F., *Scan. J. Nutr.*, **38**, 62-66 (1994). (in Norwegian)
64. Jorgensen,M.H., Lassen,A. and Michaelsen,K.F., *Scand. J. Nutr.*, **39**, 50-54 (1995).
65. Katan,M.B., *J. Am. Diet. Assoc.*, **94**, 1097-1098 (1994).
66. Katan,M.B., *Lancet*, **346** (no. 8985), 1245-1246 (Nov. 11, 1995).
67. Katan,M.B., van Staveren,W.A., Deurenberg., Barendse-Van Leeuwen,J., Germing-Nouwen,C., Soffers,A., Berkel,J. and Beyen,A.C., *Prog. Lipid Res.*, **25**, 193-195 (1986).
68. Kleemola,P., Virtanen,M. and Pietinen,P., The 1992 dietary survey of Finnish adults, *Publications of the National Public Health Institute*, (B2/1994) (Helsinki).
69. Kohlmeier,L., Simonsen,N., van't Veer,P., Strain,J.J., Martin-Moreno,J.M., Margolin,B., Huttunen,J.K., Navajas,J.F.-C., Martin,B.C., Thamm,M., Kardinaal,A.F.M. and Kok,F.J., *Cancer Epidemiol. Biomar. Prev.*, **6**, 705-710 (1997).
70. Koivistoinen,P. and Hyvônen,L., Lipidforum seminar "*Trans* fatty Acids, analysis and metabolism", 21-22 Oct. 1991, Helsinki, Finland, 1991. (as referenced by Precht and Molkentin, 1995).
71. Koletzko,B., Mrotzek,M. and Bremer,H.J., in *Human Lactation 2, Maternal and Environmental factors*, pp. 589-594 (1986) (edited by M. Hamosh and A.S.Goldman, Plenum Press, New York).
72. Koletzko,B., Mrotzek,M., Eng,B. and Bremer,H.J., *Am. J. Clin. Nutr.*, **47**, 954-959 (1988).
73. Koletzko,B., Theil,I. and Abiodun,P.O., *Z. Ernahrungswiss.*, **30**, 289-297 (1991).
74. Kromhout,D., Menotti,A., Bloemberg,B., Aravanis,C., Blackburn,H., Buzina,R., Dontas,A.S., Fidanza,F., Giampaoli,S., Jansen,A., Karvonen,M., Katan,M., Nissinen,A., Nedeljkovic,S., Pekkanen,J., Pekkarinen,M., Punsar,S., Räsänen,L., Simic,B. and Tohsima,H., *Preventive Medicine*, **24**, 308-315 (1995).
75. Kummerow,F.A., *J. Food Sci.*, **40**, 12-17 (1975).
76. Lake,R., Thomson,B., Devane,G. and Scholes,P., *J. Food Comp. Anal.*, **9**, 365-374 (1996).
77. Laryea,M.D., Leichsenring,M., Mrotzek,M., El-Amin,E.O., El Kharib,A.O., Ahmed,H.M. and Bremer,H.J., *Int. J. Food Sci. Nutr.*, **46**, 205-214 (1995).
78. Lee,R.D. and Nieman,D.C., in *Nutritional Assessment*, (Second Edition) pp. 91-145 (1996) (Mosby-Year Book, St. Louis).
79. Leichsenring,M., Hardenack,M. and Laryea,M.D., *Z. Ernahrungswiss.*, **31**, 130-137 (1992).
80. London,S.J., Sacks,F.M., Caesar,J., Stampfer,M.J., Siguel,E. and Willett,W.C., *Am. J. Clin. Nutr*, **54**, 340-345 (1991).

112 CONSUMPTION OF *TRANS* FATTY ACIDS

81. London,S.J., Sacks,F.M., Stampfer,M.J., Henderson,I.C., Maclure,M., Tomita,A., Wood,W.C., Remine,S., Robert,N.J., Dmochowski,J.R. and Willett,W.C., *J. Natl. Cancer Inst.*, **85**, 785-793 (1993).
82. Mansour,M.P. and Sinclair,A.J., *Asia Pacific J. Clin. Nutr.*, **3**, 155-163 (1993).
83. Mensink,R.P. and Hornstra,G., *Brit. J. Nutr.*, **73**, 605-612 (1995).
84. Mensink,R.P. and Katan,B., *Prog. Lipid Res.*, **32**, 111-122 (1993).
85. Mertz,W., Tsui,J.C., Judd,J.T. Reiser,S., Hallfrisch,J., Morris,E.R., Steele,P.D. and Lashley,E., *Am. J. Clin. Nutr.*, **54**, 291-295 (1991).
86. National Council for Applied Economic Research, Market Information Survey of Households, (1985-1990) (New Delhi, India).
87. Noakes,M. and Nestel,P.J., *Food Australia*, **46**, 124-129 (1994).
88. OECD (Organization for Economic Cooperation and Development), Food Consumption statistics: Statistiques de la consommation des denrees alimentaires (1979-1988) (Paris).
89. Ohlrogge,J.B., Emken,E.A. and Gulley,R.M., *J. Lipid Res.*, **22**, 955-960 (1981).
90. Ovesen,L., Leth,T. and Hansen,K., *Lipids*, **31**, 971-975 (1996).
91. Pfalzgraf,A. and Steinhart,H., *Deutsche Lebensmittel-Rundschau*, **91**, 113-114 (1995).
92. Picciano,M.F. and Perkins,E.G., *Lipids*, **12**, 407-408 (1977).
93. Precht,D. and Molkentin,J., *Die Nahrung*, **39**, 343-374 (1995).
94. Ratnayake,W.M.N. and Chen,Z.Y., in *Development and Processing of Vegetable Oils for Human Nutrition*, pp. 20-35 (1995) (edited by R. Przybylski and B.E. McDonald, American Oil Chemists' Society, Champaign, IL).
95. Ratnayake,W.M.N., Hollywood,R., O'Grady,E. and Pelletier,G., *J. Am. Coll. Nutr.*, **12**, 651-660 (1993).
96. Roberts,D.C.K., *Food Australia*, **47**, 263-265 (1995).
97. Roberts,T.L., Wood,D.A., Riemersma,R.A., Gallagher,P.J. and Lampe,F.C., *Lancet*, **345**, 278-282 (1995).
98. Roine,P. and Pekkarinen,M., *Voeding*, **25**, 383-393 (1964).
99. Samman,S., *Food Australia*, **47**, S10-S13 (1995).
100. Sanders,T.A.B., *Nutr. Res. Rev.*, **1**, 57-78 (1988).
101. Schaafsma,G., *IDF Nutr. Newsletter I*, 20-21 (1992).
102. Sebedio,J.-L., Langman,M.F., Eaton,C.A. and Ackman,R.G., *J. Am. Oil Chem. Soc.*, **58**, 41-48 (1981).
103. Senti,F.R. (editor), *Health Aspects of Dietary* Trans *Fatty Acids*. (1985) (Life Science Research Office, Federation of American Societies for Experimental Biology, Bethesda, MD).
104. Singh,R.B., Niaz,M.A., Ghosh,S., Beegom,R., Rastogi,V., Sharma,J.P. and Dube,G.K., *Int. J. Cardiol.*, **56**, 289-298 (1996).
105. Slover,H.T., Thompson,R.H., Davis,C.S. and Merola,G.V., *J. Am. Oil Chem. Soc.*, **62**, 775-786 (1985).
106. Somerset,S., Food Australia, 46, 564-565 (1994).
107. Stender,S., Dyerberg,J., Hølmer,G., Ovesen,L. and Sandström,B., *Clinical Science*, **88**, 375-392 (1995).
108. Stender,S., Dyerberg,J., Hølmer,G., Ovesen,L. and Sandström,B., Transfedtsyrers betydning for sundheden, *Ugeskr Laeger*, **156**, 3764-3769 (1994). (in Danish)
109. Steinhart,H. and Pfalzgraf,A., *Z. Ernahrungswiss.*, **31**, 196-204 (1992). (in German)
110. Steinhart,H. and Pfalzgraf,A., *Fat Sci. Technol.*, **96**, 42-44 (1994). (in German)
111. Thomas,L.H., Jones,P.R., Winter,J.A. and Smith,H., *Am. J. Clin. Nutr.*, **34**, 877-886 (1981).
112. Thomas,L.H., Olpin,S.O., Scott,R.G. and Wilkins,M.P., *Human Nutr. Food Sci. Nutr.*, **41F**, 167-172 (1987).
113. Thomas,L.H. and Winter,J.A., *Hum. Nutr. Food Sci. Nutr.*, **41F**, 153-165 (1987).
114. Thomas,L.H., Winter,J.A. and Scott,R.G., *J. Epidemiol. Commun. Health*, **37**, 16-21 (1983).
115. Toponen,T., *Nutritional importance of* trans-*fatty acids*, Masters' Thesis, Dept. of Nutrition, Univ. of Helsinki (1983). (as referenced by Precht and Molkentin, 1995).
116. Troisi,R., Willett,W.C. and Weiss,S.T., *Am. J. Clin. Nutr.*, **56**, 1019-1024 (1992).
117. United States Department of Agriculture (USDA). Civilian consumption of visible and invisible food fats per person, 1959-71. (1971) ERS/USDA Statistical Bulletin 489, U.S. Fats and Oils Statistics, 1950-1971, p. 102.
118. van den Reek,M.M., Craig-Schmidt,M.S. and Clark,A.J., *J. Am. Diet. Assoc.*, **86**, 1391-1394 (1986).
119. van den Reek,M.M., Craig-Schmidt,M.S., Weete,J.D. and Clark,A.J., *Am. J. Clin. Nutr.*, **43**, 530-537 (1986).

120. van Dokkum,W., Kistemaker,C. and Hilwig,G.N.G., *Voeding*, **50**, 214-218 (1989). (in Dutch)
121. van Staveren,,W.A., Deurenberg,P., Katan,M.B., Burema,J., de Groot,L.C.P.G.M. and Hoffmans,D.A.F., *Am. J. Epidemiol.*, **123**, 455-463 (1986).
122. Vessby, B., Becker, W. and Aro, A., *International Society for the Study of Fatty Acids and Lipids (ISSFAL) Newsletter*, **3(4)**, 6-10 (1996).
123. Wadolowska,L., Przybylowicz,K., Trzaskowska,M.A., Cichon,R., Kozikowski,W., Klobukowski,J., Stefanowicz,M., Romaszko,E. and Malarkiewicz, J., *Pol. J. Food Nutr. Sci.*, **6**, 127-136 (1997).
124. Wahle,K.W.J. and James,W.P.T., *Euro. J. Clin. Nutr.*, **47**, 828-839 (1993).
125. Wahle,K.W.J., McIntosh,G., Duncan,W.R.H. and James,W.P.T., *Eur. J. Clin. Nutr.*, **45**, 195-202 (1991).
126. Willett,W.C., Stampfer,M.J., Manson,J.E. and Colditz,G.A., *Lancet*, **341**, 581-585 (1993).
127. Wills,R.B.H., Myers,P.R. and Greenfield.H., *Food Technol. Aust.*, **34**, 240-244 (1982).
128. Wolff,R.L., *J. Am. Oil Chem. Soc.*, **71**, 277-283 (1994).
129. Wolff,R.L., *J. Am. Oil Chem. Soc.*, **72**, 259-272 (1995).
130. Won,J.S. and Ahn,M.-S., *Korean J. Nutr.*, **23**, 19-24 (1990). (in Korean)

CHAPTER 4

ANALYSIS OF TRANS *FATTY ACIDS*

W.M.N. Ratnayake

Nutrition Research Division, Food Directorate, Health Protection Branch, Health Canada, Banting Building, Postal Locator 2203C, Ottawa, Ontario, Canada K1A 0L2

A. INTRODUCTION

Trans isomers of naturally occurring *cis*-unsaturated fatty acids are produced when liquid vegetable or marine oils are partially hydrogenated to produce margarine, shortenings and other hardened fat products. Isomeric *trans* fatty acids are also formed in the intestinal tract of ruminants and they appear in small amounts in dairy products and beef fat.

Partially hydrogenated fats have complex fatty acid profiles because of the existence of both *cis* and *trans* isomers and this makes the analysis of such fats a challenging task. In partially hydrogenated vegetable oils (PHVO) more than 40 *cis* and *trans* isomers, primarily of the C_{18} chain length, are possible. The *trans* isomers of PHVO are predominantly those of oleic acid (18:1), with smaller quantities of various *trans,trans* and *cis,trans* isomers of linoleic acid (18:2) [1,2]. Partially hydrogenated fish oils (PHFO), invariably have much more complex fatty acid profiles because of the large number of possible isomers of the chain lengths ranging from C_{16} to C_{22} [3,4]. Except for three or four isomers, which are centered around the original double bond position (*e.g.* 9 in vegetable oils), most of the isomers in hydrogenated oils are present only in low concentrations [1-4] and this makes an extra burden in the analysis of hydrogenated fats. Unavailability of appropriate reference standards, particularly those of minor, unusual *cis* and *trans* isomers, impedes the analysis of hydrogenated fats further.

In nutrition work, food labeling or quality control in edible oil processing plants, fats are analysed to determine detailed fatty acid compositions which include the types and levels of various, saturated fatty acids (SAFA), monounsaturated fatty acids (MUFA), polyunsaturated fatty acids (PUFA), total *trans* content and the levels of the individual *cis-trans* isomers. Owing to the complexity of the hydrogenated fat mixtures there is no simple definitive method for simultaneous determination of both the total *trans* content (sum of all *trans* fatty acids) and fatty acid composition. Traditionally, total *trans* content was determined by infrared (IR) spectroscopic techniques that do not quantify individual fatty acids. Nowadays a complete analysis is achieved using capillary gas chromatography (GC) in combination with other techniques, particularly silver ion chromatography, IR spectroscopy and mass spectrometry (MS). Several other techniques which rely on special instrumentation, including proton (^{1}H) and ^{13}C-nuclear magnetic spectroscopy, supercritical fluid chromatography, and Raman spectroscopy are also available. However, such equipment is costly and the techniques may well be restricted to few laboratories; they remain research tools rather than analytical procedures for routine use.

In the following sections, applications of IR spectroscopy, GC, silver nitrate thin-layer chromatography (Ag-TLC) and silver ion high-performance chromatography (Ag-HPLC) for the analysis of *trans* fatty acids in partially hydrogenated fats are described. The progress made in the development of analytical techniques for *trans* fatty acids up to the latter part of 1970s was reviewed by Conacher [5] and Scholfield [6]. Firestone and Sheppard [7], and the British Nutrition Foundation [8] have reviewed some of the recent developments.

B. INFRARED SPECTROSCOPY

Trans double bonds absorb strongly at 980-965 cm-1 due to the deformation of the C-H bond adjacent to the *trans* double bond. The measurement of the intensity of the characteristic absorption band under a defined set of analytical conditions constitutes the basis for the various IR methods for the determination of total *trans* unsaturation in fats. Earlier methodology for determination of *trans* unsaturation and other applications of IR spectroscopy to lipid analyses were reviewed by Kauffman [9] and O'Connor [10] and more recent developments were reviewed by Firestone and Sheppard [7]. The basic

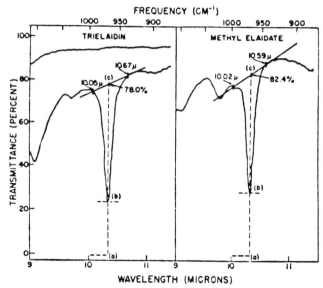

Fig. 4.1. IR absorption of *trans* unsaturation in esters (trielaidin and methyl elaidate) [11].
% *trans*, as elaidic acid, methyl elaidate, or trielaidin

$$\% \ trans = \frac{a, \text{sample (acid, methyl ester or triacylglycerol)}}{a, \text{standard (elaidic acid, methyl elaidate or trielaidin)}}$$

Where a = absorptivity = A/BC, A = absorbance = log ac/ab, B = internal cell length, C = concentration of solution in g/L. (Reproduced by kind permission of the American Oil Chemists' Society).

steps of the IR method could be illustrated using the older official method Cd 14-61 of the American Oil Chemists' Society (AOCS) [11]. In this method, triacylglycerols, fatty acids or fatty acid methyl esters (FAME) are dissolved in carbon disulphide, and the transmittance spectrum is recorded with respect to a carbon disulphide blank over the region 1110 cm^{-1} to 910 cm^{-1} (9-11 μm). The appropriate baseline is drawn and the absorbance of the *trans* analytical band at 967 cm^{-1} is measured (Figure 4.1). The % *trans* content is determined by comparison with an external standard analysed under identical analytical conditions. Trielaidin, methyl elaidate or elaidic acid are the recommended external standards.

Since the introduction of the basic procedure in 1946 [12], various improvements, ranging from minor refinements to major modifications, were suggested over the last 50 years and some of the modifications have been incorporated into the current official methods of the AOCS [13], Association of Official Analytical Chemists (AOAC) [14,15] and the International Union of Pure and Applied Chemistry (IUPAC) [16]. The modifications were aimed mainly at correcting the various background absorbances that interfere with the *trans* band at 967 cm^{-1}. Improvements were also made for calibration and reference materials. The introduction of new IR instrumentation, such as Fourier-*trans*form infrared (FTIR) and attenuated total reflectance (ATR) spectroscopy led also to development of methods tailored for these instruments. FTIR spectroscopy offers several advantages over conventional IR spectroscopy [17,18]; these include improved signal to noise ratio, due to averaging of multiple scans, high accuracy of the wavelength calibration due to a He-Ne laser, improved light through-put and speed of analysis, due to the Michelson interferometer principle, and spectra manipulation by means of computers. The ATR technique, first introduced in 1961 for analysis of organic compounds [19], offers many attractive features for analysis of *trans* fatty acids also [20]. Particularly, ATR cells allow analysis of neat samples and therefore, the need for the volatile toxic carbon disulphide solvent is eliminated. Consequently, errors due to weighing, dilution and air bubble formation within the cell, which are normally encountered with sodium chloride or potassium bromide cells, are also eliminated.

1. Background Interference

Despite the advances in IR instrumentation, errors can arise in the IR methods still, specially for samples containing lower amounts (less than 15%) of *trans* fatty acids, because of absorbances which interfere with that due to the *trans* bond per se. One such interfering absorption arises from conjugated *trans* fatty acids isomers which are very often present in small amounts in a wide variety of foods including common vegetable oils, meats, sea foods, milks, cheeses, canned foods, infants foods [21] and some seed oils [22]. Conjugated *trans,trans* bonds exhibit an absorption at 989 cm^{-1} and *cis,trans* isomers a doublet at 982 and 948 cm^{-1} that are sufficiently close to interfere with the isolated *trans* double bond absorption [9]. Therefore, the application of IR spectroscopy is limited to samples containing less than 5% conjugated *trans* fatty acids [13-16]. A band near 933 cm-1, due to the O-H out of plane deformation of carboxyl groups of the component fatty acids of fats and oils interferes also with the *trans* measurement, particularly at *trans* levels below 15% [7]. Conversion of the acids to their methyl esters partly reduces the interference.

The most serious interference to the *trans* analytical band comes from other broad bands in the spectrum which produce a strong slope in the region of the *trans* absorption that converts the *trans* band into a shoulder at low levels of *trans* unsaturation [23]. This interference is more pronounced in triacylglycerols than in methyl esters [24]. In triacylglycerols with no *trans* unsaturation, the interfering absorption is equivalent to 3-4% *trans* unsaturation as determined by spiking triacylglycerol samples with trielaidin. In contrast, non-*trans* methyl esters exhibit a negative absorption in the region of the *trans* band and as a result produce *trans* levels which are 1.5-3% lower than the actual *trans* content [23]. These errors are more severe at low *trans* levels (<15% *trans*) and differences have also been observed in fatty acids of different chain lengths [23]. Firestone and LaBouliere in 1965 [14,23] addressed this situation by proposing arithmetic correction factors for the

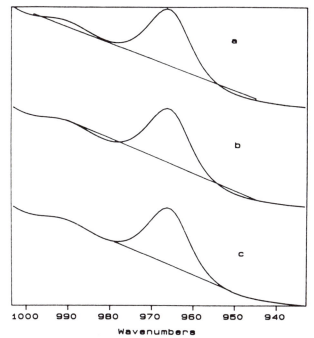

Fig. 4.2. FTIR spectra indicating three different methods of calculating the baseline. a, 998 cm⁻¹ to 944 cm⁻¹ [11]; b, 990 cm⁻¹ to 945 cm⁻¹ [28]; c, manual method by drawing a tangent [26]. (Reproduced by kind permission of the Authors [29] and *Journal of the Science of Food and Agriculture*).

positive triacylglycerol absorptions and negative methyl ester absorption. Another remedy proposed included eliminating the interferences by using a carbon disulphide solution of a non-*trans*-containing triacylglycerol or FAME as the IR reference blank [25]. This procedure gave results which were 1-2% high for both triacylglycerols and methyl esters. The high bias associated with triacylglycerols may be overcome in part by using FAME derivatives for the measurements and mixtures of methyl elaidate and either methyl linoleate [26] or oleate [27] as calibration standards. Accuracy has also been improved appreciably by the use of modern FTIR Spectroscopy linked to a data processor [18,27,28].

   Although the positive bias associated with triacylglycerols is partly resolved by analysing the sample as methyl esters, the negative bias seen with zero-*trans* methyl esters is difficult to overcome completely. The baseline procedure, used in the calculation of the absorptivities (Figure 4.1), is intended to eliminate all the possible background interferences. However, there is some ambiguity as to the selection of the two points for the construction of the linear baseline. The IUPAC official method uses a starting point of 1000 cm⁻¹ and ends at 925 cm⁻¹ for methyl esters [16] whereas in the older versions of the AOCS official methods [11,12], the baseline is between 998 and 944 cm⁻¹ (Figure 4.2a). At high *trans* concentrations, these wavelengths coincide closely in general with those of the peak minima of the *trans* band. But at low

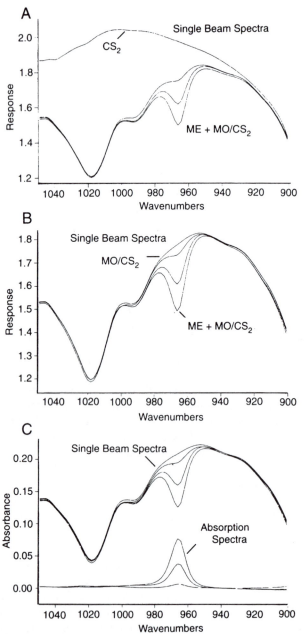

Fig. 4.3. (a) FTIR single-beam spectra of reference background material CS$_2$ (top spectrum), and of CS$_2$ solutions of 1, 4 and 7% methyl elaidate (ME) in methyl oleate (MO) (three bottom spectra). (b) Same as (a) after addition of standard MO to the CS$_2$ reference background material. (c) By "rationing" the single beam spectrum for each of these CS$_2$ solutions of 1, 4 and 7% ME in MO against that of MO in CS$_2$, the resulting FTIR absorption spectra exhibited symmetric 966 cm^{-1} bands on a horizontal background. (Reproduced by kind permission of the authors [30] and *Journal of the American Oil Chemists' Society*).

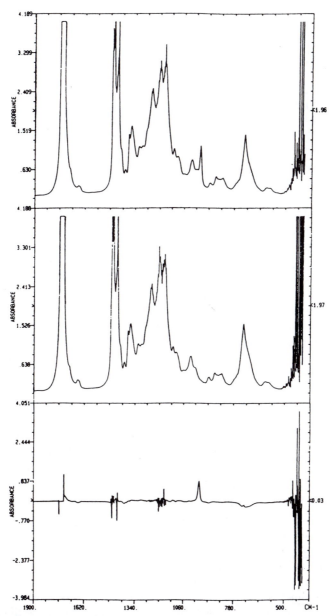

Fig. 4.4. The difference IR spectrum (bottom figure) of a *trans* fatty acid margarine fat (top figure) and unhydrogenated soybean oil (centre figure). (Reproduced by kind permission of the Authors [31] and *Journal of Food Science*).

concentration, the baseline constructed between the wavelengths could include an area not due to *trans* absorption (Figure 4.2a) [27-29]. A third suggestion (Figure 4.2b) uses 990 and 945 cm^{-1} as the limits but might include some negative areas for *trans* values less than 5% [28]. The simplest solution to this is to draw the baseline between peak minima (Figure 4.2c) [26,27]. Since the peak minima vary slightly between samples of different *trans* unsaturation, best results are obtained when the baseline of the sample is drawn exactly as the baseline in the spectrum of one of the calibration standards (mixtures of methyl elaidate with methyl oleate or methyl linoleate) having approximately the same intensity of absorption at 967 cm^{-1} [13,15,26]. An identical baseline is constructed very easily by superimposing the two spectra. This technique allows accurate analysis of *trans* contents as low as 0.5% of total fatty acids.

For convenience, all the existing IR methods assume that the baseline between 940 to 990 cm^{-1} of methyl oleate or other zero-*trans* methyl esters is linear. But, as shown in Figure 4.2, the baseline is a curve and sloping [27-30]. This means, as noted by Toschi *et al.* [29], that none of the above baseline procedures is fully satisfactory for eliminating the background noise. This problem could be overcome by making use of the special capabilities provided by FTIR spectroscopy [29-31]. With single-beam FTIR instruments the spectrum is obtained by "ratioing" the single-beam spectrum of the analyte against that of air, solvent blank or reference solution. In *trans* analysis, ratioing is performed either against pure carbon disulphide or a carbon disulphide solution of triacylglycerols, free fatty acids or methyl esters containing no *trans* bonds. Mossoba *et al.* [30] have shown that the sloping background is easily eliminated by ratioing against carbon disulphide solution of an unhydrogenated oil having only *cis* double bonds. This results in a symmetrical *trans* band on a horizontal background (Figure 4.3) and allows accurate integration of the *trans* band. Speed and convenience may be added to the method by using an ATR liquid cell instead of the conventional sodium chloride cell for IR measurement. The ATR quantification was based on integration of the *trans* band area between 990 and 945 cm^{-1}. Mossoba *et al.* [30] used mixtures of methyl elaidate and oleate or the corresponding triacylglycerols as the reference background material. This method allows detection of *trans* fatty acids as low as 0.2% of total fatty acids but because of the high noise to signal ratio at such lower levels, the limit of quantification was placed at 1%. The estimated error for this level was 1.8% [30].

Ulberth and Haider [31] and Toschi *et al.* [29] corrected the sloping baseline by a spectral subtraction technique using computer-assisted FTIR. The spectrum of the reference sample (methyl oleate or methyl esters of a non-hydrogenated oil in carbon disulphide) is directly subtracted from that of the *trans* analyte. This is possible because the *cis* isomer has a very similar spectrum on either side of the *trans* double bond peak. The subtraction results in a new spectrum with a well defined *trans* peak and flat baseline (Figure 4.4) which allows an unambiguous determination of the *trans* peak area [31]. An important advantage of this method is that the calibration curve generated by the spectrum subtraction technique is linear and passes through the zero, whereas the linear baseline technique has a negative intercept [29]. This spectrum subtraction technique introduces systematic errors of only 1-2%.

2. Trans *Calibration Plots*

In the older IR methods, the % *trans* of the analyte was calculated with respect to a single measurement of the absorbance of the *trans* calibration standard. However, as noted above on several occasions, for correcting some of the background noise and consequently improving the accuracy of the *trans* measurements, calibration plots of absorbance (alternatively peak area or peak height) versus % *trans* developed using two-component standard mixtures of a *trans* reference material (often methyl elaidate or its triacylglycerol) and a non-*trans* material (methyl stearate, methyl oleate, methyl linoleate and methyl esters from an unhydrogenated oil) are being used nowadays. The calibration plots generated by the previously mentioned subtraction technique of Toschi *et al.* [29] and the ATR technique of Mossoba *et al.* [30] are generally linear for a wide *trans* range. Direct read out of area under an absorption peak is a prerequisite for this kind of calibration. This feature is not always available with commercial software packages used for quantitative FTIR analysis.

The calibration plots developed according to the classical technique of drawing the baseline between two fixed points could be linear for most *trans* ranges, but some of the data points at the lower end of the curve could deviate from linearity. To circumvent this, the most recent official method of the AOCS (Cd 14-95) [13] recommends two calibration plots: one for *trans* levels above 10% and the other for those below 10%. Both plots are linear. As little as 0.5% *trans* unsaturation can be determined by this procedure. Also, this technique yields reproducible results. In a collaborative study, the relative standard deviation (RSDr) for a sample with 6% *trans* analysed by the AOCS two-calibration plot method was 3.7% and for a sample with 1.3% *trans* fatty acids, the RSD_r value was 13.9% [13]. In contrast, in another collaborative study, the RSD_r value for a sample with a *trans* content of 5% analysed by a similar procedure but using the single calibration plot was 34.6% [32].

3. Trans *Calibration Material*

All the current IR methods use methyl elaidate or its triacylglycerol as the calibration standard generally. However, methyl elaidate (18:1 9*t*) is not the ideal IR calibration standard, because partially PHVO contain an assortment of *trans* isomers of 18:1, 18:2 and 18:3 [1,2] in which very often 18:1 10*t* and 18:1 11*t* are the predominant *trans* components. In milk fats, 18:1 11*t* is always the major *trans* isomer and 18:1 9*t* is not present to an appreciable extent [1,33]. Furthermore, the various *trans* isomers in PHVO, particularly the *trans*-PUFA do not absorb to the same extent as methyl elaidate. For example, mono-*trans* isomers of linoleate have about 84% as much absorption as methyl elaidate, and *trans,trans*-linoleate despite having two *trans* bonds, is about 165-174% [34-36]. Because of these differences, as shown by Ratnayake and Pelletier [37], the *trans* isomers in PHVO do not absorb to the same extent as methyl elaidate or two-component mixtures of methyl elaidate and oleate. Consequently, the actual *trans* content in a PHVO will be greater than that determined using methyl elaidate as the calibration standard.

It is suggested that a fatty acid mixture of known *trans* content and well-characterized fatty acid composition, derived from a PHVO, is a more suitable calibration standard than methyl elaidate [37]. The usefulness of a PHVO-methyl ester external standard was demonstrated by analysis of a number of margarine samples. *Trans* content measured by using PHVO FAME as the external standard was in reasonable agreement ($\pm 3\%$) with values determined by Ag-TLC fractionation followed by GC analysis. The Ag-TLC/GC technique gives reliable *trans* isomer levels. Whereas the values determined by the official methods of AOAC (method 994.14)[15] and AOCS (method cd 14-610 [11], both of which use methyl elaidate as the calibration standard, were 16% and 12%, respectively, lower than the Ag-TLC/GC values. Ratnayake and Pelletier [37] proposed that a well-characterized PHVO methyl ester sample be made available from a laboratory supplier for use as an IR *trans* calibration standard. A procedure such as Ag-TLC/GC could be used for establishing the fatty acid composition and the actual *trans* content of the PHVO calibration standard.

4. IR Methods Based on Absorbance Ratio

Trans content could also be measured using other bands in the IR spectrum in conjunction with the characteristic *trans* band. Allen in 1969 [38,39] described a rapid IR method in which the ratio of absorbance at 965 cm^{-1} for the *trans* double bond to that at 1170 cm^{-1} for esters or 935 cm^{-1} for acids for the carbon-oxygen stretch of the carbonyl moiety was utilized for calibration and calculation. In practice, absorbance ratios are determined over a wide range of concentrations for calibration mixtures of trielaidin and triolein, or their methyl esters or acids in carbon disulphide. From these data, linear regression equations were calculated. To determine % *trans* in the test sample, the ratio of *trans* absorbance to ester or acid is determined and substituted in the appropriate linear equation. Since only ratios are measured, the method is independent of sample size so accurate weighing and dilution is unnecessary, a feature that might be of value with small samples from animal tissues or isolated using preparative chromatographic methods. However, this method may not give as reliable data as the alternative IR methods [25]. Errors might arise due to large differences in the areas of the two absorbances (up to 7000 times for the lower content of *trans* isomers) and to water vapour in the unpurged spectrometer absorbing near the carbonyl absorption. The latter could be reduced by purging for 5 minutes with air passed through molecular sieves to remove water and carbon dioxide [29]. Also, the method is not applicable to methyl esters or triacylglycerols containing small amounts of free fatty acids or to free fatty acids containing small amounts of methyl esters or triacylglycerols [40].

A similar internal ratioing method has been described for measuring *trans* content in margarines in which the measurements were made using FTIR equipment combined with ATR [18]. In ATR, the absorbance is a function of the extinction coefficient of the absorbing species, the efficiency of optical contact and the effective path length, which is controlled by the surface area of the crystal covered by the sample. For a given sample the first two are generally fixed but the coverage may vary. In practice, the absolute absorbance at 967 cm^{-1} might vary between replicate samples as a result of inconsistent coverage. Belton *et al.* [18] eliminated this random variation in absorbance by ratioing the absorbance at 967 cm^{-1} to that of the ester carbonyl stretching absorbance at 1743 cm^{-1}. The absorbance at 967 cm^{-1} was measured with respect to a baseline drawn between 998 and 944 cm^{-1} whilst that at 1743 cm^{-1} was measured with respect to a baseline drawn between 1600 and 1850 cm^{-1}. Linear calibration curves of absorbance (peak height) ratio versus % *trans*, with an intercept much closer to zero, were developed using standard mixtures of methyl elaidate-soybean oil methyl esters or trielaidin-soybean oil and soybean oil or its methyl ester as the reference material. Margarines are analysed directly by this method without eliminating water, because water has little effect on the ATR spectra. This is probably due to the preferential absorption of the lipid on the ATR cell. Nevertheless, the water content must be known for expressing the *trans* content relative to total margarine fatty acids. A drawback is that the method gives the number of *trans* double bonds per carbonyl ester group in the whole sample, but does not distinguish between two esters with one *trans* bond or one ester with two *trans* bonds and one with none.

5. *Partial Least Square Chemometrics - FTIR*

Chemometrics is probably an unfamiliar subject to many lipid chemists. Application of chemometrics to lipid analysis was reviewed by Kaufmann [41]. Basically, chemometrics is a mathematical approach for unraveling and elucidation of analytical features of interest from many multi-component data sets. Partial least square (PLS) is a powerful chemometric technique for quantitative analysis of complex systems and the method of choice for analysis of overlapping bands of complex infrared spectra from multicomponent matrices [42]. The power of PLS is based on its ability to make use of spectral information from broad spectral regions, rather than absorbance, peak area or peak height measurements, and correlate mathematically spectral changes to changes in the concentration of a component of interest, while simultaneously accounting for other spectral contributions that may perturb the spectrum. As such, a PLS calibration model is capable of delivering accurate and reproducible results as long as the calibration spectra contain enough

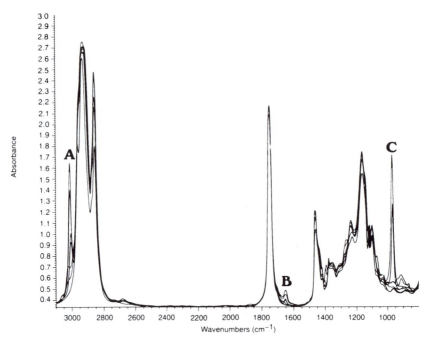

Fig. 4.5. Overlaid spectra of triacylglycerols of stearic, oleic, linoleic, linolenic, elaidic and linoelaidic acids illustrating the major bands of interest related to *cis* and *trans* analysis in edible oils: (A) *cis* C-H stretching absorption; (B) C = C stretching absorption; (C) *trans* C = C-H bending absorption. (Reproduced by kind permission of the Authors [43] and *Journal of the American Oil Chemists' Society*).

information that is representative of both the component of interest and non-related spectral variations associated with the samples to be analysed.

In 1992, Ulberth and Haider [31] described briefly the first application of a PLS method using FTIR spectroscopy for determination of *trans* content in edible fats. However, the analytical results were not validated. Recently, the McGill FTIR Group [43] developed a more versatile PLS-FTIR method for simultaneous determination of the percent *cis* and *trans* content of edible fats and oils. A generalized, industrial sample-handling platform was also designed for handling both liquid oils and solid fats and was incorporated into an FTIR spectrometer. PLS calibrations were derived for *cis* content, *trans* content, iodine value and saponification number using a calibration set that included both trielaidin and trilinoelaidin and five *cis* triacylglycerols (tripalmitoleic, triolein, trilinolein, trilinolenin and trierucin) and six saturated triacylglycerols that covered a broad range of molecular weights. This approach allowed the spectral shifts between the different *cis* and *trans* components to be accounted for in the calibrations. Furthermore, as the PLS procedure makes use of broad spectral regions in determining *cis* and *trans* contents, the spectral contribution of the weak *trans* (3025 cm^{-1}) band in the *cis* region is included, as is that of a weak *cis* (913 cm^{-1}) band in the *trans* region (Figure 4.5). Another attractive feature of this approach is that by including saturated triacylglycerols of different chain-lengths in the calibration set, the PLS model accounts for the contributions of triacylglycerols absorption that lead to errors in the classical IR methods. Other advantages include use of neat samples, therefore requiring no weighing and dilution with carbon disulphide, and there is no need for construction of baselines and other measures for correcting the background interferences. Sample handling consisted of warming the neat sample in a microwave oven, aspirating the sample into a sodium chloride cell (path-length 25 μm), recording its spectrum, evacuating the sample, and loading the subsequent sample in the same manner. The cell needed only cleaning prior to and at the end of an analytical run with isooctane. This FTIR-PLS method works well over a wide range of *trans* components, including those between 0 and 15% and measures the *cis* and *trans* content in a reproducible manner (±0.7%). The measured accuracy being 1.5% for standard addition. The McGill FTIR Group [43] automated the analysis fully by programming the spectrometer in Visual Basic (Windows) to provide a simple, prompt-based interface and to allow an operator to carry out *cis-trans* analyses without any knowledge of FTIR spectroscopy. The derived calibration is transferable between instruments, eliminating the need for recalibration. A typical analysis requires less than 2 minutes per sample. The automated system is specially designed for industrial "on line" use in an oil-processing environment. A similar procedure has not been developed to be used in a lipid research laboratory, but such a system provides an opportunity for replacing the classical IR methods, many of which are tedious and affected by environmental concerns. In addition, the FTR-PLS method is advantageous in terms of its ability to determine both *cis* and *trans* content in a single analysis. This is not possible with other IR methods.

C. GAS CHROMATOGRAPHY

GC is undoubtedly the most convenient and widely used analytical tool for analysis of fatty acids of fats, oils and lipids. The fatty acids are invariably analysed as their methyl esters. The preparation of methyl esters was reviewed by Christie [44]. For all routine lipid samples, hydrogen chloride (5%) or sulphuric acid (2%) in methanol, despite the comparatively long reaction times needed, is probably the best general purpose reagent available. Nevertheless, one of the most popular of all transesterification reagent is boron trifluoride in methanol [45]. However, caution should be exercised in using the latter reagent, which has a limited shell-life, even when refrigerated, and the use of old or too concentrated solutions might result in the production of artifacts and the loss of appreciable amounts of PUFA [44]. For mixed lipid samples with no unesterified fatty acids or single lipid classes containing ester-bound fatty acids, alkaline transesterification is recommended for speed and simplicity

TABLE 4.1

Chemical structures and maximum isothermal operating temperatures of cyanosilicone phases

Phase	Composition	Polarity	Temperature °C	Reference
SP-2300	100%A	Lowest	–	[78]
SP-2310	50%A/50%B	Low	–	[78]
SP-2330	10%A/90%B	Low-medium	250	[78]
SILAR-9CP	10%A/90%B	Low-medium	200	[78]
SP-2380	5%A/95%B	High-medium	275	[75],[78]
SILAR-10C	100%B	High	200	[78]
SP-2340	100%B	High	250	[78]
SP-2560	100%B	High	250	[75],[78]
CP-SIL 88	100%B	High	250	[48]
OV-275	100%C	Highest	–	[75]

A = Cyanopropyl phenyl polysiloxane $[C_6H_5\text{-Si(O)-CH}_2\text{-CH}_2\text{-CN}]_n$
B = Biscyanopropyl polysiloxane $[CN\text{-CH}_2\text{-CH}_2\text{-CH}_2\text{-Si(O)-CH}_2\text{-CH}_2\text{-CH}_2\text{-CN}]_n$
C = Biscyanoethyl polysiloxane $[CN\text{-CH}_2\text{-CH}_2\text{-Si(O)-CH}_2\text{-CH}_2\text{-CN}]_n$

[46]. Potassium hydroxide in methanol and sodium or potassium methoxide in anhydrous methanol are the common reagents for base-catalysed transesterification.

GC can afford excellent separation of *cis* and *trans* isomers. However, much depends on the choice of column, stationary phase, GC operating parameters and on the skill of the analyst in the interpretation of the chromatographic peaks, particularly when appropriate reference *trans* and *cis* fatty acids are not available. Highly polar cyanosilicone stationary phases, containing various polar substituents including nitrile groups marketed under trade designations such as SILAR-10C, SILAR-9CP, SP-2340, OV-275, SP-2380, SP-2560 [47] and CP-Sil 88 [48] are mandatory for *cis-trans* isomer analysis. Structures of some of these phases are given in Table 4.1. These high polarity phases enable analysts to separate components that cannot be resolved by non-polar or intermediate polarity phases. For instance, cyanosilicone phases resolve *cis*, *trans*, and their positional isomers for which small structural differences produce a change in the dipole moment, but an insignificant difference in boiling point. Thus, these phases allow separation of FAME by the degree of unsaturation and also by the geometry and position of the double bond. The earlier literature in the GC analysis of isomeric fatty aids on various types of cyanosilicone columns was reviewed comprehensively by Conacher [5], Ottenstein *et al.* [47], Christie [44] and Firestone and Sheppard [7], and more recent literature has been briefly assessed by Wolff [49], Ratnayake [50] and Precht and Molkentin [51].

In the past, *trans* fatty acid analyses were performed using packed columns. Following an international collaborative study [52], a packed column GC method employing a 6.1 m and 3 mm 15% OV-275 column was adopted as an official method for analysis of PHVO by the AOAC [53], AOCS [54] and the Technical Commission of the Oil and Fat Industry in Italy [55]. However, packed column GC is an outdated technology and it is now being replaced by

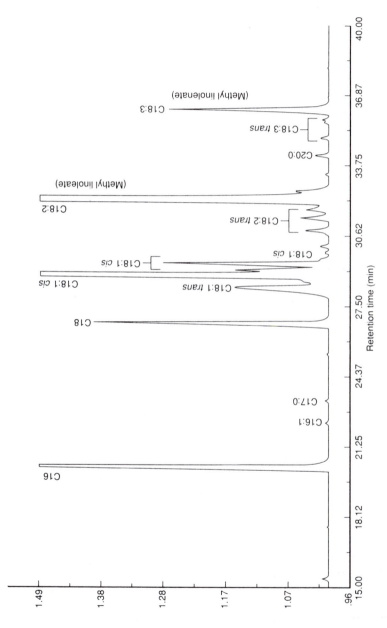

Fig. 4.6. Gas chromatogram of FAME from a partially hydrogenated soybean oil sample using a 60 m × 0.25-mm i.d. fused silica capillary column coated with SP-2340™. Oven temperature program: initial temperature, 150°C; initial hold time, 0 min; program rate 1.3°C/min; final temperature, 200°C; final hold time, 10 min. [64] (Reproduced by kind permission of the American Oil Chemists' Society).

capillary GC technology. For those who are still interested in packed column GC, the review written by Firestone and Sheppard [7] is recommended. This present review focuses only on capillary GC.

Development of capillary columns for *cis-trans* isomer separation occurred concurrently with that of the cyanosilicone packed columns. However, it was not until the late 1970s when glass capillary columns became commercially available that cyanosilicone columns for *cis-trans* isomer separation came into more common use. The availability and use of capillary columns further accelerated with the development of flexible fused silica (FFS) in the early 1980s as a column material, that allowed preparation of stable capillary columns with uniform thickness and uniformity of wall coatings. FFS capillary columns are easy to install and are now widely used for all types of GC analyses.

In the 1970s and 80s, although various types of cyanosilicone capillary columns were available, most isomeric fatty acid analyses were performed using SP-2340 or OV-275 capillary columns [27,56,57-62]. Based on an inter-laboratory study [63], a 60 m × 0.25 mm FFS capillary coated with a 0.2-μm thick film of SP-2340 was adopted by AOCS in 1990 [64] for determination of fatty acid composition and the total *trans* unsaturation in hydrogenated and unhydrogenated oils (other than marine oils) (Figure 4.6). At present, however, 100 m capillary columns coated with newer SP-2560, SP-2380 or CP-Sil 88 stationary phases are considered as most progressive in *trans* fatty acid separation.

i) Oleic acid isomers. Cyanosilicone capillary columns always give excellent baseline separation between a *cis*-18:1 isomer and its corresponding *trans* isomer. For a given double bond position, the *trans* isomer always elutes before the corresponding *cis* isomers, the difference tending to increase with the polarity. However, overlaps could occur between some *cis* and *trans* isomers with different double bond positions. The extent of overlaps may vary slightly from one liquid phase to the other. Ratnayake and Beare-Rogers [65] have observed that on the 60 m SP-2340 column a *cis*-18:1 isomer whose double bond is at the y position overlaps with a *trans*-18:1 isomer having the double bond at the y + 6 position. For example, 18:1 6c overlaps with 18:1 12t, and 18:1 9c with 18:1 15t. Slover *et al.* [59] in their analysis of US margarines noted similar overlaps on the 100 m × 0.25 mm OV-275 capillary column. That means, as observed by many workers [7,44,56-62,65,66] in the analyses of PHVO and ruminant fats using SP-2340 or OV-275 capillary columns, the 18:1 6t to 18:1 11t isomers are readily separable from the *cis*-18:1 isomers but the *trans*-18:1 isomers with double bonds from 12 to 16 positions overlap into the *cis* region. As illustrated in Figure 4.7, the main overlap takes place in the major 18:1 peak. This peak (3 in Figure 4.7) contains five *cis* isomers (18:1 6c to 18:1 10c) and four *trans* isomers (18:1 12t to 18:1 15t). The 18:1 13t, 18:1 14t and 18:1 15t isomers are completely immersed among the early eluting *cis* isomers, but the 18:1 12t isomer appears usually as a non-quantifiable shoulder on the front side of the major 18:1 peak. Column overload or the presence of

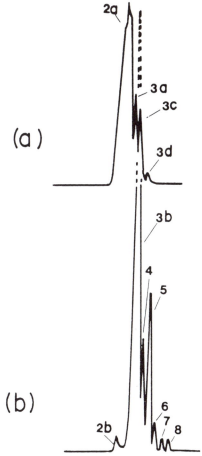

Fig. 4.7. Gas chromatogram of fractions separated by Ag-TLC (a) *trans*-18:1 and (b) *cis*-18:1 FAME fractions of a partially hydrogenated vegetable oil sample (60 m × 0.25-mm i.d. fused silica capillary column coated with SP-2340™; Oven temperature program: initial temperature, 160°C; initial hold time, 0 min; program rate 1.5°C/min; final temperature, 200°C; final hold time, 10 min.) [65]. (Reproduced by kind permission of the *Journal of Chromatographic Science*).

large amounts of *cis* isomers affects the resolution of the 18:1 12*t* isomer. Sometimes quantifiable partial separation of the 18:1 12*t* is possible in samples containing high levels of *trans* unsaturation (usually >30%) and relatively low levels of *cis* isomers. In such samples, the separation could be improved by performing the analysis using dilute samples and a smaller sample load. These *cis-trans* isomer overlaps place a constraint on accurate determination of *trans* unsaturation by GC alone using either the SP-2340 or OV-275 capillary column. In some margarines, the underestimation of *trans*-18:1 content determined using 60 m SP-2340 capillary FFS column according to the AOCS official method Ce 1c-89 can be as high as 32% [65]. In PHVO, the proportion of *trans*-18:1 isomers with double bonds further from the carboxyl group (from 18:1 12*t* to 18:1 16*t*) may depend on the hydrogenation conditions and

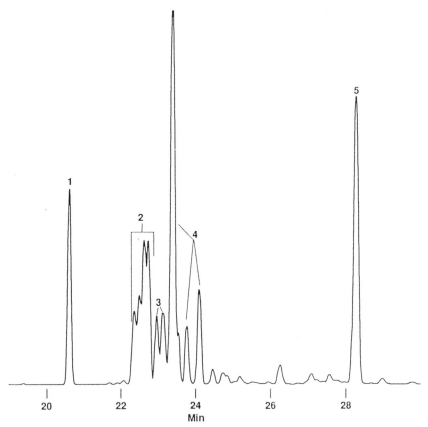

Fig. 4.8. The C$_{18}$ region of a gas chromatogram of FAME from a partially hydrogenated vegetable oil sample (100 m × 0.25-mm i.d. × 20 μm film fused silica capillary column coated with SP-2380™; operated isothermally at 175°C). Peak identification. 1, 18:0; 2, 18:1 6*t*-11*t*; 3, 18:1 12*t*-14*t*; 4, 18:1 6*c*-12*c*; 5, 18:2*n*-6. (Reproduced by kind permission of the authors [75]).

the source oil, and this will result in variation in the extent of overlaps of the isomers from one PHVO to another.

Recently several laboratories demonstrated improved *cis-trans* 18:1 isomer separation using capillary columns coated with SP-2560, SP-2380 or CP-Sil 88 liquid phases [32,33,37,48,49,51,67-78]. These phases are less polar than OV-275 [75]. SP-2380 is a relatively new liquid phase and is slightly less polar than SP-2560 due to a 5% phenyl substitution in its polymer backbone (Table 4.1) [75,78]. It is a stabilized phase and therefore offers a higher maximum temperature than SP-2560 (275°C versus 250°C)]. In these three phases, the overlaps between *cis-* and *trans-*18:1 isomers are minimized. Most importantly, as illustrated in Figures 4.8-10, the 18:1 12t isomer could be readily resolved and the 18:1 13t and 18:1 14t isomer pair, which always elute together, could also be separated from the major *cis* isomer peak. In addition to these improvements, as illustrated in Figure 4.10, a 100 m × 0.25 mm SP-2560 FFS capillary column is capable of resolving the 18:1 16t isomer from the *cis*

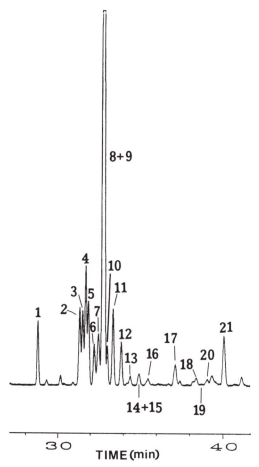

Fig. 4.9. The C$_{18}$ region of a gas chromatogram of FAME from a partially hydrogenated canola oil sample (100 m × 0.25-mm i.d. × 20 μm film fused silica capillary column coated with CP-Sil™; Operated isothermally at 170°C). Peak identification: 1, 18:0; 2,18:1 6*t*-8*t*; 3,18:1 9*t*; 4, 18:1 10*t*; 5, 18:1 11*t*; 6,18:1 12*t*; 7,18:1 13*t*+14*t*; 8,18:1 6*c*-9*c*; 9,18:1 15*t*; 10,18:1 10*c*; 11,18:1 11*c*; 12,18:1 12*c*; 13,18:1 13*c*; 14,18:1 16*t*; 15,18:1 14*c*; 16,18:1 15*c* +18:2*tt*; 17,18:2 9*c*,13*t*+18:2 8*t*,12*c*; 18, 18:2 9*c*,12*t*; 19,18:1 16*c*; 20,18:2 9*t*,12*c*; 21,18:2*n*-6 (W.M.N. Ratnayake, unpublished work).

isomers; it elutes between the 18:1 13*c* and 18:1 14*c* isomers. This separation is not possible on a 100 m CP-Sil 88 column, in which the 18 16t isomer overlaps with 18:1 14*c* (Figure 4.9). There are no data on separation of the individual *trans* isomers on SP-2380 and therefore, at present it is not known whether the 18:1 16*t* isomer could be separated on this column [75]. Clearly, at present a 100-m SP-2560 column could be considered as the most efficient column for *cis-trans* 18:1 isomer separation. The 18:1 15*t* is the only isomer that remains unresolved on the 100-m SP-2560 column (Figure 4.10). This is not a drawback, since 18:1 15*t* is a minor component in PHVO and ruminant fats. Its level normally does not exceed 1.3-1.5% % of total *trans*-18:1 isomers. The equivalent chain length (ECL) values on SP-2560 capillary

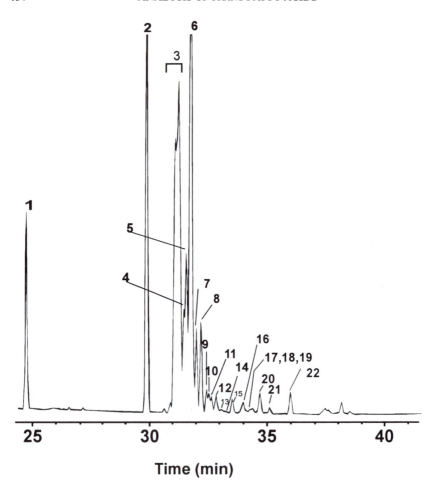

Fig. 4.10. The C₁₆, C₁₈ and C₂₀ regions of a gas chromatogram of FAME from a partially hydrogenated canola oil sample (100 m × 0.25-mm i.d. × 20 μm film fused silica capillary column coated with SP-2560™; Oven temperature program: initial temperature, 165°C; initial hold time, 0 min; program rate 1.5°C/min; final temperature, 210°C; final hold time, 25 min). Peak identification: 1, 16:0; 2, 18:0; 3,18:1 6*t*-11*t*; 4,18:1 12*t*; 5, 18:1 13*t*+14*t*; 6, 18:1 6*c*-10*c* +18:1 15*t*; 7,18:1 11*c*; 8,18:1 12*c*; 9,18:1 13*c*; 10,18:1 16*t*;11,18:1 14*c*; 12,18:1 15*c*; 13,18:2 *tt*; 14,18:2 *tt*; 15,18:2 9*c*,13*t*+18:2 8*t*,12*c*; 16, 18:2 9*c*,12*t*; 17,18:1 16*c*; 18,18:2 *ct*; 19,18:2 9*t*,12*c*; 20,18:2n-6; 21,18:2 9*c*,15*c*; 22, 20:0.(W.M.N. Ratnayake, unpublished work).

column of 18:1 isomers encountered normally in PHVO are shown in Table 4.2.

The column length appears to have only a minimal influence on the *cis-trans* isomer separation on the above cyanosilicone phases. Wolf and Bayard [77] noted that a 100 m CP-Sil capillary column produces a slightly better resolution of the individual *trans*-18:1 isomers compared to a 50 m CP-Sil 88 column. Sidisky *et al.* [75] observed that increasing the column length from 100 to 150 metres of SP-2560, SP-2380 and OV-275 resulted in only minimal improvements in *cis-trans* isomer separation, but was coupled with increased

TABLE 4.2

Equivalent chain length (ECL) values of 18:1 and 18:2 FAME isomers normally encountered in partially hydrogenated vegetable oils on an SP-2560 capillary column (100 m × 0.25 mm i.d.) at 180°C.

FAME	ECL	FAME	ECL
18:0	18:00	18:2 *tt*[a]	18.96
18:1 8*t*	18.43	18:2 *tt*[a]	18.99
18:1 9*t*	18.46	18:2 *tt*[a]	19.08
18:1 10*t*	18.48	18:2 *tt*[a]	19.09
18:1 11*t*	18.52	18:2 9*t*,12*t*	19.16
18:1 12*t*	18.55	18:2 9*c*,13*t*	19.24
18:1 13*t*	18.61	18:2 8*t*,12*c*	19.24
18:1 14*t*	18.61	18:2 9*c*,12*t*	19.36
18:1 15*t*	18.63	18:2 8*c*,13*c*[b]	19.43
18:1 16*t*	18.92	18:2 9*t*,12*c*	19.45
18:1 8*c*	18.52	18:2 9*t*,15*c*	19.46
18:1 9*c*	18.63	18:2 10*t*,15*c*[b]	19.46
18:1 10*c*	18.63	18:2 9*c*,13*c*[b]	19.49
18:1 11*c*	18.73	18:2 9*c*,12*c*	19.58
18:1 12*c*	18.80	18:2 9*c*,14*c*[b]	19.60
18:1 13*c*	18.88	18:2 9*c*,15*c*	19.63
18:1 14*c*	18.97		
18:1 15*c*	19.00		
18:1 16*c*	19.45		

[a] double bond positions unknown
[b] tentative identification
(Unpublished work of W.M.N. Ratnayake)

analysis times. Overall, 100 m columns provided the best resolution of geometrical and positional 18:1 isomers, with reasonable analysis times.

On the other hand, the column temperature affects the *cis-trans* isomer resolution greatly. In the author's laboratory (unpublished data), on 100 m SP-2560 FFS, the least overlaps of the *cis-trans*-18:1 isomers were achieved when the column temperature was programmed from an initial temperature of 165°C to a final temperature 210°C at a rate of 1.5°C per min (Figure 4.10). In this program, as pointed out earlier, all the *trans*-18:1 isomers, except the *trans*-15 isomer, are separated from the *cis*-18:1 isomers. Minor drawbacks of this analysis are that 18:1 16*c* isomer is only partially resolved from the *trans* 18:2 isomers and there is less satisfactory separation between the individual *trans* isomers; particularly of the early eluting isomers (18:1 6*t* to 18:1 11*t*). However, information regarding individual isomers is not necessary in many circumstances. Nevertheless, individual *trans* isomers could be easily separated by operating the column isothermally at 165°C (Figure 4.11). The 18:1 16*c* isomer is also well resolved from the *trans*-18:2 isomers (Figure 4.11). But this results in complete overlap of the 18:1 13*t* + 14*t* pair with the major *cis*-18:1 peak. Isothermal operation of CP-Sil 88 at 170°C gives a similar elution pattern also (Figure 4.9).

The influence of column temperature on the separation of *cis-trans*-18:1

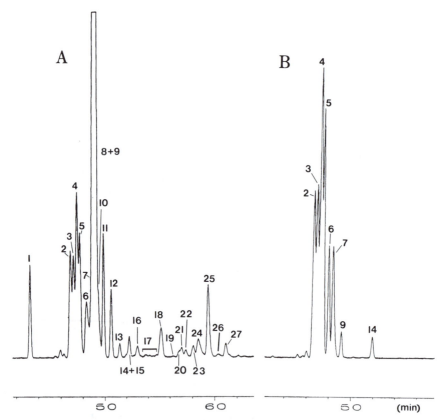

Fig. 4.11. Gas chromatogram of (A) the FAME C$_{18}$ region and (B) *trans* 18:1 FAME fraction isolated by Ag-TLC from a partially hydrogenated canola oil sample (100 m × 0.25-mm i.d. × 20 *μ*m film fused silica capillary column coated with SP-2560™; Operated isothermally at 165°C). Peak identification; 1, 18:0; 2,18:1 6*t*-8*t*; 3,18:1 9*t*; 4,18:1 10*t*; 5,18:1 11*t* 6,18:1 12*t*; 7, 18:1 13*t*+14*t*; 8, 18:1 6*c*-9*c*; 9,18:1 15*t*; ; 10,18:1 10*c*; 11,18:1 11*c*; 12,18:1 12*c*; 13,18:1 13*c*; 14,18:1 16*t*; 15,18:1 14*c*; 16,18:1 15*c*; 17,18:2 *tt*; 18,18:2 9*c*,13*t*+18:2 8*t*,12*c*; 19,18:2*tt*; 20,18:2 9*c*,12*t*; 21,18:2 8*c*,13*c*; 20,18:2 9*c*,12*t*; 22,18:1 16*c*; 23,18:2 9*t*,12*c*; 24,18:2 10*t*,15*c* + 18:2 9*t*,15*c*; 25,18:2n-6; 26,18:2 9*c*,15*c*; 27, 20:0 (W.M.N. Ratnayake, unpublished work).

isomers of milk fat FAME on 100 m CP-Sil 88 was evaluated by Molkentin and Precht [71,72] and the best separation of 18:1 isomers of milk fat is obtained by isothermal operation at 175°C. This operating temperature also gives good separation of the 18:1 isomers of PHVO origin, but still the *trans* isomers with remote double bonds of 18:1 overlap with the *cis* isomers [79].

ii) Linoleic and α-linolenic acid isomers. In the analysis of isomeric fatty acids, the main focus has always been on the *cis* and *trans* isomers of oleic acid, which are the most abundant isomeric group in many dietary fats. However, for accurate determination of the fatty acid profile of dietary fats, the isomers of linoleic and linolenic acids, which are frequently present in low concentrations in both partially hydrogenated and non-hydrogenated dietary fats [80-85], should also be taken into consideration. The PUFA isomers found in non-hydrogenated fats or in many common dietary fats are the result of exposure of linoleic or linolenic acids to some form of heat treatment; such as steam deodorization or stripping during refining of oils [80,84,85] or simple heating in deep fat frying (see Chapter 2) [86]. In these processes, the double bonds do not shift in position, but isomerize from *cis* to *trans*, resulting in the formation of small amounts of geometrical *trans* isomers. This isomerization usually occurs at temperatures higher than 200°C. α-Linolenic acid is more prone to isomerization than linoleic acid, whereas monoenoic acids do not isomerize at all [80-88]. Among the three double bonds in α-linolenic acid, the two external double bonds (*i.e.* 9 and 15) are more susceptible to isomerization than the 12 central double bond [88]. *Trans* isomer formation during refining of oils is temperature- and time-dependent, but under severe conditions, total levels up to 3.5% have been reported [86].

The three possible *trans*-geometrical isomers of linoleate, *i.e.* *trans*-9,*trans*-12; *cis*-9,*trans*-12 and *trans*-9,*cis*-12 are readily separated in the order stated on capillary columns coated with polar cyanosilicone columns [47,57,89]. The 18:2 isomer group of PHVO is more complex than that of non-hydrogenated fats. In addition to the above three geometric isomers of linoleic acid, a PHVO may contain at least 16 other 18:2 isomers [2,34,90]. As shown in Figures 4.9-11 and Table 4.2, the elution patterns and the equivalent chain length (ECL) values of the major isomers and many of the minor isomers have been established [2]. In general, the isomers are eluted in the order *trans,trans* < *trans,cis* < *cis,trans* followed by *cis,cis*. ECL values on FFS capillary columns of different stationary phases for a number of unusual synthetic *cis,cis*-18:2 isomers with more than one methylene group were also recently compiled by Christie [44]. Fatty acids of this type are not encountered in common vegetable oils or dietary fats, but can be present in trace levels in natural sources, especially lipids of marine invertebrates.

Except for the three geometrical isomers, 18:2 reference standards are not commercially available and therefore the analyst may find difficulties in correctly identifying the 18:2 isomers in PHVO. In this situation, the elution patterns shown in Figures 4.9-11 should be useful, but caution should be exercised particularly in identifying the *trans*-9,*trans*-12 isomer. In some cyanosilicone capillary columns (*e.g.* SP-2340), this isomer overlaps with 18:2 9*c*,13*t*, which is very often the major 18:2 isomer in PHVO [2] and ruminant fats [author's unpublished work]. Because of this overlap, in the past, the major 18:2 isomer in Canadian margarines [91] and blood samples of patients with coronary artery disease [92] was misidentified as *trans*-9,*trans*-12-18:2. Correct identification of this isomer is rather important [93], because it has been suggested that this isomer interferes with the metabolism of linoleic acid and the biosynthesis of prostaglandins [94-97]. An expert committee of Health and Welfare Canada in 1980 recommended that the *trans,trans*-octadecadienoic isomer content in Canadian margarines should be less than 1% of the total fatty acids [98]. As shown by Ratnayake and Pelletier [2], preparative Ag-TLC followed by GC and GC/MS analyses allow unambiguous identification and

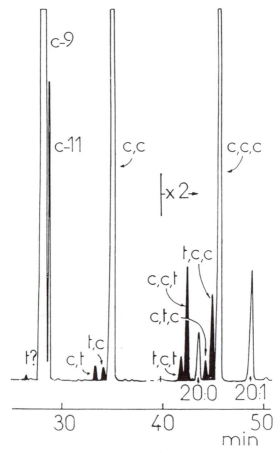

Fig. 4.12. Partial chromatogram of FAME prepared with a sample of commercial deodorized rapeseed oil. Analysis on a CP-Sil 88™ fused silica capillary column (50 m × 0.33 mm i.d., 0.24 μm film; operated isothermally at 160°C). The configurations of the double bonds are given in the order 9 and 12 for 18:2*n*-6 geometrical isomers, and in the order 9, 12 and 15 for 18:3*n*-3 geometrical isomers. (Reproduced by kind permission of the author [83] and *Journal of the American Oil Chemists' Society*).

quantification of *cis*-9,*trans*-13 and *trans*-9,*trans*-12 as well as other 18:2 isomers in PHVO.

α-Linolenic acid has eight geometric isomers, but usually only four are present in industrially deodorized vegetable oils [80,82,83]. They have been identified as 18:3 9*t*,12*c*,15*t* (*tct*; minor), 18:3 9*c*,12*c*,15*t* (*cct*; major),18:3 9*c*,12*t*,15*c* (*ctc*, minor) and 18:3 9*t*,12*c*,15*c* (*tcc*, major) [80]. These isomers are also common in margarines and other common foods prepared from partially hydrogenated canola oil or soybean oil [99,100]. These isomers give peaks that can be recognized readily in GC analyses on cyanosilicone capillary columns [49,65,80,88,101,102] and are eluted from the column in the order stated above (Figure 4.12). All of these isomers precede the all *cis* α-linolenic

TABLE 4.3

Chromatographic characteristics of geometrical isomers of 18:3
FAME on a CP Sil 88 capillary column (50 m × 0.25 mm i.d.;
isothermal operation at 150°C). Adapted from Wolff [49].

18:3 isomer	Rf (Ag-TLC)	ECL (CP-Sil 88)
9t,12t,15t	0.65	19.71
9c,12t,15t	0.47	19.84
9t,12c,15t	0.47	19.87
9t,12t,15c	0.47	19.96
9c,12c,15t	0.35	19.91
9c,12t,15c	0.35	20.06
9t,12c,15c	0.35	20.11
9c,12c,15c	0.11	20.15

acid. Minor amounts of other geometric isomers of α-linolenic acid have been reported in oils heated at very high temperatures for very long periods but are not encountered in common dietary fats. Wolff [49,83] recently demonstrated that the eight geometric isomers of α-linolenic acid are resolved into seven peaks and elute in the order *ttt, ctt, tct, ttc, cct, ctc, tcc* and *ccc* on a CP-Sil 88 capillary column. Their ECL and retardation values (Rf) on Ag-TLC are shown in Table 4.3.

Although the elution order of the geometric isomers of α-linolenic acid has been established, care should be taken not to confuse them with 20:0 or 20:1. Wolff [102] demonstrated that, depending on the temperature of the column, the 20:1 11c isomer, which is present in appreciable amounts in some vegetable oils (*e.g.* canola oil and peanut oil), may elute before, with or after α-linolenic acid in GC on a CP Sil 88 column (Figure 4.13). Generally the lower temperatures (<165°C) are best suited to separate the *cis*-20:1 acid. At column temperatures above 180°C, the 20:1 acid co-elutes with the *tcc* isomer. SP-2560 capillary columns give similar overlap problems also (unpublished data). If not aware of this problem, this might lead to overestimation of the *trans* geometrical isomers of α-linolenic acid. It is imperative to check the behaviour of 20:1 11c relative to α-linolenic acid and its geometric isomers each time a new cyanosilicone capillary column is installed or when the column temperature or other operating parameters are changed. Dutchateau *et al.* [79] recently proposed that resolution between 20:1 and 18:3 isomers should be considered as one of the suitability criteria in the selection of the GC column and conditions for analysis of PHVO and refined oils by a single step direct GC method. Using a computer-assisted optimization procedure, Dutchateau *et al.* [79] have determined that for 50 m × 0.25 mm CP-Sil, 60 m × 0.25 mm SP-2340 and 50 m × 0.22 mm BPX70 FFS columns, isothermal operations at 175, 192 and 198°C, respectively, give optimal separation of the various *cis* and *trans* isomers of mono-, di- and triethylenic fatty acids.

iii) Trans *isomers of arachidonic, eicosapentaenoic and docosahexaenoic*

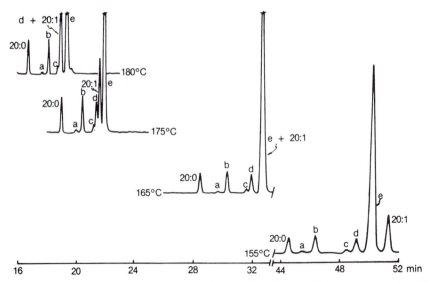

Fig. 4.13. Influence of temperature on the elution order of *cis*-11 20:1 and *α*-linolenic acid geometrical isomers. Selected chromatograms obtained with a CP-Sil 88™ capillary column (50 m × 0.25 mm i.d., 0.20 *μ*m film) of FAME prepared with a sample of commercial deodorized rapeseed oil. Temperature of the column as indicated on the chromatograms. All injections at the same load. Identification of peaks: (a) 18:3 9*t*,12*c*,15*t*; (b) 18:3 9*c*,12*c*,15*t* (c) 18:3 9*c*,12*t*,15*c*; (d) 18:3 9*t*,12*c*,15*t*; (e) 18:3 9*c*,12*c*,15*c*. 20:1 is *cis*-11 isomer. (Reproduced by kind permission of the author [102] and *Journal of the American Oil Chemists' Society*).

acids. Trans isomers of arachidonic acid were reported in rodent tissues after feeding diets containing *trans*-18:2 isomers [68,69,95]. Several *trans* isomers of eicosapentaenoic acid [101,103] and a 19-*trans* isomer of docosahexaenoic acid [101] were found in rats fed *trans*-18:3 isomers. Long-chain *trans*-PUFA have also been found in human platelets [104] and human breast milk [73]. These C_{20} and C_{22} unusual *trans*-PUFA have been satisfactory separated on cyanosilicone capillary columns [68,69,73,101,104]. Figure 4.14 illustrates the order of elution of some of these long-chain PUFA relative to other fatty acids on a 100 m SP-2560 FFS column.

   iv) Isomeric fatty acids of hydrogenated fish oils. The fatty acid profile of PHFO are far more complex than that of PHVO. In PHFO, long-chain fatty acids (C_{20}-C_{24}) constitute up to 40% or even more of total fatty acid content [105,106], whereas in PHVO the figure is usually less than 1%. Although *trans*-monoenoic acids of chain length C_{16} to C_{22} comprise the major proportion of *trans* fatty acids in commercially PHFO, variable quantities of *trans,trans*-, *cis,trans*- and *trans,cis*-dienoic acids are also present [3,4,105,106]. *Trans* isomers of trienoic and higher polyunsaturated fatty acids may be present in minor quantities [105]. It is possible to separate these complex mixtures to some extent on cyanosilicone capillary columns [107] (Figure 4.15). However, as for PHVO, the main obstacle is the incomplete separation of the *trans* from *cis*-monoethylenic isomers of the various chain-lengths. In addition, some chain-length overlaps (*e.g.* 18:3 and 20:1 isomers) may occur on very polar

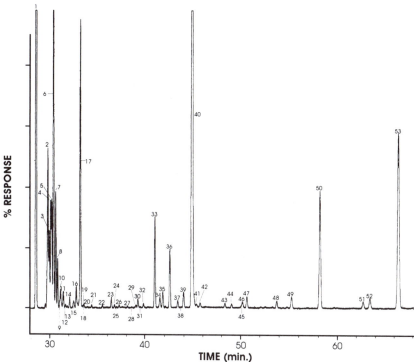

Fig. 4.14. The C$_{18}$, C$_{20}$ and C$_{22}$ regions of a gas chromatogram obtained on a SP-2560™ capillary column (100 m × 0.25-mm i.d. × 20 μm film; operated isothermally at 180°C) of FAME from liver phospholipids of rats fed partially hydrogenated canola oil. Peak identification: 1, 18:0; 2,18:1 8*t*-9*t*; 3,18:1 10*t*; 4, 18:1 11*t*; 5,18:1 12*t*; 6, 18:1 13*t*+14*t* + 18:1 6*c*-10*c* +18:1 15*t*; 7,18:1 11*c*; 8,18:1 12*c*; 9, 18:2*tt*; 10,18:1 13*c*; 11,18:1 16*t* + 18:1 14*c*; 12, 18:2*tt* + 18:1 15*c*; 13,18:2 9*t*,12*t*; 14,18:2 9*c*,13*t* + 18:2 8*t*,12*c*; 15, 18:2 9*c*,12*t* + 18:2 8*c*,13*c* + 18:2 9*t*,13*c*; 16,18:1 18:2 9*t*,12*c* + 18:2 10*t*,15*c* +18:2 9*t*,15*c* + 18:2 9*c*,13*c*; 17, 18:2*n*-6; 18,18:2 9*c*,14*c*; 19, 18:2 9*c*,15*c*; 20, unknown; 21,18:2 12*c*,15*c*; 22, 20:0; 23, 20:1 11*c*; 24, 20:1 12*c*; 25, 18:2 conjugates; 26, 20:1 13*c*; 27, 18:2 conjugates; 28, 18:2 conjugates; 29, 20:2 11*c*,14*t*; 30, 20:2 8*c*,14*c*; 31, 20:2 11*t*,14*c*; 32, 20:2*n*-6; 33, 20:3*n*-9; 34, 20:3 5*c*,8*c*,14*c*; 35, 20:3 5*c*,11*c*,14*c*; 36, 20:3*n*-6; 37, 20:4 5*c*,8*c*,11*c*,15*t*; 38, unknown; 39,20:4 5*c*,8*c*,11*c*,14*t*; 40, 20:4*n*-6; 41, 22:1; 42, 22:1; 43, 22:2; 44, 22:2; 45, 20:5*n*-3; 46, unknown; 47, unknown; 48, 24:0; 49, 24:1; 50, 22:5*n*-6; 51, 22:5*n*-3; 52, *t*-22:6*n*-3; and 53, 22:6*n*-3 [69] (Reproduced by kind permission of *Lipids*).

capillary columns [102]. Partial analysis of PHFO is possible by combining GC with other separation techniques, particularly Ag-TLC [105,107]. The combined techniques are described below.

D. GAS CHROMATOGRAPHY-INFRARED SPECTROSCOPY

From the discussion presented previously, it is evident that almost quantitative fatty acid data for PHVOs can be obtained by direct GC analysis employing a 100 m capillary column coated with either SP-2560 or CP-Sil88 or a liquid phase of equivalent efficiency and selecting the correct column temperature program. Direct GC analysis may produce minor overlaps of *cis* and *trans*-18:1 isomers (particularly 18:1 15*t* overlaps with *cis*-18:1 isomers), but the data should be accurate enough for many applications, including food labeling and quality control work. However, if higher accuracy is required GC

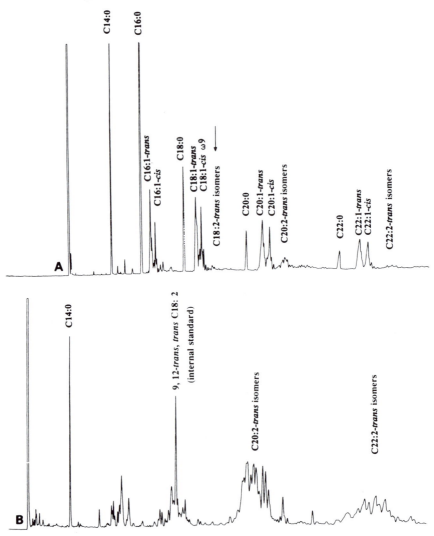

Fig. 4.15. Gas chromatogram of FAME of partially hydrogenated fish oil before (A) and after (B) separation by Ag-TLC of the diene fraction. GLC conditions; SP-2560™ (100 m × 0.25 mm i.d.) Oven temperature profile; initial temperature 175°C, initial hold time 0 min; program rate, 2°C/min; final temperature 190°C. The arrow in (A) indicates the position of 9,12 *trans,trans* 18:2 used as an internal standard. In (B) the internal standard is seen in the bunch of *trans,trans* 18:2 isomers. All the diene isomer fractions contain *trans,trans*; *trans,cis*; *cis,trans* isomers. Retention times: C16 26.50 min; C18:0, 33.14 ; internal standard 9,12 *trans,trans* 18:2, 38.47 min: C20:2 *trans* isomers approximately 50 min, and C22:2 *trans* isomers approximately 72 min. (Reproduced by kind permission of the authors [107] and Journal of Lipid Research).

analysis should be used in conjunction with other analytical techniques. Ratnayake *et al.* [35] in 1990 proposed use of a combined capillary GC and IR method for accurate determination of total *trans*-18:1, total *cis*-18:1 and general fatty acid composition in PHVO. In this method, the total *trans*-

unsaturation determined by IR was correlated to the capillary GC weight percentage of the component *trans* FAME by a linear equation:

$$IR\ trans = \%trans\text{-}18{:}1 + 0.84(\%trans\text{--}18{:}2) + 1.74(\%trans,trans\text{--}18{:}2) + 0.84(\%trans\text{--}18{:}3)$$

where 0.84, 1.74 and 0.84 are correction factors relating GC weight percentages to the IR *trans*-equivalents for mono-*trans* octadecadienoic (*trans*-18:2), *trans,trans*-octadecadienoic (*trans,trans*-18:2) and mono-*trans*-octadecatrienoic (*trans*-18:3) acids respectively. The formula forms the basis for determining the total *trans*- and *cis*-18:1 in margarines. GC provides the proportions of *trans*-18:2, *trans,trans*-18:2 and *trans*-18:3, because these isomers are easily separated on GC without any serious overlaps or interference from other isomers (Figures 4.6,8-11). Whereas IR gives the total *trans* content and therefore, total *trans*-18:1 content is calculated from the mathematical formula. The total *cis*-18:1 obtained as the difference between total 18:1 fatty acids (which is obtained directly from GC) and *trans*-18:1. This combined GC/IR procedure after an international collaborative study [32] was adopted as an official method of AOAC (Method 994.15) [108] and AOCS (method Cd 14b-93) [109] for determination of the fatty acid composition of dietary fats made from PHVO. This method is routinely used in the author's laboratory and it was used to determine the fatty acid composition of margarines [99] and common bakery foods [100] available in Canada. However, the method should be used with caution. This is because of the difficulty of obtaining reproducible results with IR (discussed previously), particularly with samples containing low (<5%) levels of *trans* fatty acids. Furthermore, the method is applicable only to samples containing C18 unsaturated fatty acids, such as those derived from PHVO. The method cannot be used for fats containing large amounts of unsaturated fatty acids of chain-lengths other than C18, as for example PHFO.

E. SILVER NITRATE THIN-LAYER CHROMATOGRAPHY

The basic principle of silver ion chromatography is straight forward and involves formation of a coordination complex by interaction of *pi* electrons of carbon-carbon double bonds (or triple bonds) in the carbon chain of the fatty acid molecule with silver ions. The greater the number of double bonds the stronger the complexation. Because of steric hindrance, *trans* double bonds form weaker complexes compared to *cis* double bonds. Thus silver ion chromatography (thin-layer and high-performance liquid) is an effective means of fractionation of fatty acids into simpler fractions on the basis of the number, the configuration and to some extent the positions of the double bonds. The methodology and applications to fatty acids and other lipids have been reviewed extensively by Nikolova-Damyanova [110]. Ag-TLC fractionation followed by analysis of the fractions by capillary GC or GC/MS allows complete and accurate analysis of *trans* fatty acids, especially the mono- and diethylenic isomers.

The first step in Ag-TLC is the preparation of the TLC plate with a uniform layer of silver nitrate. This could be achieved using any of the several techniques given in the literature [110]. In the author's laboratory, Ag-TLC plates are prepared by the following simple procedure. Commercial pre-coated silica gel G TLC plates (thickness 0.2 to 0.3 mm for analytical work and 0.5 mm for preparative work) are initially cleaned by developing in a TLC tank containing ethyl acetate. This helps to remove impurities such as phthalates, dust particles and other air borne-impurities that might be present on the plate. After briefly drying in an oven at 100°C, the plate is placed, horizontally, the side containing the silica gel facing downwards, in a glass tray (such as a cake baking tray) containing a 10% solution of silver nitrate in acetonitrile, for 30 minutes. The glass tray is covered with aluminium foil to minimize the evaporation of acetonitrile. The plate is activated by heating in an oven at 110°C for about one hour. If the plate is not used immediately, it should be stored in a desiccator over drying agents in a dark place. Plates produced by this procedure have a uniform layer of silver nitrate.

Usually, the fatty acids are subjected to Ag-TLC in the form of methyl ester derivatives. The FAMEs are applied to the plate as a solution in hexane or other non-polar organic solvent. 100% Toluene or mixtures of toluene-hexane, hexane-diethyl ether and hexane-chloroform in different proportions are frequently used as developing solvent. The plates are normally developed at room temperature, but the resolution of some positional 18:1 [111] and 18:2 [2,69] isomers may be improved by developing at temperatures at about -20°C or -25°C. When analysing fish oils or other highly unsaturated oils, to minimize possible oxidation, the plates should be developed in a dark place. The separated bands, made visible under ultraviolet light by spraying the plate with an 0.1% solution of 2',7'-dichlorofluorescein in ethanol, are scraped off and extracted with diethyl ether or any other suitable solvent. A 1:1 mixture of hexane-chloroform is particularly effective when extracting highly unsaturated FAME. This extracted material is very often contaminated with trace amounts of silver nitrate and dichlorofluorescein, which can be removed by washing the extract with bicarbonate, ammonia or sodium chloride solutions [112]. A slightly different technique is used in the author's laboratory [69]; 1% solution of sodium chloride in 90% methanol is added to the silica gel band until the silver-dichlorofluorescein complex disappears. Water is added and the methyl esters extracted with hexane.

An important application of Ag-TLC is for the determination of *trans*-and *cis*-monoene contents of PHVO. As shown in Figure 4.16 Ag-TLC gives a good separation of *trans*-18:1 as a group from the *cis*-18:1 fatty acid and other unsaturated fatty acids in PHVO. Incorporation of a known amount of methyl penta- or heptadecanoate as internal standard to the *trans*-18:1 or *cis*-18:1 fractions permits quantification by GC. Such a procedure was standardized by IUPAC for determination of *trans*-18:1 in natural and hydrogenated animal and

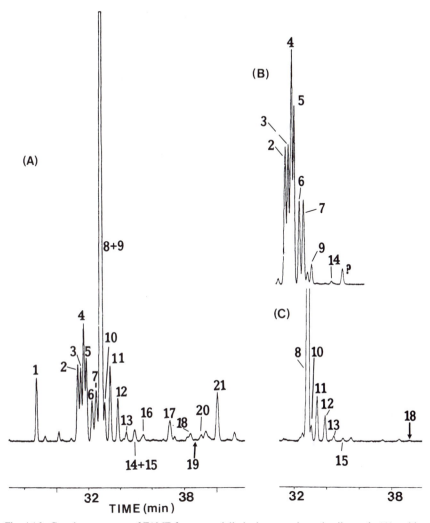

Fig. 4.16. Gas chromatogram of FAME from a partially hydrogenated canola oil sample (A) and its *trans* (B) and *cis* (C) 18:1 isomer fractions isolated by Ag-TLC (developed in toluene at room temperature). The GC conditions and peak identifications are as in Fig. 4.9 (W.M.N. Ratnayake, unpublished work).

vegetable oils and fats (IUPAC method 2.302, [113]). In practice, however, as proposed by Christie and co-workers [29,114,115], a convenient means of quantification of *trans*-18:1 involved collection of both the saturated and *trans*-18:1 fatty acids together as a single fraction. This fraction is analysed by GC to determine the ratio of *trans*-18:1 to SFAs and then the absolute amount of the former is calculated from the unfractionated sample (determined prior to Ag-TLC); the SFA components of the sample serve as an internal standard. This technique eliminates the errors resulting from sample application, scraping losses, incomplete extraction, weighing of small quantities of internal standard and isolated bands. The simplest method, however, is to treat methyl palmitate [54] or methyl stearate [116] instead all the SFA, in the sample as the internal standard.

Another approach is to treat the *trans*-18:1 isomers with double bonds close to the carboxyl group (from 6 to 11) as the internal standard [73,74]. This is possible because, as discussed earlier, the 18:1 6*t* to 18:1 11*t* isomers are always well resolved from the *cis*-18:1 isomers on polar cyanosilicone capillary GC columns (Figures 4.11,8-11,16). In this method, the proportion of the *trans*-18:1 isomers from 12 to 16 (the isomers with remote double bonds) that overlapped with the *cis*-18:1 isomer peak on capillary GC is calculated by comparing the 18:1 region of the GC chromatogram of the isolated *trans*-18:1 fraction with that of the parent FAME prior to Ag-TLC fractionation. This calculation is done with respect to the well separated *trans*-18:1 isomers with double bonds close to the carboxyl group. The total *trans*-18:1 content is then calculated by summing the proportion of the *trans*-18:1 isomers (18:1 12*t* to 18:1 16*t*) that overlapped with the *cis* isomers with the well-separated *trans*-18:1 (from 18:1 6*t* to 18:1 11*t*) isomers. This also allows calculation of the total *cis*-18:1 content. An attractive feature of this approach, is that it requires isolation of only the *trans*-18:1 band and it is not necessary to include any SFAs in the *trans*-18:1 fraction. This procedure was used to determine the fatty acid composition in Canadian human milk [33,67,73] and adipose tissue samples [74]. The plates were developed in toluene at -20°C and GC analyses of the isolated fractions were performed on a 100 m SP-2560 capillary column.

Most of the Ag-TLC separations reported so far have been performed with glass plates coated with silica gel. However, Ulberth and Henninger [116] have demonstrated that pre-coated plastic-backed sheets with 0.2-mm layer of silica gel (Merck Cat. No. 5748) impregnated dynamically with 10% (w/v) silver nitrate in acetonitrile [117] give equally good separation of *trans* fatty acids in PHVO. Ulberth and Henninger [116] developed the sheets in a similar way to glass plates with n-hexane-diethyl ether (9:1) and sprayed with 0.05% rhodamine B in ethanol. For isolation of the separated bands, the regions containing the relevant fractions were cut off with a pair of s*cis*sors and the cuttings were placed in a small bottle containing 5 mL diethyl ether. Extraction was complete in 30 min. No differences in chromatographic behaviour between conventional Ag-TLC and pre-coated Ag-TLC sheets were noted [116]. Advantages of the TLC-sheets include i) cutting out the relevant fractions with a pair of s*cis*sors is easier than scraping off the plates, thereby avoiding possible cross contamination; ii) the lengthy extraction procedure is substituted by equilibration of the cuttings in diethyl ether and iii) rhodamine B is insoluble in diethyl ether, thus rendering washing steps unnecessary.

Ag-TLC, to a limited extent, has the ability to separate different positional isomers [26,118-120]. The positional isomers of *cis* and *trans*-18:1 migrate in the form of a sinusoidal curve (Figure 4.17) [118]. The positional isomer separation may be important in certain applications (*e.g.* collection of sub-

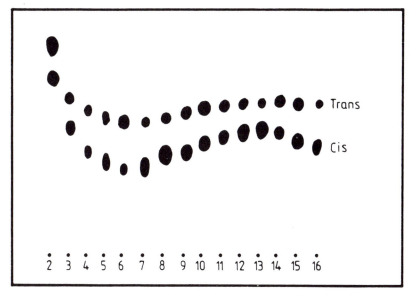

Fig. 4.17. Ag-TLC of methyl *cis-* and *trans-*octadecenoates. The plate was impregnated with 15% (w/w) silver nitrate and developed with dibutyl ether-hexane (40:60 v/v); spots were visualized by heating with a glass-blower's torch. The numbers at the bottom refer to the position of the double bond in the fatty acid molecule. (Reproduced by kind permission of the authors [118]).

fractions enriched with particular positional isomers for structural analysis) but can be a minor drawback in the determination of total *trans*-18:1 or total *cis*-18:1 content. This is because, if not careful, the collected fractions may not include all the positional isomers; some isomers may be left behind on the plate or cross-contamination of bands may occur. The latter is most likely to occur with the *cis*-18:1 and 18:2 + 18:3 fractions. Fortunately, the positional isomer separations are more pronounced in plates developed at low temperatures, -20°C [110,121]. The separation of the positional isomers could be minimized by developing the Ag-TLC plate at room temperature.

Among the other important applications of Ag-TLC in the area of hydrogenated fats has been for the isolation of preparative quantities of PUFA isomers for structural identification work. Ratnayake and Pelletier [2], used combined techniques of reversed phase HPLC, Ag-TLC, GC/MS, hydrazine reduction and oxidative ozonolysis for identification of 18:2 isomers in PHVO. Initially, the 18:2 isomer group was isolated using HPLC. Then the 18:2 group was fractionated by Ag-TLC (developed in toluene at -25°C) into five sub-fractions, which eluted in the order, methylene-interrupted *trans,trans*-18:2 > methylene-interrupted *trans,cis-* and *cis,trans*-18:2 + non-methylene-interrupted *trans,trans*-18:2 > methylene-interrupted *cis,cis*-18:2 > non-methylene-interrupted *cis,trans-* and non-methylene-interrupted *cis,cis*-18:2. This elution order shows that geometry and position, as well as the relative distance between the two double bonds, could influence the 18:2 isomer separation. The 18:2 isomers whose two double bonds are separated by one

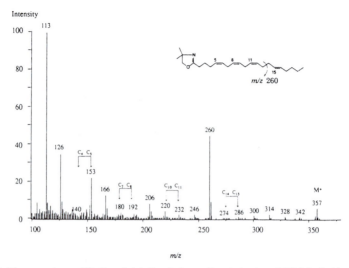

Fig. 4.18. Mass spectrum of 2-alkenyl-4,4-dimethyloxazoline derivative of 20:4 5*c*,8*c*,11*c*,15*t* [69] (Reproduced with kind permission of *Lipids*).

methylene group generally have a higher mobility than those 18:2 isomers whose double bonds are separated by two or more methylene groups. The structures of the 18:2 isomers were elucidated by subjecting each of the five fractions to GC/MS analysis as their picolinyl ester derivatives. The picolinyl derivatives (as well as other nitrogen containing ester derivatives such as pyrrolidides and 2-alkenyl-4,4-dimethyloxazolines (DMOX)) of unsaturated fatty acids have distinctive mass spectra with characteristic ions that can be used to locate the positions of the double bonds. In addition to GC-MS analysis, each 18:2 FAME fraction was subjected to partial hydrazine reduction to obtain 18:1 isomers representative of the ethylenic bonds in the original 18:2 isomers, Ag-TLC of the 18:1 isomers, oxidative ozonolysis of the isolated *cis*- and *trans*-18:1 bands and capillary GC of the dimethyl ester ozonolysis products [2]. This led to the identification of *cis*-9,*trans*-13-18:2 as the major 18:2 isomer in PHVO. Other important isomers characterized were *trans*-9,*trans*-12, *cis*-9,*trans*-12, *trans*-9,*cis*-12 and *cis*-9,*cis*-15. A number of minor isomers were detected and some structures identified were *trans*-8,*cis*-12, *trans*-8,*cis*-13, *cis*-8,*cis*-13, *trans*-9,*cis*-15, *trans*-10,*cis*-15 and *cis*-9,*cis*-13.

   A similar combined procedure of Ag-TLC was used by Ratnayake *et al.* [69] to identify C_{20} *trans*-PUFA isomers of arachidonic acid in rats fed partially hydrogenated canola oil. In this work GC-MS analysis was performed on DMOX derivatives of the fatty acids. The DMOX derivative, in contrast to picolinyl esters or pyrrolidides, are easily separated on capillary cyanosilicone capillary columns. The elution pattern is very much similar to fatty acid methyl esters. In addition mass spectra of DMOX derivatives of unsaturated fatty acids can be more clear and informative than those of picolinyl or pyrrolidide derivatives. A typical mass spectral pattern of a DMOX derivative of a polyunsaturated fatty acid is shown in Figure 4.18. This mass spectral pattern (DMOX of 20:4 5*c*,8*c*,11*c*,15*t*) gave very intense ions at *m/z* 113 and 126, which are characteristic of the DMOX derivative of unsaturated fatty acids. The positions of the double bonds at 8, 11 and 15 were clearly located by ions (*m/z* 180 versus 192, *m/z* 220 versus 232 and *m/z* 274 and 286) which differed by 12 atomic mass units (amu). In DMOX and other nitrogen containing ester derivatives, a mass interval of 12 amu instead of regular 14 amu between two neighbouring even mass homologous fragments containing n-1 and n carbon atoms in the acid moiety indicate a double bond between n and n + 1 in the chain. The 5 bond of 20:4 5*c*,8*c*,11*c*,15*t* was located by the appearance of the prominent ion at *m/z* 153. This ion is diagnostic for the 5 position. The geometry of the double bonds in 20:4 5*c*,8*c*,11*c*,15*t* was deduced by comparison of the experimental and calculated ECL values. For more details of this approach for tentative identification of double positions in unsaturated fatty acid the reader is referred to a review on GC of fatty acids by Christie [44].

Mossoba *et al.* [122,123] used the combined technique of Ag-TLC and GC/matrix isolation (MI)/FTIR for identification and quantification of FAME isomers in hydrogenated soybean oil and margarines. MI/FTIR was helpful in differentiating *trans,trans* and *cis,trans* positional isomers based on the =C-H stretching vibrations at 3010 and 3018 cm⁻¹ for *cis*, and 3035 and 3005 cm⁻¹ for *trans* configurations, as well as the intensity of the 2935 cm⁻¹ band relative to that of the carbonyl band at 1745 cm⁻¹. The GC/MI/FTIR technique revealed the presence of four *trans,trans*-18:2 isomers including the *trans*-9,*trans*-12 isomer in a sample of hydrogenated soybean oil. The amounts of this isomer in the two brands of US margarines analysed by Mossoba *et al.* were low (0.3%) [122].

A limitation of Ag-TLC is that complex mixtures of FAME containing a range of chain-lengths and higher degree of unsaturation, such as those encountered in fish oils, produce very complex chromatograms [124-126]. In PHFO, numerous mixed bands are expected due to separation according to chain length superimposed on the separation of geometric and positional isomers. The cross overlap of bands occurs especially in FAME isomers containing more than three double bonds. Although Ag-TLC may not resolve all the isomers, it is still helpful for measuring at least the levels of mono- and diethylenic isomers in PHFO [107,124]. Almendingen *et al.* [107] determined the *trans*-monoene content by the SFA internal standard procedure described previously for PHVO; *i.e.* collecting the two zones from Ag-TLC (developed in hexane-diethyl ether, 9:1 v/v) corresponding to SFA + *trans*-monoenes as one fraction and determining the *trans*-monoene content with respect to the total SFA by GC. Figure 4.19 illustrates the GC chromatograms of FAME from a PHFO-margarine before and after Ag-TLC fractionation. The *trans*-diene in PHFO, was determined by GC analysis of the Ag-TLC separated total diene fraction with and without addition of 9-*trans*,12-*trans*-18:2 as the internal standard [107] (Figure 4.15). The sum of *trans*-monoene and *trans*-

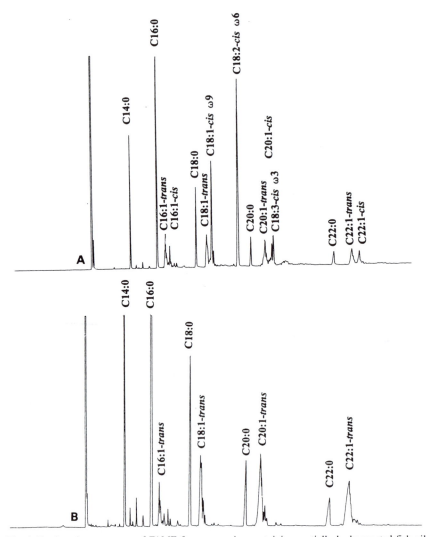

Fig. 4.19. Gas chromatogram of FAME from margarine containing partially hydrogenated fish oil before (A) and after (B) separation of the *trans* monoene and the saturated fatty acid fraction by Ag-TLC. GLC conditions are as described in Fig. 4.15. The main areas of *trans* C20:1 and C22:1 in (B) were calculated to be 78% and 93% of total *trans* monoene area, respectively. Retention times (A) C16:0, 26.46 min; *trans* 18:1, 35.38 min; *trans* C18:1, 35.38 min; C20:1 *trans*, 46.22 min; *trans* C22:1, 62.0 min (Reproduced by kind permission of the authors [107] and *Journal of Lipid Research*).

diene of PHFO margarine sample analysed by Almendingen *et al.* [107] accounted only for about 65% of the total *trans* content measured by IR, however. This shows that there is large amount of *trans*-trienes and *trans*-tetraenes in PHFO undetected by the Ag-TLC/GC technique.

A complete analysis of all the isomers in PHFO has not been reported so far. A substantial amount of information regarding *trans*-isomers might be possible to obtain by using an alternative preliminary separatory technique, such as methoxy-bromo mercuric-adduct fractionation in conjunction with Ag-TLC [105]. An important feature of methoxy-bromo mercuric-adduct fractionation is that it separates FAME strictly according to the number of double bonds and there is no sub-fractionation due to configuration or position of the double bonds. In the early 1980s, Sebedio *et al.* [105] used this technique to measure the fatty acid classes (*i.e.* monoenes, dienes, trienes etc.) in PHFO. Unfortunately, further fractionation of each fatty acid class for determination of *cis-trans* isomers was not pursued.

F. SILVER ION HIGH-PERFORMANCE LIQUID CHROMATOGRAPHY

The first attempt to apply HPLC in the silver ion mode for analysis of *cis-trans* isomers was reported in 1973 by Mikes *et al.* [127] who partially separated methyl elaidate from methyl oleate on a Corasil silica gel column coated with 0.8% silver nitrate and 1.75% ethylene glycol. Since then great improvements have been made in silver ion-HPLC (Ag-HPLC) especially in the laboratories of Battaglia [128], Christie [29,129-139] and Adlof [140,141]. With the recent availability of a commercial Ag-HPLC column, this technique is gaining acceptance as an alternative technique to Ag-TLC for the separation of *cis* and *trans*-18:1 isomers in PHVOs [29,79,139,140]. The reader is referred to reviews by Nikolova-Damyanova [110], Christie [142] and Firestone and Sheppard [7] for more detailed information. This section on Ag-HPLC is focused primarily with recent applications for *cis-trans* isomer analysis.

The main problem with Ag-HPLC has been the development of a stable and reproducible column with a controlled silver content and good shelf life. The approach used in the earlier days was to impregnate HPLC grade silica gel with silver nitrate and pack it into columns [128,143,144]. Good resolution of isomers was achieved with such columns. For example, Battaglia and Frohlich [128] using Spherisorb S5W columns (240 mm × 5 mm i.d.) impregnated with silver nitrate of amounts ranging from 2 to 30%, with a mobile phase of 1% tetrahydrofuran in hexane, using a ultraviolet (UV) detector, separated 18:1 FAME from margarine into *cis-trans* and positional isomers. In this system, the optimum separation of *cis-trans* and positional isomers is reached at a silver nitrate load of 20%. The heavily loaded silver ion columns (*e.g.* 30%) allow a faster chromatography and are therefore more efficient in routine work, however. Unfortunately such columns are not available commercially, and their preparation requires much practice and skill. A major problem is that the silver ions bleed from them continuously into the mobile phase. Silver ions are corrosive and have the potential to contaminate the HPLC system and damage the detector. Furthermore, because of this leaching, the operational life of columns is short; the column performance tend to decrease after about 20 to 50 analyses [128,145].

A more versatile and practical approach to the preparation of Ag-HPLC

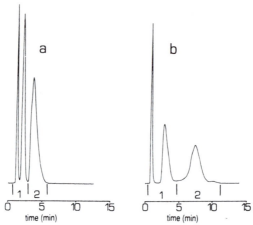

Fig. 4.20. Silver ion HPLC separation of methyl esters of partially hydrogenated soybean oil. a, ChromSpher Lipids™ column, mobile phase, dichloromethane/1,2-dichloroethane/acetonitrile (50:50:0.0005 v); b, Spherisorb S5SCX™ column, mobile phase, same solvents in ratio 50:50:0.001. Fraction 1, saturated and *trans*-monoenoic esters; 2, *cis*-monoenoic esters [29]. (Reproduced by kind permission of the authors and *Journal of the Science of Food and Agriculture*).

column is to bind a silica-based ion-exchange medium (chemically bonded sulfonic acid groups) with silver ions [129,146]. The approach of Powell was to pump 1 M aqueous solution of silver nitrate (150 mL) through a RSilCAT (5μ particles) HPLC column, followed by large volumes of water, methanol, acetone, ethyl acetate, chloroform and hexane [146]. The procedure adopted by Christie [129] involved a standard prepacked ion exchange column (Nucleosil™ 5SA) and introducing 20% aqueous silver nitrate (1 mL) via the HPLC injector system (in 50 μL aliquots at 1 min intervals) while pumping water through the column. Finally the column is washed with methanol for 1 hour followed by a further 1 hour with 1,2-dichloroethane. Only 50-80 mg of silver ions are bound to the stationary phase, but this is sufficient for many FAME separations. These columns are stable for long periods of time and there is no leaching of silver ions. In addition, the samples separated are much cleaner than those obtained with Ag-TLC [129]. A commercial silver column of this type is available from Chrompack Ltd.

Excellent separations of geometrical and positional have been achieved with the above described silica-based ion-exchange columns loaded with silver nitrate. By using a short HPLC column (5 cm × 4.6 mm i.d.) of Spherisorb™ S5SCX in the silver ion form or a similar commercial silver ion column (ChromSpher Lipids™) and isocratic elution with a mobile phase consisting of 1,2-dichloroethane-dichloromethane (1:1, v/v) and small amounts of acetonitrile (0.01 to 0.025%) at a flow rate of 1 mL min^{-1}, utilizing an evaporative light-scattering detector (ELSD) (also termed mass detector), Toschi *et al.* [29] were able to achieve a base-line separation of *trans*-18:1 FAME isomers as a group from that of the *cis*-18:1 isomers. The nature of separation achieved with methyl esters of hydrogenated soybean oil is shown in Figure 4.20 [29]. Saturated, *trans*- and *cis*-18:1 were clearly resolved within

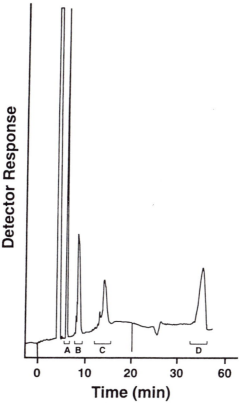

Fig. 4.21. Analysis of partially hydrogenated vegetable oil fatty acid methyl esters by silver-ion high-performance liquid chromatography. Sample size, 20 µg; flow rate 1.0 mL/min 0.15% acetonitrile in hexane; RI detector. Fractions: A, saturates; B, *trans*-18:1; C, *cis*-18:1; D, 18:2. (Reproduced by kind permission of the authors [141] and *Journal of the American Oil Chemists' Society*).

5-10 minutes on both columns with somewhat stronger retention by the Spherisorb™ S5SCX column. The shorter analysis time and the complete separation of *cis*- and *trans*-18:1 (no *cis*- and *trans* overlaps) makes the Ag-HPLC technology a good alternative to Ag-TLC for preliminary *trans*-isomer fractionation before GC analysis. Quantification of *trans*-18:1 isomers could be performed using a procedure similar to that adopted for Ag-TLC/GC (discussed previously), *i.e.* analysis of PHVO methyl ester sample by capillary GC first to determine the relative proportions of saturated fatty acids, and then to isolate the saturated esters with *trans*-monoenes as a fraction for re-analysis by GC [29]. This methodology has given excellent results with hydrogenated soybean and rapeseed oils [29]. The results are reproducible and the precision is better than by FTIR spectroscopy. A drawback is that this method gives information only on the content of *trans*-monoenes in the sample. Dienoic and polyenoic methyl esters (including those with *trans* bonds) are not eluted under the HPLC conditions described above. Toschi *et al.* [29] cleared the

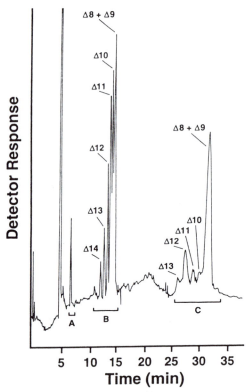

Fig. 4.22. Analysis of partially hydrogenated vegetable oil 18:1 FAME positional isomers by silver-ion high-performance liquid chromatography. Column: ChromSpher Lipids™ (250 mm × 4.6 mm i.d., 5μL particle size). Sample size: 0.4 μg; flow rate 1.0 mL/min, 0.08% acetonitrile in hexane; ultraviolet detector at 206 nm. Fractions: A, saturated; B, *trans*-18:1; C, *cis*-18:1. (Reproduced by the kind permission of the authors [141] and *Journal of the American Oil Chemists' Society*).

PUFAs from the column by elution with acetonitrile-methanol (9:1, v/v) at the end of each day and with acetonitrile-methanol (1:1, v/v) every second day. Adlof *et al.* [141] developed an alternative mobile phase to elute PUFA of vegetable oils within a reasonable time. By eluting with 0.15% acetonitrile in hexane at a flow rate of 1.0 mL min^{-1}, methyl linoleate was eluted in about 45 minutes using a ChromSpher Lipids™ column (4.6 mm × 250 mm i.d., 5μ particle size) (Figure 4.21).

A further modification of the solvent system permits the separation of the individual positional *cis* and *trans*-18:1 isomers with no *cis-trans* overlaps [140,141]. By eluting isocratically with 0.08% acetonitrile in hexane, positional isomers of both *cis* and *trans*-18:1 of hydrogenated vegetable oil were successfully resolved (Figure 4.22) [141]. The elution orders of the positional isomers are the reverse of those obtained by capillary GC (Figure 4.16). Use of small sample quantities (0.5 μg or less) improved the isomer resolution. However, 18:1 8 and 18:1 9 positional isomer pairs (in both *cis* and *trans*) could not be separated and the 18:1 10c was poorly separated from the 18:1 8c and 18:1 9c isomer pair. These results are consisted with the Ag-TLC patterns of 18:1 isomers (Figure 4.17). A UV detector is required for the

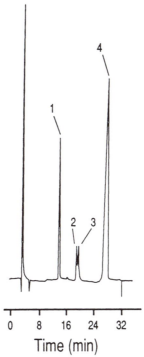

Fig. 4.23. Separation of isomerized methyl linoleate. Sample size: 20 μg. Flow-rate: 1.0 mL/min, 0.125% acetonitrile in hexane. UV detection at 210 nm. Peaks: 1 = *trans*-9,*trans*-12: 2 = *trans*-9, *cis*-12; 3 = *cis*-9,*trans*-12; 4 = *cis*-9,*cis*-12. (Reproduced by kind permission of the authors [140] and *Journal of Chromatography*)

optimum positional FAME isomer analyses due to its greater sensitivity when compared with refractive index (RI) or flame ionization detectors (FID).

Adlof *et al.* [141] examined the reproducibility and quantification of data by HPLC. Good reproducibility (±4% of peak area or better) was achieved with an FID detector, but positional isomer resolution was reduced because a larger sample size was required for detection. The reproducibility with the RI detector was 5-6% of the peak area. Comparison of results obtained by HPLC-FID with GC-FID for the saturated, *trans*-18:1, *cis*-18:1 and 18:2 FAME varied from 1 to 10% of peak area. A similar relationship was obtained with RI detection but the RI response for 18:2 was almost twice that observed with Ag-HPLC-FID or GC. Correction factors would be required for an RI detection system.

HPLC with the ChromSpher Lipids™ column using a solvent system of containing 0.125 to 0.15% acetonitrile in hexane can also provide good resolution of the geometrical isomers of FAME containing up to four double bonds in a reasonable time [140]. The separation of the geometrical isomers of methyl linoleate (4 isomers), methyl α-linolenate (8 isomers) and methyl arachidonate (16 isomers) are illustrated in Figures 4.23-25. Up to 200 μg of sample can be applied to the column without severe peak distortion or

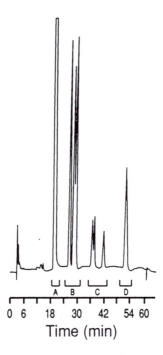

Fig. 4.24. Separation of isomerized methyl linolenate. Sample size: 20 μg. Flow-rate: 1.0 mL/min, 0.125% acetonitrile in hexane. UV detection at 210 nm. Groups: A = three *trans;* B = two *trans*, one *cis;* C = one *trans*, two *cis;* D = three *cis* double bonds. (Reproduced by kind permission of the authors [140] and *Journal of Chromatography).*

significant loss of retention with resolutions often exceeding the capabilities of capillary GC. The order of the elution of methyl linoleate isomers (Figure 4.23) was *trans*-9,*trans*-12-; *cis*-9,*trans*-12-; *cis*-9,*trans*-12- and *cis*-9,*cis*-12- and differed from that obtained with cyanosilicone capillary GC, in which *cis*-9,*trans*-12-18:2 elutes before *trans*-9,*cis*-12-18:2. The separation of the eight geometrical isomers of methyl α-linolenate (Figure 4.24) was similar but with better resolution than the separation on capillary GC. Also, the resolution of 15 of 16 isomers of methyl arachidonate (Figure 4.25) far exceeded the capabilities of current GC.

The Ag-HPLC method has several advantages, including rapid analysis time (15-45 min) and complete separation of the *cis*- and *trans* isomers with no *cis*-*trans* overlap. The complete separation of *cis* and *trans* isomers can be used to provide quantitative data for positional isomers not separated by capillary GC. However, Ag-HPLC is more limited as a stand alone method for the study of hydrogenated fats. Although the resolution of geometrical isomers of linoleic, linolenic and arachidonic acids has been demonstrated [140,147], the separation of the 18:2 isomers in partially hydrogenated vegetable oils and the various *trans*-PUFA isomers in PHFO has yet to be demonstrated. The major problem remaining appears to be in reproducing the analytical results. In the

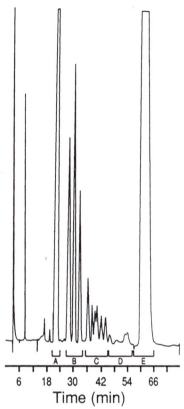

Fig. 4.25. Separation of isomerized methyl arachidonate. Sample size: 100 μg. Flow-rate: 1.0 mL/min, 0.125% acetonitrile in hexane. UV detection at 210 nm. Groups: A = four *trans;* B = three *trans*, one *cis;* C = two *trans,* two *cis*; D = one *trans*, three *cis*; E = four *cis* double bonds. (Reproduced by kind permission of the authors [140] and *Journal of Chromatography*).

author's laboratory a rapid loss of resolution (using a ChromSpher Lipids™ column) was noted after only 4 weeks of daily use. In contrast a capillary GC column could be used at least up to one year without any significant loss of resolution. Ag-HPLC could be a good alternative to Ag-TLC for preliminary isomer fractionation before GC analysis. Such a procedure was recently recommended by AOCS for *trans* isomer analysis [148].

G. CONCLUSIONS

From the above it can be seen that by selecting the proper operating parameters and polar cyanopolysiloxane capillary column, detailed fatty acid composition including *trans* isomers and other isomeric fatty acids in partially hydrogenated vegetable oils, can be obtained by GC analysis alone. For applications requiring more precise *trans* fatty acid data and detailed information of individual isomers, it is necessary to couple GC with either Ag-TLC or Ag-HPLC. The current analytical tools are, however, inadequate for complete analysis of hydrogenated fish oils.

ABBREVIATIONS

Ag-HPLC, silver ion high performance-liquid chromatography; Ag-TLC, silver nitrate thin-layer chromatography; AOAC, Association of Official Analytical Chemists; AOCS, American Oil Chemists' Society; ATR, attenuated total reflectance; c, *cis*; DMOX, 2-alkenyl-4,4-dimethyloxazoline; ECL, equivalent chain length; FAME, fatty acid methyl esters; FID, flame ionization detector; FFS, flexible fused silica; FT, Fourier *trans*form; GC, gas chromatography; IR, infra-red; IUPAC, International Union of Pure and Applied Chemists; MI, matrix isolation; MS, mass spectrometry; PHFO, partially hydrogenated fish oils; PHVO, partially hydrogenated vegetable oils; PLS, partial least square; PUFA, polyunsaturated fatty acids; RI, refractive index; RSD, relative standard deviation; SFA, saturated fatty acids; t, *trans*; UV, ultra-violet.

REFERENCES

1. Emken,E.A., *Am. J. Clin. Nutr.*, **62**, 659S-669S (1995).
2. Ratnayake,W.M.N. and Pelletier,G., *J. Am. Oil Chem. Soc.*, **69**, 95-105 (1992).
3. Sebedio,J-L. and Ackman,R.G., *J. Am. Oil Chem. Soc.*, **60**, 1986-1991 (1983).
4. Sebedio,J-L. and Ackman,R.G., *J. Am. Oil Chem. Soc.*, **60**, 1992-1996 (1983).
5. Conacher,H.B.S., *J. Chromatogr. Sci.*, **14**, 405-411 (1976).
6. Scholfield,C.R., in *Geometrical and Fatty Acid Isomers*, pp. 17-52 (1979) (edited by E.A. Emken and H.J. Dutton, *AOCS Press*, Champaign).
7. Firestone,D. and Sheppard,A., in *Advances in Lipid Methodology-One*, pp. 273-322 (1992) (edited by W.W. Christie, The Oily Press., Ayr, Scotland).
8. British Nutrition Foundation, *Trans* Fatty acids, The Report of the British Nutrition Foundation Task Force, July 1995. Holborn House, London.
9. Kauffman,F.L., *J. Am. Oil Chem. Soc.*, **41**, 4,6,21,38,42 (1964).
10. O'Connor,R.T., *J. Am. Oil Chem. Soc.*, **38**, 648-659 (1961).
11. American Oil Chemists' Society, *Official Methods and Recommended Practices*, Official Method cd 14-61, Reapproved, 1989 (edited by D. Firestone, AOCS, Champaign, IL).
12. Official and Tentative Methods of the American Oil Chemists' Society, AOCS, Chicago, IL, 1946, Tentative Method cd 14-61 (edited by V.C. Mehlenbacher).
13. American Oil Chemists' Society, Official Methods and Recommended Practices, Official Method cd 14-95, 1995 (edited by D. Firestone, AOCS, Champaign, IL).
14. Official Methods of Analysis of the Association of Official Analytical Chemists International, 16th Edition, Official Method 965.34, p. 27A (Supplement March 1995) (edited by P. Cunniff, AOAC International, Arlington, VA).
15. Official Methods of Analysis of the Association of Official Analytical Chemists International, 16th Edition, Official Method 994.14, p. 27C (Supplement March 1995) (edited by P. Cunniff, AOAC International, Arlington, VA).
16. International Union of Pure and Applied Chemistry, *Standard Methods for the Analysis of Oils, Fats and Derivatives*, 7th Edition, Method 2.207 (1987) (edited by C. Paquot and A. Hautfenne, Blackwell Scientific Publications, Oxford).
17. Chenery,D.H. and Sheppard,N., *Appl. Spectr.*, **32**, 79-89 (1978).
18. Belton,P.S., Wilson,R.H., Sadeghi-Jorabchi,H. and Peers,K.E., *Lebensm. Wiss. Technol.*, **21**, 153-157 (1988).
19. Fahrenfort,J., *Spectrochim. Acta*, **17**, 698-709 (1961).

20. Dutton,H.J., *J. Am. Oil Chem. Soc.*, **51**, 407-409 (1974).
21. Chin,S.F., Liu,W., Storkson,J.M., Ha,Y.L. and Pariza,M.W., *J. Food Comp. Anal.*, **5**, 185-196 (1992).
22. Hilditch,T.P. and Williams,P.N., *The Chemical Constitution of Natural Fats*, Fourth Edition, (Chapman and Hall, London) (1964).
23. Firestone,D. and LaBouliere,P., *J. Assoc. Off. Anal. Chem.*, **48**, 437-443 (1965).
24. Firestone,D. and DeLaluz Villadelmar,M., *J. Assoc. Off. Anal. Chem.*, **44**, 459-464 (1961).
25. Huang,A. and Firestone,D., *J. Assoc. Off. Anal. Chem.*, **54**, 47-51, 1971.
26. Madison,B.L., Depalma,R.A. and D'Alonzo,R.P., *J. Am. Oil Chem. Soc.*, **59**, 178-181 (1982).
27. Lanser,A.C. and Emken,E.A., *J. Am. Oil Chem. Soc.*, **65**, 1483-1487 (1988).
28. Sleeter,R.T. and Matlock,M.G., *J. Am. Oil Chem. Soc.*, **66**, 121-127 (1989).
29. Toschi,T.G., Capella,P., Holt,C. and Christie,W.W., *J. Sci. Food Agric.*, **61**, 261-266 (1993).
30. Mossoba,M.M., Yurawecz,M.P. and McDonald,R.E., *J. Am. Oil Chem. Soc.*, **73**, 1003-1009 (1996).
31. Ulberth,F. and Haider,H.-J., *J. Food Sci.*, **57**, 1444-1447 (1992).
32. Ratnayake,W.M.N., *J. Assoc. Off. Anal. Chem. Internat.*, **78**, 783-802 (1995).
33. Ratnayake,W.M.N. and Chen,Z.Y., *Lipids,* **31**, S279-S282 (1996).
34. Kuemmel,D.F. and Chapman,L.R., *Anal. Chem.*, **38**, 1611-1614 (1966)
35. Ratnayake,W.M.N., Hollywood,R., Grady,E. and Beare-Rogers,J.L., *J. Am. Oil Chem. Soc.*, **67**, 804-810 (1990).
36. Scholfield,C.R., Jones,E.P., Butterfield,R.O. and Dutton,H.J., *Anal. Chem.*, **35**, 1588-1591 (1963).
37. Ratnayake,W.M.N and Pelletier,G., *J. Am. Oil Chem. Soc.*, **73**, 1165-1169 (1996).
38. Allen,R.R., *J. Am. Oil Chem. Soc.*, **46**, 552-553 (1969).
39. Allen,R.R., *Lipids,* **4**, 627-628 (1969).
40. O'Connor,R.T., Report of the Instrumental Techniques Committee, 1968-1969, *J. Am. Oil Chem. Soc.*, **46**, 602A, 604A (1969).
41. Kaufmann,P., in *Advances in Lipid Methodology - One*, pp. 149-180 (1992) (edited by W.W. Christie, The Oily Press, Ayr, Scotland).
42. Fuller,M.P., Ritter,G.L. and Draper,C.S., *Appl. Spectr.*, **42**, 217-227 (1988).
43. van de Voort,F.R., Ismail,A.A. and Sedman,J., *J. Am. Oil Chem. Soc.*, **72**, 873-880 (1995).
44. Christie,W.W., *Gas Chromatography and Lipids*, The Oily Press, Ayr, Scotland (1989).
45. Morrison,W.R. and Smith,L.M., *J. Lipid Res.*, **5**, 600-608 (1964).
46. Christopherson,S.W. and Glass,R.L., *J. Dairy Sci.*, **52**, 1289-1290 (1969).
47. Ottenstein,D.M., Witting,L.A., Silvis,P.H., Hometchko,D.J. and Pelick,N., *J. Am. Oil Chem. Soc.*, **61**, 390-394 (1984).
48. Chrompack News , 23 (2), 8 (1996)
49. Wolff,R.L., in *New Trends in Lipid and Lipoprotein Analyses*, pp. 147-180 (1995) (edited by J.-L. Sebedio and *E.G.* Perkins, AOCS Press, Champaign, IL).
50. Ratnayake,W.M.N., in *New Trends in Lipid and Lipoprotein Analyses*, pp. 181-190 (1995) (edited by J.-L. Sebedio and *E.G.* Perkins, AOCS Press, Champaign, IL).
51. Precht,D. and Molkentin,J., *Die Nahrung,* **39**, 343-374 (1995).
52. Gildenberg,L. and Firestone,D. *J. Assoc. Off. Anal. Chem.*, **68**, 46-51 (1985).
53. Association of Official Analytical Chemists, Official Method of Analysis, 15th Edition, Official Method 985.21, p.970 (1990) (edited by K. Helrich, AOAC, Arlington, VA).
54. American Oil Chemists' Society, Official Methods and Recommended Practices, Official Method Cd 17-85, Reapproved, 1989 (edited by D. Firestone, AOCS, Champaign, IL).
55. Technical Commission of the Oil and Fat Industry, Subcommission on Animal and Vegetable Fats, *Riv. Ital. Sostanze Grasse,* **63**, 301-303 (1986).
56. Ottenstein,D.M., Witting,L.A., Walker,G., Mahadevan,V. and Pelick,N., *J. Am. Oil Chem. Soc.*, **54**, 207-209 (1977).
57. Lanza,E. and Slover,H.T., *Lipids,* **16**, 260-267 (1981).
58. Enig,M.G., Pallansch,L.A., Sampugna,J. and Keeney,M., *J. Am. Oil Chem. Soc.*, **60**, 1788-1795 (1983).
59. Slover,H.T., Thompson,R.H., Davis,C.S. and Merola,G.V., *J. Am. Oil Chem. Soc.*, **62**, 775-786 (1985).
60. Slover,H.T. and Lanza,E., *J. Am. Oil Chem. Soc.*, **56**, 933-943 (1979).
61. Lin,K.C., Marchello,M.J. and Fischer,A.G., *J. Food Sci.*, **49**, 1521-1524 (1984).
62. Sampugna,J., Pallansch,L.A., Enig,M.G. and Keeney,M., *J. Chromatogr.*, **249**, 245-255 (1982).
63. Firestone,D., *J. Assoc. Off. Anal. Chem.*, **73**, 105-110 (1990).

64. American Oil Chemists' Society, Official Methods and Recommended Practices, Official Method Ce 1c-89 (1990) (edited by D. Firestone, AOCS, Champaign, IL).
65. Ratnayake,W.M.N. and Beare-Rogers,J.L., *J. Chromatogr. Sci.*, **28**, 633-639 (1990).
66. Ratnayake,W.M.N., *J. Am. Oil Chem. Soc.*, **69**, 192 (1992).
67. Ratnayake,W.M.N. and Chen,Z.Y., in *Development and Processing of Vegetable Oils For Human Nutrition*, pp. 20-35 (1995) (edited by R. Przybylski and B.E. McDonald, AOCS Press, Champaign, IL).
68. Beyers,E.C. and Emken,E.A., *Biochim. Biophys. Acta*, **1082**, 275-284 (1991).
69. Ratnayake,W.M.N., Chen,Z.Y., Pelletier,G. and Weber,D., *Lipids*, **29**, 707-714 (1994).
70. Wolff,R.L., *J. Am. Oil Chem. Soc.*, **71**, 277-283 (1994).
71. Molkentin,J. and Precht,D., *Chromatographia*, **41**, 267-272 (1995).
72. Molkentin,J. and Precht,D., *Z. Ernahrungswiss*, **34**, 314-317 (1995)
73. Chen,Z.Y., Pelletier,G., Hollywood,R. and Ratnayake,W.M.N., *Lipids*, **30**, 15-21 (1995).
74. Chen,Z.Y., Ratnayake,W.M.N., Fortier,L., Ross,R. and Cunnane,S.C., *Can. J. Physiol. Pharmacol.*, **73**, 718-723 (1995).
75. Sidisky,L.M., Kiefer,K.H. and Doghty,E.I., Paper presented at the Pittcon '96 Conference, March 3-8, 1996, Chicago, IL.
76. Precht,D. and Molkentin,J., *Int. Dairy J.*, (in press) (1997).
77. Wolff,R.L. and Bayard,C.C., J. Am. Oil Chem., **72**, 1197-1201 (1995).
78. Sidisky,L.M., Stormer,P.L., Nolan,L., Keeler,M.J. and Bartram,R.J., *J. Chromatogr. Sci.*, **26**, 320-324 (1988).
79. Duchateau,G.S.M.J.E., van Oosten,H.J. and Vasconcellos,M.A., *J. Am. Oil Chem. Soc.*, **73**, 275-282 (1996).
80. Ackman,R.G., Hooper,S.N. and Hooper,D.L., *J. Am Oil Chem. Soc.*, **51**, 42-49 (1974).
81. Wolff,R.L. and Sebedio,J.L., *J. Am. Oil Chem. Soc.*, **68**, 719-725 (1991).
82. Wolff,R.L., *J. Am Oil Chem. Soc.*, **69**, 106-110 (1992).
83. Wolff,R.L., *J. Am. Oil Chem. Soc.*, **70**, 219-224 (1993).
84. Wolff,R.L., *Sci. Alim.*, **13**, 155-163 (1995).
85. O'Keefe,S.F., Gaskins-Wright,S., Wiley,V. and Chen,I.C., *J. Food Lipids*, **1**, 165-176 (1994).
86. Wolff,R.L., *J. Am. Oil Chem. Soc.*, **70**, 425-430 (1993).
87. O'Keefe,S.F., Wiley,V.A. and Wright,D., *J. Am. Oil Chem. Soc.*, **70**, 915-917 (1993).
88. Wolff,R.L., Nour,M. and Bayard,C.C., *J. Am. Oil Chem. Soc.*, **73**, 327-332 (1996)
89. Kobayashi,T., *J. Chromatogr.*, **194**, 404-409 (1980).
90. Houtsmuller,U.M.T., *Fette Seifen Anstrichm.*, **80**, 162-169 (1978).
91. Sahasrabudhe,M.R. and Kurian,C.J., *Can. Inst. Food Sci. Technol. J.*, **12**, 140-145 (1979).
92. Siguel,E.N. and Lerman,R.H., *Am. J. Cardiol.*, **71**, 916-920 (1993).
93. Siguel,E.N. and Ratnayake,W.M.N., *Am. J. Cardiol.*, **75**, 424 (1995).
94. Anderson,R.L., Fullmer,C.S. and Hollenbach,E.J., *J. Nutr.*, **105**, 393-400 (1975).
95. Privett,O.S., Stearns,E.M. and Nickell,E.C., *J. Nutr.*, **92**, 303-310 (1967).
96. Hwang,D.H. and Kinsella,J.E., *Prostaglandins Med.*, **1**, 121-130 (1978).
97. Kinsella,J.E., Bruckner,G., Mai,J. and Shimp,J., *Am. J. Clin. Nutr.*, **34**, 2307-2318 (1981).
98. Davignon,J., Little,J.A., Holub,B., McDonald,B.E. and Spence,M., *Report of the ad hoc committtee on the composition of special margarines*, Ministry of Supply and Services of Canada, Ottawa, 1980, pp. 1-70.
99. Ratnayake,W.M.N., Hollywood,R. and O'Grady,E., *Can. Inst. Sci. Technol. J.*, **24**, 81-86 (1991).
100. Ratnayake,W.M.N., Hollywood,R., O'Grady,E. and Pelletier,G., *J. Am. Coll. Nutr.*, **12**, 651-660 (1993).
101. Grandgirard,A., Piconneaux,A., Sebedio,J.L., O'Keefe,S.F., Semon,E. and Le Quéré,J.L., *Lipids*, **24**, 799-804 (1989).
102. Wolff,R.L., *J. Am. Oil Chem. Soc.*, **71**, 907-909 (1994).
103. Chardigny,J.M., Sebedio,J.L., Grandgirard,A., Martine,L., Berdeaux,O. and Vatele,J.M., *Lipids*, **31**, 165-168 (1996)
104. Chardigny,J.M., Sebedio,J.L., Juaneda,P., Vatele,J.M. and Grandgirard,A., *Nutr. Res.*, **13**,1105-1111 (1993).
105. Sebedio,J.L. and Ackman,R.G., *Lipids*, **16**, 461-467 (1981).
106. Ackman,R.G., in *Nutrition Evaluation of Long-Chain Fatty Acids in Fish Oil*, pp. 25-88 (1982) (edited by S.M. Barlow and M.E. Stansby, Academic Press, London).
107. Almendingen,K., Jordal,O., Kierulf,P., Sandstad,B. and Pedersen,J.I., *J. Lipid Res.*, **36**, 1370-1384 (1995).

108. Official Methods of Analysis of the Association of Official Analytical Chemists International, 16th Edition, Official Method 994.15, p.26 (Supplement March 1995)(edited by P. Cunniff, AOAC International, Arlington, VA).
109. American Oil Chemists' Society, Official Methods and Recommended Practices, Official Method Cd 14b-93, Revised 1995 (edited by D. Firestone, AOCS, Champaign, IL).
110. Nikolova-Damyanova,B., in *Advances in Lipid Methodology - One*, pp.181-237 (1992) (edited by W.W. Christie, Oily Press, Ayr, Scotland).
111. Morris,L.J., Wharry,D.M. and Hammond,E.W., *J. Chromatogr.*, **31**, 69-76 (1976).
112. Christie,W.W., *Lipid Analysis*, Pergamon Press, Oxford (1982).
113. International Union of Pure and Applied Chemistry, Standard Methods for the Analysis of Oils, Fats and Derivatives, 7th Edition, Method 2.302 (1987) (edited by C. Paquot and A. Hautfenne, Blackwell Scientific Publications, Oxford).
114. Christie,W.W. and Moore,J.H., *J. Sci. Food Agric.*, **22**, 120-124 (1971).
115. Christie,W.W., *Lipid Technology*, **7**, 113-115 (1995).
116. Ulberth,F. and Henninger,M., *J. Am Oil Chem. Soc.*, **69**, 829-831 (1992).
117. Aitzetmuller,K. and Goncalves,L.A.G., J. Chromatogr., **519**, 349-358 (1990).
118. Gunstone,F.D., Ismail,I.A. and Lie Ken Jie,M.S.F., *Chem. Phys. Lipids*, **1**, 376-385 (1967).
119. Christie,W.W., *J. Chromatogr.*, **34**, 405-406 (1968).
120. Lie Ken Jie,M.S.F. and Lam,C.H., *J. Chromatogr.*, **124**, 147-151 (1976).
121. Breuer,B., Stuhlfauth,T. and Fock,H.P., *J. Chromatogr. Sci.*, **25**, 302-306 (1987).
122. Mossoba,M.M., McDonald,R.E., Chen,J.-Y., Armstrong,D.J. and Page,S.W., *J. Agric. Food Chem.*, **38**, 86-92 (1990).
123. Mossoba,M.M., McDonald,R.E., Armstrong,D.J. and Page,S.W., *J. Agric. Food Chem.*, **39**, 695-699 (1991).
124. Sebedio,J-L., Langman,M., Eaton,C.A. and Ackman,R.G., *J. Am. Oil Chem. Soc.*, **58**, 41-48 (1981).
125. Ackman,R.G., Hooper,S.N. and Hingley,J., *J. Chromatogr. Sci.*, **10**, 430-436 (1972).
126. Sebedio,J-L. and Ackman,R.G., *J. Chromatogr. Sci.*, **19**, 80-85 (1981).
127. Mikes,F., Schurig,V. and Gil-Av,E., *J. Chromatogr.*, **83**, 91-97 (1973).
128. Battaglia,R., and Frohlich,D., *Chromatographia*, **13**, 428-431 (1980).
129. Christie,W.W., *J. High Resolut. Chromatogr. Chromatogr. Commun.*, **10**, 148-150 (1987).
130. Christie,W.W., Brechany,E.Y. and Shukla,V.K.S., *Lipids*, **24**, 116-120 (1989).
131. Christie,W.W., Brechany,E.Y. and Stefanov,K., *Chem. Phys. Lipids*, **46**, 127-135 (1988).
132. Christie,W.W. and Breckenridge,G.H.M., *J. Chromatogr.*, **469**, 261-269 (1989).
133. Stefanov,K., Konaklieva,M., Brechany,E.Y. and Christie,W.W., *Phytochemistry,* **27**, 3495-3497 (1988).
134. Christie,W.W., *J. Chromatogr.,* **454**, 273-284 (1988).
135. Christie,W.W., *Fat. Sci. Technol.,* **93**, 65-66 (1991).
136. Laakso,P. and Christie,W.W., *J. Am. Oil Chem. Soc.*, **68**, 213-223 (1991).
137. Laakso,P., Christie,W.W. and Petersen,J., *Lipids*, **25**, 284-291 (1990).
138. Nikolova-Damyanova,B., Christie,W.W. and Herslof,B., *J. Am. Oil Chem. Soc.*, **67**, 503-507 (1990).
139. Nikolova-Damyanova,B., Herslof,B. and Christie,W.W., *J. Chromatogr.*, **609**, 133-140 (1992).
140. Adlof,R.O., *J. Chromatogr. A*, **659**, 95-99 (1994).
141. Adlof,R.O., Copes,L.C. and Emken,E.A., *J. Am. Oil Chem. Soc.*, **72**, 571-574 (1995).
142. Christie,W.W., in *New Trends in Lipid and Lipoprotein Analyses*, pp. 59-74 (1995) (edited by J.-L. Sebedio and E.G. Perkins, AOCS Press, Champaign, IL).
143. Heath,R.R., Tomlinson,J.H. and Doolittle,R.E., *J. Chromatogr. Sci.,* **15**, 10-13 (1997).
144. Smith,E.C., Jones,A.D. and Hammond,E.W., *J. Chromatogr.,* **188**, 205-212 (1980).
145. Scholfield,C.R., *J. Am. Oil Chem. Soc.*, **56**, 510-511 (1979).
146. Powell,W.S., *Anal. Biochem.,* **115**, 267-277 (1981).
147. Juanéda,P., Sebedio,J-L. and Christie,W.W., *J. High Resolut. Chromatogr.,* **17**, 321-324 (1994).
148. American Oil Chemists' Society, Official Methods and Recommended Practices, Official Method Ce 1g-96, Revised 1997 (edited by D. Firestone, AOCS, Champaign, IL).

CHAPTER 5

BIOCHEMISTRY OF TRANS-MONOENOIC FATTY ACIDS

Gunhild Hølmer,

Department of Biochemistry and Nutrition, Technical University of Denmark, Building 224, 2800 Lyngby, Denmark.

A.   Introduction
B.   Digestion and Absorption
C.   Metabolism of *Trans* Octadecenoic Acids *in Vitro*
    1.   Activation
    2.   Desaturation and elongation
    3.   Incorporation in lipid classes
    4.   Effects of C18:1 isomers on desaturation and elongation
    5.   Oxidative degradation
D.   Metabolism of *Trans* Octadecenoic Fatty Acids *in Vivo*
    1.   Influence on tissue lipid classes
    2.   Regiospecific distribution of total *trans*-C18:1
    3.   Deposition of positional isomers in different lipid classes
    4.   Distribution of geometrical and positional isomers between *sn*-1 and *sn*-2 positions in phosphoacylglycerols (PC and PE)
    5.   Effects of *trans* fatty acids on polyunsaturated fatty acid metabolism
    6.   Desaturation studies with human cells
    7.   Isomeric fatty acids and eicosanoid formation
E.   Human Studies on *Trans* Fatty Acids
    1.   Deposition in tissues
    2.   Metabolism of octadecenoates *in vivo*
    3.   Isomeric fatty acids in fetal and neonatal metabolism
F.   Concluding Remarks

A. INTRODUCTION

Trans monoenoic fatty acids comprise the major part of the *trans* fatty acids present in fats and oils after a partial hydrogenation, either industrial or biological, and they are therefore interesting from a nutritional point of view. Due to the energetically favourable formation of the *trans* geometry, a higher amount of *trans* than *cis* isomers is formed during the hydrogenation process. The hydrogenation of the polyunsaturated fats and oils leads, because of the original scattering of double bonds along the carbon chains, to a great variety of *trans* monoenoic isomers with an almost Gaussian distribution. The *cis*-isomer fraction is in contrast dominated by the monoenes originally present. It is therefore important to stress that that elaidic acid often used for experiments intending to elucidate *trans* fatty acid metabolism is not necessarily representative for the fate of other monoenoic *trans* fatty acids, and in the natural dietary fat mixtures several different competitive metabolic reactions may take place.

In the present chapter the metabolism of monoenoic *trans* fatty acids will be discussed but also the related *cis* positional isomers will be dealt with. As most of the *trans* fatty acids present in our diet are mixed with polyunsaturated fatty acids, special attention is given to the influence of isomeric fatty acids on the formation of long-chain polyenoic fatty acids, which are important components of membranes and precursors for regulatory lipids.

B. DIGESTION AND ABSORPTION

The *trans* fatty acids in the diet are either present in triacylglycerols (margarines and dairy fats) or in phosphoacylglycerols (meat) and the absorption is dependent on the digestion. In early studies, Deuel *et al.* [18] examined the digestibility and absorption of hydrogenated fats and oils and reported values from 79% to 98% depending on the melting points of the fats.

The absorption of ^{14}C-elaidic acid in rats has been studied and compared with oleic, palmitic and stearic acids, all randomly incorporated into soybean oil [15], and the appearance in lymph was comparable for all fatty acids tested. Similar results were obtained by Ono and Frederickson [83] and Clement [13]. Studies with partially hydrogenated fats, mainly triacylglycerols, underline that these absorption studies can be extended to other octadecenoic isomers [26].

In a recent paper, the influence of *trans* fatty acids on the intestinal absorption of other components, such as cholesterol and glucose, was examined in rats. No significant differences were observed compared to a diet with exclusively *cis* fatty acids [99A].

In vitro studies on pancreatic lipase digestion of triacylglycerols with C18:1 positional isomers [32] revealed an inhibition for acids with double bonds in positions ranging from the $\Delta 2$ to $\Delta 7$, but this might not be of great importance as fatty acids with such structures are nearly absent from partially hydrogenated fats. The pancreatic digestion of triacylglycerols containing elaidic acid showed no discrimination compared to other fatty acids or to the position in the triacylglycerol molecule [48,29].

Extensive human studies with deuterated fatty acids have been carried out by Emken and coworkers [22-24,27,28]. They examined the absorption of corresponding pairs of *trans* and *cis* positional isomers and found that even if the melting points were above body temperature both isomers were absorbed as well as oleic acid. This underlines the *in vitro* findings on pancreatic lipase activity against *trans*- and *cis*-C18:1 isomers.

After the absorption and transfer to the lymph the isomers are transported with the bloodstream to various tissues for deposition or catabolism. The fate in the blood including the impact on lipoproteins and blood lipids will be covered in Chapter 7.

C. METABOLISM OF *TRANS* OCTADECENOIC FATTY ACIDS *IN VITRO*

The metabolism of *trans* isomeric fatty acids has been studied intensively both *in vitro* and *in vivo*, mostly in rat experiments, and especially the influence on the metabolism of polyunsaturated fatty acids (PUFA) has been addressed because adverse effects have been postulated. In this context, it has to be pointed out that many of the trials have been carried out with essential fatty acid (EFA) deficient systems, which are not representative for the situation *in vivo*. Furthermore the use of minor amounts of fatty acids (radio-labelled) may give a distorted picture.

Most of the reports have dealt with rat liver and subcellular fractions prepared from this organ.

1. Activation

Before further metabolism, the fatty acids need to be activated to their CoA-esters. Lippel *et al.* [67,68] reported on a higher activation rate for *trans*-9- and *trans*-11-octadecenoate than for the corresponding *cis* isomers. The fatty acids with double bond positions from $\Delta 8$ to $\Delta 12$, the most abundant isomers in partially hydrogenated oils, were the least reactive. However, when albumin was added, no differences were observed. Changes in temperature, pH or buffer did not influence the activation profile [69].

Later Norman *et al.* [77] found for a number of corresponding *cis*- and *trans*-monoenoic isomers ranging in chain length from C16 to C22 that neither the position nor the configuration had a major impact on the activation, except for the C22:1 isomers (11 versus 13), and they concluded that the activation reaction was of minor importance compared to those introduced by the following enzyme reactions.

2. Desaturation and Elongation of Isomeric Trans and Cis Isomers

Trans monoenoic fatty acids isomers and the corresponding *cis* compounds may undergo desaturation and elongation. Normally a C18:1 acid such as oleic acid is desaturated by the $\Delta 6$-enzyme in competition with C18:2(*n*-6) and

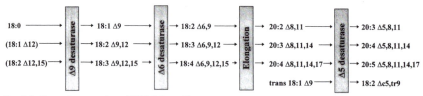

Fig. 5.1. General conversion of C18-fatty acids.

C18:3(n-3) in liver microsomes (Figure 5.1), but with essential fatty acid (EFA) deficiency the Δ9 desaturation is optimized.

Using rat liver microsomes from EFA-deficient rats, a number of C18:1 isomers were subjected to the Δ9-desaturation system under non-saturated conditions, and it was shown that most of the *trans* fatty acids could be desaturated by the Δ9-desaturase normally acting on stearic acid [87]. Acids with Δ7t- and Δ11t-unsaturation produced the corresponding conjugated dienoic acids Δ7t,9c and Δ9c,11t, respectively, but the *trans*-8 and the *trans*-10 were unable to react. For the C18 *cis*-monoenoic acids the Δ9 desaturase only acted on the Δ14- and Δ15-isomers. Some of the C18:1 *trans* fatty acids were also to some extent substrates for the Δ6 and Δ5 desaturases, the formation of 18:2 Δ6c10t from 18:1 Δ10t and 18:2 Δ5c7t from 18:1 Δ7t has been reported [87].

Further desaturation of the dienoic acids formed has also been suggested as shown in Figure 5.2.

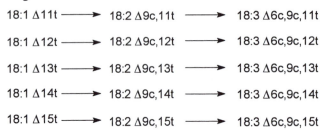

Fig. 5.2. Conversion of *trans* monoenoic acids.

The Δ5 desaturation of monoenes with chain lengths varying between C16 and C20 showed a preference for the longer chain. A seven-fold increase was found for 20:1t compared to 18:1t, the conversion being of the same order of magnitude as shown for 20:2 c11,c14 and 20:3 8c,11c,14c desaturation [88].

The desaturation of twelve different C18 *trans*-monoenes were studied extensively by Holman, Mahfouz *et al.* [70,35,36]. Besides the expected *trans* isomers, *cis,cis*-components were formed as a result of Δ9-desaturation as shown in Figure 5.3. Whether the isomerization happens before the desaturation is not clear, but it is obvious that C18:1 *trans* isomers present in partially hydrogenated oils are potential precursors for a number of *cis,trans*- and *cis,cis*-18:2 acids and thereby long-chain PUFA. Holman [36] even speculated on the probable formation of eicosatetraenoic acids from *trans* monoenoic acids and thus an influence on eicosanoids.

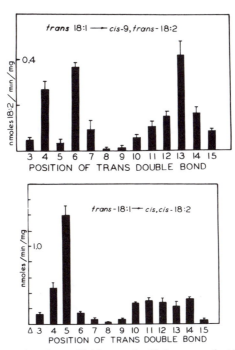

Fig. 5.3. Rates of desaturation of *trans*-18:1 isomers by Δ9 desaturase of rat liver microsomes to *cis, cis*-(upper) and *cis, trans*-18:2 (lower). Reproduced from [36] by permission.

However, it is important to note that the experimental conditions for the conversions *in vitro* are far from the *in vivo* situation, and the *trans* ratios applied may be much higher than normally found *in vivo*. Finally, the competition from other dietary fatty acids is important, and especially the presence of ample amounts of EFA changes the conversions completely.

Besides desaturation, chain elongation occurs. This process is generally believed to be a faster reaction than desaturation [6], but dependent on chain-length and number as well as position of double bonds, with maximal rates for C18 acyl chains. This implies that the *trans* monoenoic isomers could more easily be elongated than desaturated. Kameda *et al.* [50] studied with rat liver microsomes the elongation of *trans* C18:1 with Δ4 to Δ15 unsaturation and found varying rates, with Δ7t and Δ9t isomers as the most reactive, but still far from the reactivity of oleic acid (30-40% decrease). Somewhat lower rates were seen for the Δ8, Δ10, Δ11 and Δ12 isomers, whereas the fatty acids with bonds in the outer positions Δ4, 5, 6 and Δ13,14,15, respectively, were practically not chain elongated.

3. *Incorporation into Lipid Classes*

The specificity of acyltransferases catalysing the acylation leading to either triacylglycerols or phosphoacylglycerols is of great importance for obtaining membrane structures with the proper fluidity [59]. The preference of 16:0 and

18:0 for the sn-1-position is evident but not limited to these long-chain saturated acids. The incorporation of *trans* fatty acid isomers of C18:1 has been studied too [82], and all isomers except the $\Delta10$ and $\Delta12$ were incorporated preferentially into position 1. As the most abundant fatty acids in the partially hydrogenated vegetable oils are the $\Delta8$, 9, 10 and 11 isomers, they exert a strong competition with the saturated fatty acids. The positional *cis* isomers with $\Delta8$, 10 and 12 unsaturation are also rapidly incorporated [91,60,61], whereas the natural $\Delta9$ and $\Delta11$ are poor substrates for acylation in position 1.

For position 2, there is a clear preference for polyunsaturated fatty acids, but it seems that there also is a preponderance of certain monoenoic fatty acids, both *cis* and *trans*, and those with bonds at position 5, 9 and 12 are preferred. *Trans* $\Delta9$ is therefore the most abundant geometric isomer in position 2 of the phosphoacylglycerols, but still the major part of the *trans* acids is found in position 1. For further details, see reference [61].

4. Effects of C18:1 Isomers on Desaturation and Elongation

The recognition of decreased deposition of polyunsaturated fatty acids after the intake of *trans* fatty acids has prompted investigations on the influence of the specific *trans* and *cis* C18:1 isomers on the desaturation and elongation of various fatty acids important for cell structures.

In vitro studies with microsomal fractions of rat liver have been used most frequently and to enhance the desaturation activities ($\Delta9$ and $\Delta6$), EFA-deficient animals were introduced.

In early experiments by Brenner and his group [10] it was concluded that the effect of elaidic acid and vaccenic acid on the $\Delta6$ desaturation in liver microsomes was negligible compared to the influence of α–linolenic acid. Later Mahfouz *et al.* [71] made extensive examinations of the influence of double bond position in series of *cis*- and *trans*-C18-monoenes on the desaturation processes. To ascertain the inhibitory effect on the $\Delta9$, $\Delta6$ and $\Delta5$ desaturases, palmitic, linoleic and eicosa-8,11,14-trienoic acids were used as substrates in rat liver microsomal preparations. The ratios of inhibitor to fatty acid substrate were 3:1 for $\Delta6$ or 6:1 for $\Delta5$, respectively, and for $\Delta9$ both ratios were tested. A strong inhibition of the $\Delta6$ desaturation was found with the $\Delta3$, $\Delta4$, $\Delta7$ and $\Delta15$ *trans*-18:1 isomers, whereas the $\Delta9$ desaturase was inhibited by $\Delta3$, $\Delta5$, $\Delta7$, $\Delta10$, $\Delta12$, $\Delta13$ and $\Delta16$ *trans* isomers. The $\Delta5$ desaturase was most influenced by the $\Delta3$, $\Delta9$, $\Delta13$ and $\Delta15$ *trans* isomers. As a control, 18:0 was tested in the same system and only a slight inhibition of the $\Delta9$ desaturase was seen; no effects were observed on $\Delta6$ and $\Delta5$ desaturases.

A corresponding experiment was carried out with the *cis*-18:1 isomers [72]. Inhibitory effects were seen also for some of the *cis* isomers, pointing again at the importance of the *cis* isomer fraction present in the partially hydrogenated oils, which is often neglected in nutritional discussions, where the *trans* fatty acids are in focus. The strongest inhibitor for the Δ6 desaturase was *cis*-Δ8 18:1 and this acid was also a potent inhibitor for the Δ5 desaturase. On the contrary, *cis*-Δ3 18:1 was the weakest inhibitor for both desaturases. The Δ9 desaturase was especially inhibited by the *cis*-Δ10 and *cis*-Δ11 isomers, both present in fairly high amounts in partially hydrogenated edible fats.

5. Oxidative Degradation

The oxidative degradation of fatty acids takes place first of all in the mitochondrial matrix by the normal β-oxidation pathway, but additionally the long-chain fatty acids may be shortened in a peroxisomal oxidation pathway in which the first step is catalysed by a flavoprotein, acyl-CoA oxidase. In contrast to mitochondrial oxidation, this reaction produces hydrogen peroxide. The remaining steps are analogous to the mitochondrial oxidation.

i. Mitochondrial oxidation. In the classical experiments by Coots [15], the β-oxidation of 1-^{14}C labelled elaidic, oleic, palmitic and stearic acids was compared. The C18:1-isomers and palmitic acid were degraded similarly but stearic acid somewhat less. On the contrary, Ono & Fredrickson [83] found equal oxidation when comparing oleic and elaidic acids after intravenous injections as albumin complexes. Later Anderson [3] and Anderson and Coots [4] compared the degradation of 1-^{14}C with the oxidation of 10-^{14}C or uniformly labelled *cis*- and *trans*-Δ9-18:1 and found that the *cis* isomer was catabolized faster than *trans*, when the labelling was on the methyl side of the double bond, probably due to the fact that the isomerization of *cis*-Δ3 12:1 formed during the degradation is a normal reaction in β-oxidation whereas the *trans* configuration demands other enzymatic reactions.

Lawson and Kummerow used rat heart mitochondria [63] for their β-oxidation studies of CoA-esters of elaidic acid, oleic acid and the partial degraded products of elaidic acid: *trans*-Δ7 16:1, *trans*-Δ5 14:1 and *trans*-Δ3 12:1. They found a much slower oxidation rate for elaidic acid than for oleic and the chain-shortened *trans* fatty acids. These results do favour the theory of the problems being centred around the double bond.

An extensive study on the β-oxidation of geometrical and positional octadecenoic acids isomers using both heart and liver mitochondria has been published by Lawson and Holman [64]. Their data are illustrated in Figure 5.4. They prepared *cis* and *trans* isomers ranging from Δ4 to Δ16, all components which could be present in partially hydrogenated oils and fats. The oxidation of corresponding CoA-esters measured by oxygen uptake was recorded. The *cis* isomers were catabolized in a similar way by heart and liver mitochondria. The fatty acids with even numbered *cis* bonds were catabolized slower than oleic acid, while the *cis* fatty acids with uneven numbered bonds reacted nearly as rapid as oleic acid. The *trans* fatty acid catabolism was different from the *cis* isomers. Liver mitochondria oxidized the even-positioned *trans* isomers more rapidly than the odd-numbered. The same pattern was observed for the heart mitochondria, but only when the double bond was near the

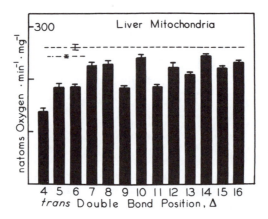

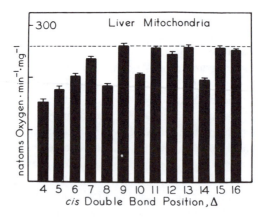

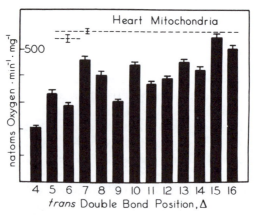

Fig. 5.4. Effect of double bond position and geometry on oxidation rates of C18:1 CoA-esters in liver and heart mitochondrial preparations with added carnitine. Bars represent means of 5-8 preparations ± SE. Dashed lines indicate results for oleyl-CoA, and for heart mitochondria for stearoyl-CoA also (short line). Reproduced by permission from [64].

middle of the chain. Both heart and liver oxidized the *cis* isomers and especially the naturally occurring Δ9 and Δ11, more rapidly than the *trans* isomers, with three exceptions: Δ8, Δ10 and Δ14. This is in accordance with the findings of Anderson and Coots [4] for *in vivo* oxidation of fatty acids with uniform labelling.

In contrast to these results, it has been reported [75] that homogenates of rat hearts oxidized oleic and elaidic acids at equal rates. Interestingly, it was also found that rats fed diets rich in *trans* acids showed a higher oxidation rate.

Human heart homogenate has also been used for oxidation experiments with elaidic and oleic acids, and equal oxidation rates were recorded. The tissue was obtained from patients undergoing by-pass surgery. A control experiment with rat homogenates prepared identically gave faster degradation of oleic than elaidic acid [62].

ii. Perfusion studies. Oxidation of *cis* and *trans* $10\text{-}^{14}\text{C-}\Delta9\text{-}18\text{:}1$ was also studied in a rat (Wistar) liver perfusion model [45], and the *trans* isomer was found to be oxidized more readily than the *cis* isomer. The difference from some of the previously reported results may be explained by the more intact model system in which the release in lipoproteins or incorporation into liver phospholipids is also possible. Based on these findings, it is suggested that the *trans* acids are more efficient energy sources than the *cis* analogues.

However, when such studies were performed with another rat strain (Sprague-Dawley), a lower *trans* oxidation rate was observed. This finding emphasizes the problems in transferring results from one species to another.

iii. Peroxisomal Oxidation. As discussed by Beare-Rogers [5], peroxisomal oxidation can be induced in a number of ways and it has certainly a function in the metabolism of very-long-chain fatty acids, including geometrical and positional isomers in partially hydrogenated marine oils. Even for the more abundant C18:1 isomers, some of which may be difficult to transfer to the mitochondrial matrix, there is supposed to be an effect. Feeding of partially hydrogenated marine oils to rats increased the peroxisomal chain shortening capacity in perfused hearts four fold compared to peanut oil-fed controls [78]. Corresponding results were obtained with hepatocytes [12] isolated from animals fed either partially hydrogenated marine oil, rapeseed oil or peanut oil. The chain shortening was increased up to three fold with the marine oil and corresponding increases in catalase and urate oxidase activities indicated peroxisomal proliferation. The induction of peroxisomal activity has also been related simply to high-fat diets [5] and in particular when this fat contained C22:1, as found in rapeseed oil (old type) and partially hydrogenated marine oils. In a feeding experiment with rats [99] given 25 % of either rape seed oil, partially hydrogenated rape seed oil, marine oil, partially hydrogenated marine oil, soybean oil and partially hydrogenated soybean oil, it was shown that the induction of peroxisomal β-oxidation could be ranked as follows: PHMO > PHRO > MO> RO > PHSO > SO. This seems to imply that the isomers formed during hydrogenation are potent activators. The specificity according to chain length as studied with acyl-CoA esters of monoenes incubated with isolated peroxisomal fractions showed preference for the oxidation of *trans* isomers, especially in rats fed a partially hydrogenated marine oil diet [84].

The mechanism behind the induction of peroxisomal activity has more recently been investigated by Flatmark *et al.* [31]. Partially hydrogenated fish oil was able to induce synthesis of peroxisomal enzymes whereas diets with soybean oil supplemented with either elaidic acid (9t-18:1), erucic acid (13c-22:1) or brassidic acid (13t-22:1) did not result in proliferation of peroxisomes. Whether this is due to other fatty acids in partially hydrogenated fish oil remains to be clarified.

The combined effect of mitochondrial and peroxisomal oxidation has been investigated in liver, heart and skeletal muscle of rats fed 40% fat with either saturated, *cis*-monoenoic or *trans*-monoenoic fatty acids as major components [101]. The total oxidation rate measured both with [1-^{14}C]- and [16-^{14}C]-palmitate was not influenced significantly by the dietary fats. The group fed saturated fat showed a higher proportion of peroxisomal oxidation, but the dietary C18 fatty acids did not result in higher activity than seen for the other groups. The conclusion seems to be that only high fat diets with C20 and C22 fatty acids induce peroxisomal activity.

Ide *et al.* [46] also measured the activities of liver mitochondrial and peroxisomal fatty acid oxidation in two strains of rats (Wistar and Sprague-Dawley) fed either *trans* fatty acids (partially hydrogenated corn oil) or olive oil. Only slight dietary dependence was observed and they concluded that the *in vivo* modulation of oxidation by changing from *cis* to *trans* is marginal. The differences found in experiments with cell-free systems, and to some extent also in perfusion studies, seem to be due to the simplified model systems.

D. METABOLISM OF TRANS OCTADECENOIC FATTY ACIDS *IN VIVO* (ANIMAL STUDIES)

1. *Influence on Tissue Lipid Classes*

Dietary *trans* fatty acids are incorporated into nearly all lipid classes of various organs and numerous experiments have shown that the influence on tissue phospholipid distribution is negligible when essential fatty acids are supplied and no tissue degeneration is observed [106,20,26,41,42,108].

The concentration of *trans* fatty acids found in lipid classes varies with the dietary level and is highest in the triacylglycerols. Alfin-Slater and coworkers have reported on the effects of feeding rats diets with 15% fats containing varying amounts of C18:1 *trans*-monoenoic isomers [2] and the incorporation of *trans* C18:1 in phosphoacylglycerols of plasma, liver, kidney, testes, heart and adipose tissue showed the greatest preference for the liver with levels increasing from 12 to 19% in parallel with increasing dietary levels from 12 to 50%. Less was deposited in other tissues, although heart and plasma were enriched also, especially after a high *trans*-containing dietary regimen, whereas *trans* fatty acids were practically absent in testes phosphoacylglycerols. The triacylglycerols reflected more or less the dietary conditions. Cholesterol esters contained lower amounts of *trans* than found for phospholipids in all organs except testes. The values found here for cholesterol esters were comparable to those in other tissues. The absence of *trans* fatty acids in testis from rats fed C20 and C22 fractions from partially hydrogenated fish oil was reported by Jensen [47]. Reichwald-Hacker *et al.* found, after feeding 12.3% of *trans*-18:1 originating from soybean oil, about 3% *trans* in the total fatty acids from total testes lipid [90].

Increasing levels of dietary *trans* monoenes were also reflected in erythrocyte membranes [2] and high levels of *trans*-18:1 were detected in the cholesterol ester fraction in adrenals [19]. The incorporation of *trans* fatty acids into lipoprotein classes was studied by Egwim and Kummerow in rats fed 20% partially hydrogenated soybean oil [20,21]. High levels of *trans*-18:1 were found in the phosphoacylglycerols of very low density lipoproteins, VLDL (18.3%), low density lipoproteins, LDL (23.3%), and high density lipoproteins, HDL (11.7%), and in the VLDL cholesterol esters (16.2%), whereas much lower amounts were present in cholesterol esters from LDL and HDL (2-4%).

There seems to be a limit for the incorporation of *trans* fatty acids, as similar values are reported after various dietary supplementations [106,107,76,74].

Brain phosphoacylglycerols did not accumulate *trans* fatty acids, but the sphingomyelins contained relatively high amounts [106]. Wood [107] also reported on the distribution among the major phospholipid classes and found for organs such as liver, kidney, muscle and especially heart, higher values for *trans* incorporation in phosphatidylethanolamine (PE) and phosphatidylserine (PS) plus phosphatidylinositol (PI) than in phosphatidylcholine (PC). The degree of incorporation of *trans* does not seem to be influenced by the linoleic content, when EFA-sufficient diets are fed [33,7,41].

2. Regiospecific Distribution of Total Trans-C18:1

The incorporation of *trans*-18:1 into triacylglycerols according to position is not equal. Reichwald-Hacker *et al.* [89] found a preferential esterification in the 1- and 3- positions. Corresponding data for incorporation in PC showed a definite preference for *trans* deposition in position sn-1 for liver, heart and serum. This is in agreement with findings by others [105,106]. Wood and Chumbler [105] also described a corresponding distribution for liver PE.

3. Deposition of Positional Isomers in Different Lipid Classes

The incorporation of *trans* 18:1 isomers into different lipid classes has been examined in a number of studies [106,89,90,104].

Wood *et al.* [106] maintained rats on a diet with 15% partially hydrogenated safflower oil (50%) for 5 weeks and examined the distribution of positional *trans* isomers in triacylglycerols from rat tissues (kidney, muscle, heart, liver, lung, spleen and adipose tissue). Almost identical incorporation was obtained for all tissues examined, with a *trans* isomer distribution comparable with the dietary fat except for an increase in especially the $\Delta 8$ and to some extent the $\Delta 9$ isomers. The presence of all the isomers underlines the previous mentioned absorption ratios of over 90% for partially hydrogenated fats. Among the *cis* isomers the $\Delta 10$-18:1 is discriminated from a relative percentage in the diet of 10.7 to values ranging from 2-4% in triacylglycerols of various tissues. Also the *cis*-$\Delta 12$ isomer is somewhat lower in the tissue triacylglycerols. Furthermore *trans*-hexadecenoates were found in triacylglycerols, and especially *trans*-$\Delta 8$ 16:1, indicating a chain shortening of the *trans*-$\Delta 10$ 18:1. The higher contents of $\Delta 8$ and $\Delta 9$ isomers in triacylglycerols and absence of the corresponding chain-shortened 16:1 may suggest that these double bond positions may pose difficulties in β-oxidation. As will be mentioned later the incorporation into phospholipids is also retarded. In experiments with rats fed 18% partially hydrogenated soybean oil with a lower *trans* content (12.3% *trans*) for 12 weeks, less deposition of isomers occurred and the above mentioned discriminations were less evident.

In contrast, the distribution of geometric isomers in the phospho-acylglycerols differed markedly from the dietary composition and varied in different organs. For the *trans* 18:1 in PC, PE and PI+PS, an enrichment of the *trans*-$\Delta 12$, $\Delta 13$ and $\Delta 14$ isomers was found in all tissues, but to varying degrees [106]. Especially, the increase in *trans*-$\Delta 12$ was notable. For *trans*-$\Delta 9$, some discrimination was seen for lung and spleen, but for the *trans*-$\Delta 10$ isomer a very tissue-dependent distribution was observed. Generally there was an extended discrimination in PE, and remarkable low values for the *trans* isomer content were found for heart and liver PC. Lung tissue possessed the highest content of the *trans*-$\Delta 10$ isomer, and only a slight discrimination compared to the dietary content was seen. For the *cis* isomers, enrichment of the $\Delta 11$, $\Delta 12$ and $\Delta 13$ isomers is evident for nearly all phospholipid classes and tissues examined. In heart, liver, spleen and muscle, the *cis*-$\Delta 9$, $\Delta 11$ and $\Delta 12$ isomers are by far the most abundant components. The *cis*-$\Delta 10$ 18:1 was not in contrast incorporated to any significant amount in the tissues examined [106].

In accordance, it was found for total organ lipids that a higher proportion of *cis*-$\Delta 11$ (*cis*-vaccenic acid) than present in the diet was deposited in liver and heart, whereas dietary distribution was reflected in adipose tissue, testes and adrenals [90]. For the *trans* isomers, the $\Delta 14$ was preferentially incorporated into liver, heart and serum, whereas discrimination occurred for the *trans*-$\Delta 10$ and $\Delta 11$ isomers. Generally, *trans* isomers were more abundant in liver and heart, whereas *cis* isomers were preferred in the testes. The remaining organs reflected more the composition found in the dietary fat [90].

The incorporation of the positional isomers into mitochondrial total membrane lipids (almost exclusively phospholipids) was examined in four groups of rats fed 15, 20 and 25% partially hydrogenated peanut oil (with 54% *trans*) for 3, 6 or 10 weeks. Supplementation with sunflower seed oil assured EFA except in one of the groups fed exclusively 25% partially hydrogenated peanut oil [41]. Total deposition of *trans* ranged from 15 to 19% after 10 weeks. The relative distribution of isomers did not depend on the level of partially hydrogenated peanut oil and varied only slightly with time. The *cis-*Δ10 18:1, one of the major *cis* isomers of the partially hydrogenated oil, was almost completely discriminated in accordance with the above mentioned data for whole organs. Likewise, the content of *trans-*Δ10 18:1 was considerably lower than in the diet whereas the deposition of *trans-*Δ12 18:1 was increased. In the group without linoleic acid supplement, isomers of linoleic and arachidonic acids were observed.

4. Distribution of Geometrical and Positional Isomers of C18:1 between the sn-1 and 2 Positions in Phosphoacylglycerols (PC and PE).

It was mentioned previously that the major part of the dietary *trans* fatty acids was incorporated into the *sn*-1 position of phosphoacylglycerols. The percentage distribution of the positional isomers of *cis* and *trans* 18:1 in the specific positions has also been examined. In PC, the major *trans* isomers present in the *sn*-1 position in liver, heart and serum were Δ11, Δ12, Δ13 and Δ14 with a definite discrimination against Δ10. On the contrary the *sn*-2 position, although only small amounts were present, favoured the *trans* Δ9, Δ11 and Δ12 isomers. The *cis* isomers were easily incorporated into the *sn*-2 position and the endogenous Δ9 and Δ11 were the major components [89]. The deposition of *cis* 18:1 Δ8 and Δ6 compared to oleic acid in rat liver mitochondrial membranes and adipose tissue was examined by Høy and Hølmer [42]. The rats were fed 10% fat and half of this was a C18:1 isomer; all diets were supplemented with sunflower seed oil to avoid EFA deficiency. The isomeric octadecenoates were primarily incorporated into the 1-position of both PC and PE at the expense of saturated fatty acids. The highest incorporation was seen in PE with 8.9% for the Δ8 isomer and 4.8% for the Δ6 isomer. No effects on the contents of PUFA in position 2 were observed. The isomers were readily incorporated into adipose tissue, 23% of the *cis* 18:1 Δ8 and 13% of the *cis* 18:1Δ6, in exchange for oleic acid.

5. Effects of Trans Fatty Acids on Polyunsaturated Fatty Acid Conversion

The *in vitro* experiments have been mentioned previously. In this section, the influence of feeding experiments with partially hydrogenated fats and the *in vivo* effect of specific ingested isomers will be addressed.

The partially hydrogenated vegetable oils contain mainly C18-monoenoic fatty acids, positional as well as geometrical isomers, and only low levels of *cis,trans-*, *trans,cis-* and negligible amounts of *trans,trans-*C18-dienoic acids. The partially hydrogenated marine oils have additionally the corresponding long-chain C20 and C22 isomers, again mostly monoenoic acids. As such fats have been used for margarine production, at least in Europe, experiments with partially hydrogenated marine oils will be discussed also.

In early feeding studies [19,20], the influence of partially hydrogenated vegetable oil on PUFA concentrations in various tissues was examined. However, the rats were not supplied with adequate linoleate and it was concluded that the decrease in incorporation of (*n*-6)-PUFA could be due to EFA deficiency. It was shown many years ago by Aaes-Jørgensen that partially hydrogenated oils accentuated the EFA deficiency symptoms, as for instance testicular degeneration [1]. Hill *et al.* [33] compared the effect of partially hydrogenated soybean oil and hydrogenated coconut oil and found decreased Δ6 and Δ9 desaturation in liver microsomes of the rats fed isomers, whereas the Δ5 desaturation was not affected. Arachidonic acid was somewhat lowered, and correspondingly linoleic acid was increased, but again it cannot be excluded that some of the effect was due to low dietary C18:2 intake.

In later experiments, the necessity for adequate linoleic acid supplementation has been considered and the influence of the level of linoleic acid studied in detail.

De Schrijver and Privett [17] compared the effect of *trans* fatty acids on the Δ6 and Δ9 desaturase activities in liver microsomes with and without sufficient linoleic acid. In both cases, a decrease in Δ6 desaturase activity lead to lower PUFA deposition, whereas the Δ9 desaturase activity was stimulated especially in the EFA-deficient animals. It must, however, be stressed that the *trans*-diet used also contained rather high amounts of *trans*-18:2 isomers.

Studies on the influence of *trans* isomers from partially hydrogenated peanut and marine oils on the PUFA deposition in rat liver and heart showed a reduced conversion of linoleic acid into arachidonic acid with a more pronounced effect of the partially hydrogenated marine oil [39]. The latter having appreciable amounts of 20:1 and 22:1 in addition to the 18:1 isomers. The very moderate deposition of 20:1 and 22:1 observed in the tissues seems to imply that the effect of the long-chain isomers may be exerted through the wide range of C18:1 isomers produced by peroxisomal chain shortening. The total *trans* fatty acids deposited in PC from rat mitochondrial membranes were equal after feeding partially hydrogenated peanut oil or marine oil [39].

The effect of partially hydrogenated vegetable and marine oils on the Δ6 and Δ5 desaturase activities was studied subsequently with rat liver microsomes at two substrate levels reflecting saturated and non-saturated conditions [52], the latter being more comparable to physiological conditions. In both cases, a significant decrease in the Δ6-desaturase activity was found for groups fed *trans* fatty acids and additionally the Δ5-desaturase activity for the group fed partially hydrogenated marine oil was much lower than the control group. Detailed analysis of the dietary fat showed that the total *trans* content was the same for both groups and was present as *trans* monoenoic acids, for the vegetable oil exclusively as 18:1 isomers and for the marine oil as equal amounts of C18, 20 and 22-monoenes.

Corresponding results on the Δ6 desaturase were obtained by Svensson [97], whereas the Δ5 desaturase activity was reported slightly increased for the group fed partially hydrogenated peanut oil.

The dietary effect of partially hydrogenated soybean oil on $\Delta 6$, $\Delta 5$ and $\Delta 9$ desaturases was also studied by Mahfouz *et al.* [73], who concluded from a comparison with lard that *trans*-18:1 was more inhibitory towards EFA conversions than saturated fatty acids. This is in accordance with decreased deposition of long-chain PUFA after *trans* intake, which has been reported in a number of studies [7,8,39]. The effect of specific isomers on the desaturation process has been addressed by Lawson *et al.* [65] who calculated the correlation coefficients between arachidonic acid present in liver phospholipid of rats and the *cis-* and *trans*-18:1 isomers after feeding partially hydrogenated soybean oil. The most prominent negative correlation was found for three *cis* fatty acids with the $\Delta 12$ isomer as the most effective. More equal correlation figures were found for the *trans* isomers, the $\Delta 12$ also being here the most effective. In a subsequent experiment [66] concentrates of *trans* and *cis* isomers from partially hydrogenated soybean oil were compared in diets with adequate linoleic acid. The *trans* isomers increased the levels of 18:2(*n*-6) and 20:5(*n*-3) in liver PC and PE, whereas the *cis* isomers had less effect in PE than PC, perhaps due to a larger incorporation of 18:1 in PC. It has been suggested that the *trans* isomers may exert their modifying effect on the desaturation, whereas the *cis* 18:1 isomers rather compete in the acylation of the *sn*-2 position of the phosphoacylglycerols [66].

There are conflicting results reported on the effect of *trans* fatty acids on the conversion of linoleic acid and the deposition of long-chain PUFA. It seems that the presence of adequate linoleic acid as well as an additional dietary intake of n-3 fatty acids is critical for the outcome of a study and has been overlooked in many experiments.

A thorough investigation of the linoleic acid requirement of rats fed *trans* fatty acids was made by Zevenbergen *et al.* [108]. Diets with rather high amounts of *trans*-octadecenoate (20 en%) from partially hydrogenated soybean oil with increasing amounts of linoleic acid (from 1 to 17% of the total fatty acids) and balanced to give nearly the same levels of other dietary fatty acids were fed to rats for 13-14 weeks. Groups without *trans* supplementation and dietary levels of 5 and 12.5 % 18:2(*n*-6) served as controls. The composition of liver and heart mitochondria showed no change in phosphoacylglycerol distribution in relation to linoleic acid supply and compared to controls. *Trans* fatty acids were found in all phospholipids investigated. The linoleic acid content was increased and correspondingly the arachidonic acid decreased when *trans* fatty acids were fed. However, with increasing dietary linoleic acid levels the arachidonic acid concentration reached the same values as found for the controls for both liver and heart mitochondrial PC. For liver PE, the same trend was observed, whereas the heart PE for all groups receiving *trans* fatty acids had higher 20:4(*n*-6) compared to controls, mainly due to a lower content of 22:5(*n*-6). This might again be a consequence of an influence on the $\Delta 6$ desaturation, which according to the newer pathway for formation of C22 PUFAs presented by Sprecher's group [96] would favour this distribution. Despite these changes the mitochondrial function in both tissues, measured as

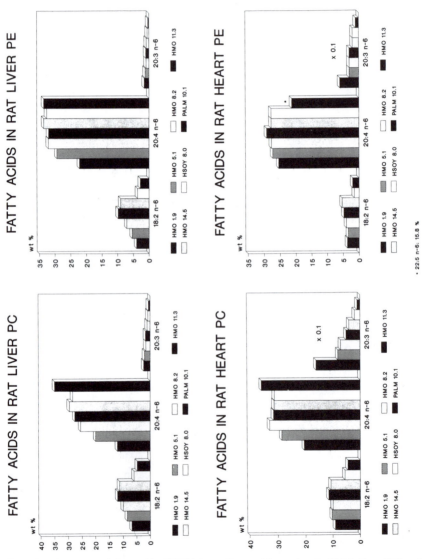

Fig. 5.5. Polyunsaturated fatty acids (PUFA) in rat liver and heart phosphatidylcholines (PC) and phosphatidylethanolamines (PE). Abbreviations: HMO, partially hydrogenated marine oil; HSOY, partially hydrogenated soybean oil; PALM, palm oil. Numbers given indicate dietary linoleic acid in %. All groups fed partially hydrogenated fats received 33% *trans* fatty acids.

ATP synthesis, was not impaired and it was concluded that a linoleate level of 5% of the dietary fatty acids is sufficient to prevent any undesirable effect of even high levels of *trans*-octadecenoates from partially hydrogenated soybean oil.

A corresponding experiment with partially hydrogenated marine oil (HMO) was also conducted [43,44]. Rats were fed for 10 weeks diets with 16% HMO + 4% vegetable oil providing 33% of *trans* fatty acids with chain lengths varying from 16 to 22. The composition of the vegetable oil was chosen to give linoleic acid contents ranging from 1.9 to 14.5% of the total fatty acids.

A group fed partially hydrogenated soybean oil, furnishing the same amount of *trans* fatty acids and 8% 18:2, and a palm oil group, without dietary *trans* acids and a 18:2 content of 10%, served as controls. Increasing levels of linoleic acid from 1.9 to 8.2% resulted in increased conversions and PUFA formation, but then a plateau was reached and at a lower level than found in the palm oil group without *trans*. For both liver and heart PC and PE, the linoleic acid content was much higher than found for the controls. For heart PE, the deposition of 22:5(n-6) was decreased in all groups fed *trans* fatty acids compared to the palm oil control indicating an inhibition of the second Δ6 desaturase step. Generally, the PE composition was less affected by the linoleic acid level than the PC and the heart less than the liver. Whether this reflects a hierarchy in the utilization of the metabolic pool or is due to organ specificities is not known. The results are shown in Figure 5.5.

All groups fed HMO had similar contents of *trans* fatty acids deposited in the PC and PE fractions irrespective of dietary 18:2 levels, and these were even lower than found in the group fed partially hydrogenated soybean oil. The C20:1 and C22:1 isomers present in the diet were only deposited in very minor amounts in the PC and PE.

Although increased contents of linoleic acid did not overcome the inhibitory effect of the rather high amount of dietary *trans* fatty acids present, no adverse effects were observed in the rats. For cardiolipins, normally very rich in linoleic acid, it was found that feeding HMO resulted in even higher values than found for groups fed partially hydrogenated soybean oil and palm oil. The decreased conversion to longer chain PUFA's and deposition in other phosphoacylglycerols resulted apparently in a greater pool of linoleic acid for cardiolipin synthesis in liver, heart and kidney. For testis, a reflection of the dietary level of linoleic acid was not seen. The deposition of total *trans* was below 2.5% of the fatty acids in spite of a dietary level of 33%, and only 18:1 isomers were deposited [44].

When different studies are compared it is probably important also to make an evaluation of the remaining dietary regimen apart from the fat component. An implication of other dietary factors cannot be fully excluded, as discussed by Koga *et al.* [53] who reported that linoleic acid metabolism was influenced by the type of dietary protein. The deposition of arachidonic acid in liver phospholipids was lower when the protein source was casein instead of soybean protein.

6. Desaturation Studies with Human Cells

Investigations with human tissue are very scarce. In studies with human fibroblasts, it was reported that both elaidate and linolelaidate are potent inhibitors of the Δ6 desaturase [94]. Also, inhibitor effects on the Δ5 desaturation have been examined and in this reaction elaidic acid was found more active than linolelaidic acid. In contrast, Δ11-*trans* 18:1, vaccenic acid, one of the major *trans* fatty acids in butter, was less inhibitory than the Δ9 isomer. The Δ11-*trans* 20:1 had an effect between that of the two C18-isomers. In contrast to these findings it has been reported [14] that all *trans*-monoenoic octadecenoates with double bonds in positions 8 to 15 inhibited the Δ5 desaturation in glioma cells when linoleic acid acted as substrate, whereas only the Δ11- and 12-*trans* were effective with 18:3(*n*-3) as substrate. The *cis*-isomers were less potent inhibitors.

7. Isomeric Fatty Acids and Eicosanoid Formation

As already mentioned, the monoenoic isomers can be incorporated into the *sn*-2 position of phosphoacylglycerols, where normally the PUFA are acylated also. This means that the isomeric fatty acids may influence the production of eicosanoids.

Holman and Johnson have in their review on EFA deficiency [37] stressed that the early findings [34] have shown that partially hydrogenated oils accentuate the EFA deficiency symptoms, and they suggested that a supply of at least 4% of linoleic acid would be necessary to fulfil the EFA requirement. A marked decrease in eicosanoid production has been reported during EFA deficiency [51,98].

Later Zevenbergen and Haddeman [109] found in experiments with rats fed partially hydrogenated vegetable oil supplemented with 2% linoleic acid that neither the production of prostacyclin by aorta pieces nor the amounts of eicosanoids formed in platelets were changed, although a slight decrease in arachidonic acid concentration was observed in the precursor phospho-acylglycerols. For the aorta, a linear relationship between PGI_2 and arachidonic acid was observed and independent of whether *trans* fatty acids were present. For platelet lipids the linoleic acid present in the diet was reflected in the arachidonic acid content. The biosynthesis of eicosanoids was normalized at 5% (weight) of dietary linoleic acid, which is in accordance with findings by others. [8,95].

A direct influence of the dietary isomers on the enzymes involved in eicosanoid production was therefore not obvious. These results were in line with the findings in tumor-bearing rats fed *trans* octadecenoates with an adequate supply of linoleic acid [103]. Plasma values for thromboxane A_2 and prostaglandins E_1, E_2 and I_2 were not affected [103].

In contrast to these findings, Chiang *et al.* [11] reported that *trans*-octadecenoic acid reduced the renal prostaglandin $F_{2\alpha}$ in hypertensive rats, and an influence on the phospholipase A_2 activity was suggested as the reason. In line with this, Koga *et al.* [53] found a significant reduction in arachidonic acid and leukotriene C_4 release in spleen from rats fed *trans* octadecenoates; the plasma prostaglandin E_2 was in contrast unaffected.

Wahle *et al.* [102] have examined the effects of the individual *cis* and *trans* isomers on porcine platelet function. Aggregation of platelets stimulated with either collagen or thrombin was studied after preincubation with $\Delta 9$, $\Delta 11$, $\Delta 12$ and $\Delta 13$ *cis-* and *trans*-C18:1 isomers. The response was measured with a sensitive impedance method. The *cis*-isomers generally inhibited the induced aggregation by 85-90%, with the $\Delta 11$ being the less effective in collagen-induced stimulation and with a parallel decreased formation of TXB_2. In contrast, *trans* isomers inhibited only the collagen aggregation by about 30-50% and correspondingly more TXB_2 was observed. In the thrombin-induced aggregation, the *cis* isomers decreased the effect also, but the *trans* isomers increased the aggregation with $\Delta 12$ as the most potent isomer. However, the *trans* acids were still able to reduce the TXB_2 production to some extent. The implications of these findings for *in vivo* conditions seem difficult to evaluate and need further investigations.

E. HUMAN STUDIES ON DEPOSITION AND METABOLISM OF MONOENOIC *TRANS* FATTY ACIDS

1. Deposition in Human Tissues

Information on incorporation of *trans* fatty acids in human organs, apart from the adipose tissue, are only to be obtained from samples taken at autopsy or during surgery. The data are therefore limited.

For the adipose tissue, results on total *trans* fatty acids present in samples obtained in UK, Germany, Israel and US are in the range of 1-7 %, the major part being 18:1 monoenes and the composition reflecting dietary intake [30].

Extensive studies have been made by Ohlrogge *et al.* [79,80,81]. The distribution of *cis-* and *trans*-C18:1 isomers was determined in total lipid extracts from adipose tissue, heart, liver, aorta and brain and compared to the corresponding isomers found in partially hydrogenated oils (average of margarines and cooking oils) and in butter. The tissues were obtained within 4 hours of death from 8 adults. Percentages of total *trans* determined for heart (1.2-4.1), aorta (1.4-3.9) and liver (1.5-2.5) were slightly lower than found for adipose tissue (2-5.8), whereas brain only contained from 0.23%-0.88% total *trans* [79,81]. The distribution of positional isomers in the *trans* octadecenoate fraction was similar to that present in the partially hydrogenated vegetable oil, although discrimination of certain fatty acids could be found. In accordance with animal experiments, the *trans* $\Delta 10$ isomer was discriminated in total liver lipids. The *trans* isomer distribution was far from that found for butter.

In contrast, the *cis* isomers in human tissues were not dominated by those found in the partially hydrogenated oil, but by the $\Delta 9$ and $\Delta 11$ isomers which can be synthesised endogenously. These fatty acids are also the major *cis* octadecenoates in butter. The only reflection was a deposition of the *cis*-Δ12, which was abundant in partially hydrogenated fats. On the contrary, the *cis*-Δ10 was effectively discriminated, also consistent with the findings in animal studies.

For liver and heart, the distribution of *cis* and *trans* octadecenoate in some of the major lipid classes was also reported [80]. A comparison with the dietary intake of *trans* octadecenoate isomers showed that the composition of human triacylglycerols was very much the same as in the diet, but again a preference for the *cis*-Δ12 was obvious.

A comparison of the *trans* isomer intake with the isomer distribution in liver triacylglycerols showed a great coincidence, except for the Δ12 isomer which was relatively increased. The phosphatidylcholines from both liver and heart had a definite preference for the *trans*-Δ11 and Δ12 isomers and even the Δ14 isomer was increased. For the *cis* fraction, the dominating isomers were the natural Δ9 and Δ11, whereas Δ10 was nearly absent in the phospholipids.

The fatty acid composition including the presence of *trans*-18:1 of human heart was also reported by Roquelin *et al.* [93]. Samples from 36 surgery patients, representing both sexes and with ages between 40 and 70 years, showed low incorporation of *trans*-octadecenoic acid in PC, PE and sphingomyelin and absence from cardiolipin. The isomer distribution examined in a combined PC and sphingomyelin fraction showed a preferential incorporation of the Δ9 and Δ11 isomers, both for the *cis* and *trans* configurations, in contrast to the findings of Ohlrogge *et al.* Such differences could possibly be attributed to dietary intakes.

The relevance of using the content of isomeric fatty acids in human adipose tissue as a clinical risk factor for cardiovascular disease has been discussed by Hudgins *et al.* [38], who correlated 19 geometrical and positional fatty acid isomers and 10 risk factors for cardiovascular disease in 76 adult males. No evidence of strong correlations between the risk factors and isomeric *cis* and *trans* fatty acids present in the adipose tissue was found, although the dietary composition was reflected in the depot fat.

2. Metabolism of Octadecenoates in Man (in vivo Studies)

The presence of *trans* fatty acids in human membrane lipids and the findings in animal experiments that some of the isomers may even prefer the *sn*-2 position of phospholipids, normally occupied by PUFA conferring fluidity to the membrane and being precursors for important cellular regulators as the eicosanoids, has intensified the need for studying the metabolism in man.

Emken and coworkers have performed extensive studies [22-29] with the use of deuterated octadecenoates and determined absorption, oxidation rates and deposition of the isomers in lipid classes of human plasma [29].

Mixtures of triacylglycerols with deuterium-labelled 9-*cis*-18:1 for comparison and equal amounts of pairs of deuterated *cis*- and *trans*-18:1 positional isomers were given to two young adult males. Blood samples were

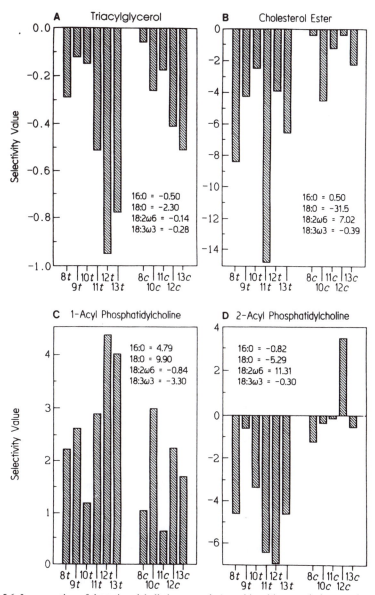

Fig. 5.6. Incorporation of deuterium-labelled *trans* and *cis* positional isomers in human plasma lipid classes. Selectivity values are related to oleic acid. For comparison, values for palmitic acid (16:0), stearic acid (18:0), linoleic acid (18:2), and α-linolenic acid (18:3) are included. Reproduced from [29A] by permission of the author.

drawn at various intervals and the composition of the plasma lipids, cholesterol esters, triacylglycerols, phosphatidylcholines (both for the *sn*-1 and *sn*-2 positions) and phosphatidylethanolamines, was analysed by GC-MS. For comparison also, the distribution of 16:0, 18:0, 18:2 and 18:3 was determined. As previously discussed, the absorption was greater than 90%. A selectivity value relative to oleic acid was calculated for the distribution of *trans*-Δ8, 9, 10, 11, 12 and 13 and *cis*-Δ8, 10, 11, 12 and 13 in plasma lipid fractions.

The plasma cholesterol esters, normally characterized with high amounts of linoleic acid, showed in accordance with this a high positive selectivity value, whereas all *trans*-octadecenoate isomers were incorporated less than oleic acid, thus leading to negative figures. Especially, a discrimination against *trans* Δ11, Δ8 and Δ13 was observed. The *cis* isomers were more readily esterified. It is noteworthy to state that compared to stearic acid having a selectivity value of -31.5, the most discriminated isomer, the Δ11 *trans*, had -15. Palmitic acid is not discriminated at all compared to oleic acid (see Figure 5.6).

The plasma triacylglycerols showed generally much lower discrimination figures. This lipid class has a function as a reservoir supplying all kinds of fatty acids for energy production as well as other metabolic functions. There is a distinct avoidance of *trans*-Δ12 and to some extent of the *trans*-Δ11 and the Δ13 isomers in this fraction, which may also be interpreted as a faster turn-over of these fatty acids. Also, some of the *cis* isomers are less readily incorporated than oleic acid into plasma triacylglycerols [25].

For the phosphatidylcholines, the most abundant class of phospholipid in plasma lipids, the selectivity for incorporation in the two *sn*-positions is interesting. As expected, the preference for the saturated fatty acids is high in the *sn*-1 position and on the contrary the PUFA's, 18:2 and 18:3, are discriminated against in this position. For all the *trans* isomers, there is a preferential incorporation compared to *cis*-18:1 and with selectivity values between 4.8 (corresponding to 16:0) and 9.9 (corresponding to 18:0) in the sn-1 position. The *trans*-Δ10 isomer shows the lowest figure, but still higher than oleic acid. For the *cis*-octadecenoates, the Δ10 isomer is preferred especially in the 1-position, but all other isomers have a higher incorporation than 9Δ18:1.

For the *sn*-2 position, the opposite picture is seen, with all isomers except elaidic acid discriminated. The *cis*-octadecenoates all show about the same incorporation as oleic acid except for a definite preference for the *cis* Δ12-isomer. The position of a double bond at Δ12 as found in linoleic acid being one of the preferred substrates for position 2, may be a prerequisite for the optimal function of the acyltransferase. The discrimination of the *trans*-Δ10 isomer is also consistent with this deposition pattern.

As seen from the selectivity values for saturated fatty acids (16:0 and 18:0) there are differences between the saturated acids in various lipid classes, but the selectivity of the geometrical and positional isomers is much more complicated than often discussed in epidemiological studies. In such studies, the *trans* fatty acids are mostly considered as synonymous with elaidic acid and the effects of positional isomers are hardly mentioned. Furthermore, the level of dietary polyunsaturated fatty acids, especially linoleic acid, being a very potent competitor in all acylations, plays a definite role for the incorporation of C18:1-isomers. The amount of isomers directed to oxidative degradation may therefore vary considerably, due both to the amount of dietary fat and to its fatty acid composition.

The influence of long-chain C20 and C22 isomers present in partially hydrogenated fish oil on the *trans* fatty acid depositions has not been examined directly in humans. However, the presence of only traces or even absence of long-chain monoenoic fatty acids in depot fat reflects a chain shortening. Investigations on pigs [85,86] and rats [43] showed only trace amounts of C20:1 and C22:1 in liver and heart phosphoacylglycerols, indicating effective chain shortening before acylation. The much lower amounts of isomers found in triacylglycerols compared to the dietary content also seem to prove the preferential oxidation of isomeric fatty acids. Even in adipose tissue, the deposition of *trans*-20:1 is less than one fifth of the amount present in the dietary fat and *trans*-C22:1 is only found as traces [43].

3. Isomeric Fatty Acids in Fetal and Neonatal Metabolism

The placental transfer of *trans* fatty acids was verified years ago in rats by comparing elaidic acid and *trans,trans*-18:2(*n*-6) with the corresponding *cis* isomers [49]. However, the amount transferred to the young is small (0.5% in extracted fat) compared to the presence of about 25% in the carcass fat of the mother. An increase is observed in contrast when the young receives mother's milk. Corresponding results have been found for swine fed partially hydrogenated fish oil by Pettersen and Opstvedt [86].

The general conclusion from animal experiments was that *trans* fatty acids did not exert any harmful effects when dietary linoleic acid was sufficient [30,92,2].

In contrast to these findings, Koletzko *et al.* [54-58] have reported that there could be a PUFA-deficiency in infants due to the presence of *trans* fatty acids. This theory is built on the finding of an inverse relation between *trans* fatty acids and contents of the essential PUFA in blood lipid classes. In both the preterm and term infants, *trans* fatty acids correlate with low birth weight and to some extent with lower contents of the important *n*-6 and *n*-3 membrane fatty acids. As the *n*-3 family is especially important for the development of brain and nerve tissues, much attention has been paid to such findings.

Also for children of age 1-15 years, the total *trans* and the *trans*-18:1 in plasma phospholipids were negatively correlated with arachidonic acid [16].

Van Houvelingen and Hornstra [100] determined the content of *trans* fatty acids in fetal tissue after abortion and compared it with the *trans* content of the plasma of the mother. A direct correlation between the *trans*-18:1(*n*-9) in mother and fetus was observed. The ratio long-chain PUFA:18:2(*n*-6) in umbilical arterial walls showed an inverse relation to *trans*-18:1(*n*-9), that was explained as a result of decreased conversion of 18:2(*n*-6).

Especially for the fetus and the preterm babies, a decreased conversion may be critical. Furthermore the deficiency could be worsened after birth when intake of mother's milk may supply additional *trans* fatty acids. The influence of *trans* fatty acid isomers in mother's milk on long-chain PUFA deposition in infant erythrocyte lipids was also studied [40]. Inverse correlations were found between total *trans* and *trans*-18:1(n-9) present in milk and the contents of 20:4(*n*-6), 22:6(*n*-3), Σ*n*-6 PUFA and Σ*n*-3 PUFA in erythrocyte PC but not PE. No significant relations were found for *trans*-18:1(*n*-7).

Whether these findings may influence the development through a lifespan is really questionable. An epidemiological model suggesting *trans* and *cis* isomers being the key to development of atherosclerosis has been proposed by Booyens [9], but must await further verifications. A recommendation of decreased intake of *trans* fatty acids for pregnant and lactating women may on the other hand be appropriate.

F. CONCLUDING REMARKS

Much attention has been paid to the effect of *trans* fatty acids in general metabolism, and especially the role of the major dietary fraction the *trans*-octadecenoic acids, over the years and recently a renewed interest has arisen because of reports on the possible correlation to coronary heart disease.

It must be emphasized, however, that much of the literature gives results obtained under conditions which are far from the physiological reality and mainly obtained in animal models. The data obtained with EFA-deficient systems and with subcellular fractions may give important clues to the possible reactions of the *trans* fatty acids, but the implication of the major dietary fatty acids and the cooperative balance between cellular components and compartments seem to overrule most of the effects found *in vitro*.

Therefore, the intake of smaller amounts of *trans* fatty acids when adequate linoleic acid is fed do not seem to pose any harmful effect to mammals and the epidemiological evidence for unfavourable characteristics of the isomeric fatty acids has to be supported by direct evidence.

ABBREVIATIONS

EFA, essential fatty acids; HMO, partially hydrogenated marine oil; PC, phosphatidylcholine; PE, phosphatidylethanolamine; PI, phosphatidylinositol; PS, phosphatidylserine; PUFA, polyunsaturated fatty acids.

REFERENCES

1.   Aaes-Jørgensen,E. and Hølmer,G., *Lipids,* **4**, 501-06 (1969).
2.   Alfin-Slater,R.B. and Aftergood,L. in *Geometrical and Positional Fatty Acid Isomers*, pp. 53-74 (1979) (edited by E.A. Emken and H.J. Dutton, Am. Oil Chem. Soc., Champaign, Ill.)
3.   Anderson,R.L., *Biochim. Biophys. Acta,* **144**, 18-24 (1967).
4.   Anderson,R.L. and Coots,R.H., *Biochim. Biophys. Acta,* **144**, 525-531 (1967).
5.   Beare-Rogers,J.L., *Adv. Nutr. Res.,* pp. 171-200 (1988).
6.   Bernert,J.T. and Sprecher,H., *Biochim. Biophys. Acta,* **398**, 344-353 (1975).
7.   Blomstrand,R. and Svensson,L., *Lipids,* **18**, 151-170 (1983).
8.   Blomstrand,R., Diczfalusy,U., Sisfontes,L. and Svensson, L., *Lipids,* **20**, 283-295 (1985).
9.   Booyens,J., *Med. Hypothesis,* **21**, 323-333 (1986).
10. Brenner,R.R. and Peluffo,R.O., *Biochim. Biophys. Acta,* **176**, 471-479 (1969).
11. Chiang,M.T., Otomo,M.I., Itoh,H. Furukawa,Y. Kimura,S. and Hujimoto,H., *Lipids,* **26**, 46-52 (1991).

12. Christiansen,R.Z., Christiansen,E.N. and Bremer,J., *Biochim. Biophys. Acta,* **573**, 417-429 (1979).
13. Clement,J., *J. Am. Oil. Chem. Soc.,* **42**, 1035-1037 (1965).
14. Cook,H.W. and Emken,E.A., *Biochem. Cell. Biol.,* **68**, 653-60 (1990).
15. Coots,R.H., *J. Lipid Res.,* **5**, 468-478 (1964).
16. Decsi,T. and Koletzko,B., *Ann. Nutr. Metab.,* **39**, 36-41 (1995).
17. De Schrijver,R. and Privett,O.S., *Lipids,* **17**, 27-34 (1982).
18. Deuel,H.J., in *The Lipids Vol. III*, pp. 218-221 (1955) (Interscience Publishers).
19. Egwim,P.O. and Sgoutas,D.S., *J. Nutr.,* **101**, 315-321 (1971).
20. Egwim,P.O. and Kummerow,F.A., *J. Nutr.,* **102**, 783-792 (1972).
21. Egwim,P.O. and Kummerow,F.A., *J. Lipid Res.,* **13**, 500-510 (1972).
22. Emken,E.A., Rohwedder,W.K., Dutton,H.J., DeJarlais,W.J., Adlof,R.O., Mackin,J., Dougherty,R. and Iacono,J.M., *Metabolism,* **28**, 575-583 (1979).
23. Emken,E.A., Dutton,H.J., Rohwedder,W.K., Rakoff,H. and Adlof,R.O., *Lipids,* **15**, 864-871 (1980).
24. Emken,E.A., Adlof,R.O., Rohwedder,W.K. and Gulley,R.M., *J. Lipid Res.,* **24**, 34-46 (1983).
25. Emken,E.A., Rohwedder,W.K., Adlof,R.O., DeJarlais,W.J. and Gulley,R.M., *J. Am. Oil Chem. Soc.,* **61**, 678 (1984).
26. Emken,E.A., *Ann. Rev. Nutr.,* **4**, 339-376 (1984).
27. Emken,E.A., Rohwedder,W.K., Adlof,R.O., DeJarlais,W.J. and Gulley,R.M., *Lipids,* **21**, 589-595 (1986).
28. Emken,E.A., Adlof,R.O., Rohwedder,W.K. and Gulley,R.M., *Lipids,* **24**, 61-69 (1989).
29. Emken,E.A., in *Health Effects of Dietary Fatty Acids,* pp. 245-263 (1990) (ed. G.J. Nelson, Am. Oil Chem. Soc., Champaign, Ill).
29A. Emken,E.A., *Am. J. Clin. Nutr.,* **62**, 659S-669S (1995).
30. Federation of American Societies for Experimental Biology (FASEB), *Health Aspects of Dietary Trans Fatty Acids,* USA, FASEB. (1985).
31. Flatmark,T., Nielsson,A., Kvannes,J., Eikhom,T.S., Fukami,M.H., Kryvi,H. and Christiansen,E.N., *Biochim. Biophys. Acta ,***962**, 122-30 (1988).
32. Heimerman,W.H., Holman,R.T., Gordon,D.T., Kowalyshyn,D.E. and Jensen,R.G., *Lipids,* **8**, 45-47 (1973).
33. Hill,E.G., Johnson,S.B., Lawson, L.D., Mahfouz, M.M. and Holman, R.T., *Proc. Natl. Acad. Sci. USA,* **79**, 953-957, (1982).
34. Holman,R.T. and Aaes-Jørgensen,E., *Proc. Soc. Exp. Biol. Med.,* **93**, 175-179 (1956).
35. Holman,R.T. and Mahfouz,M.M., *Progr. Lipid Res.,* **20**, 151-156 (1981).
36. Holman,R.T., Mahfouz,M.M., Lawson,L.D. and Hill,E.G., in *Dietary Fats and Health,* pp. 320-340 (1983) (edited by E.G. Perkins and W.J. Visek, Am. Oil Chem. Soc. Champaign, Ill).
37. Holman,R.T. and Johnson,S.B. in *Dietary Fats and Health,* pp. 247-266 (1983) (edited by E.G. Perkins and W.J. Visek, Am. Oil Chem. Soc. Champaign, Ill).
38. Hudgins,L.C., Hirsch,J. and Emken.E.A., *Am. J. Clin. Nutr.,* **53**, 474-482 (1991).
39. Hølmer,G., Høy,C.-E. and Kirstein,G., *Lipids,* **17**, 585-593 (1982).
40. Hølmer,G., Jørgensen,M., Lund,P. and Fleischer Michaelsen,K., *35th ICBL,* Aberdeen, Abstr. II P8. (1994).
41. Høy,C.-E. and Hølmer,G., *Lipids,* **14**, 727-733 (1979).
42. Høy,C.-E. and Hølmer,G., *Lipids,* **16**, 102-108 (1981),
43. Høy,C.-E. and Hølmer,G., *Lipids,* **23**, 973-980 (1988).
44. Høy,C.-E. and Hølmer,G., *Lipids,* **25**, 415-418 (1990).
45. Ide,T. and Sugano,M., *Biochim. Biophys. Acta,* **794**, 282-291 (1984).
46. Ide,T., Watanabe,M., Sugano,M. and Yamamoto,I., *Lipids,* **22**, 6-10 (1987).
47. Jensen,B., *Lipids,* **11**, 179-188 (1976).
48. Jensen,R.G., Sampugna,J. and Pereira,R.L., *Biochim. Biophys. Acta ,***84**, 481-83 (1964).
49. Johnston,P.V., Johnson,O.C. and Kummerow,F.A., *Proc. Soc. Exp. Biol. Med.,* **96**, 760-762 (1957).
50. Kameda,K., Valicenti,A.J. and Holman,R.T., *Biochim. Biophys. Acta ,***618**, 13-17 (1980).
51. Kinsella,J.E., Bruckner,G., Mai,J. and Shimp,J., *Am. J. Clin. Nutr.* **34**, 2307-2318 (1981).
52. Kirstein,D., Høy,C.E. and Hølmer,G., *Brit. J Nutr.* **50**, 749-56 (1983).
53. Koga,T., Yamato,T., Gu,J.Y., Nonaka,M., Yamada,K. and Sugano,M., *Biosci. Biotech. Biochem.,* **58**, 384-387 (1994).
54. Koletzko,B., Schmidt,H.J., Haug,M. and Harzer,G., *Eur. J. Pediatr.,* **148**, 669-675, (1989).
55. Koletzko,B. and Müller,J., *Biol. Neonate,* **57**, 172-178 (1990).

56. Koletzko,B., *Nahrung,* **35**, 229-283 (1991).
57. Koletzko,B., *Acta Pediatr.,* **81**, 302-306 (1992).
58. Koletzko,B., *World Rev. Nutr. Diet.*, **75**, 82-85 (1994).
59. Lands,W.E.M., Blank,M.L., Nutter,L.J. and Privett,O.S., *Lipids,* **1**, 224-229 (1966).
60. Lands,W.E.M., in *Geometrical and Positional Isomers*, pp. 194-96 (1979) (edited by E.A. Emken and H.J. Dutton, Am. Oil Chem. Soc., Champaign, Ill).
61. *Ibid,* pp. 181-212.
62. Lanser,A.C., Emken,E.A. and Ohlrogge,J.B., *Biochim. Biophys. Acta,* **875**, 510-515 (1986).
63. Lawson,L.D. and Kummerow,F.A., *Biochim. Biophys. Acta,* **573**, 245-254 (1979).
64. Lawson,L.D. and Holman,R.T., *Biochim. Biophys. Acta,* **665**, 60-65 (1981).
65. Lawson,L.D., Hill,E.G. and Holman,R.T., *J. Nutr.,* **113**, 1827-1835 (1983).
66. Lawson,L.D., Hill,E.G. and Holman,R.T., *Lipids,* **20**, 262-67 (1985).
67. Lippel,K., *Lipids,* **8**, 111-118 (1973).
68. Lippel,K., Carpenter,D., Gunstone,F.D. and Ismail,I.A., *Lipids,* **8**, 124-128 (1973).
69. Lippel,K., Gunstone,F.D. and Barve,J.A., *Lipids,* **8**, 119-123 (1973).
70. Mahfouz,M.M., Valicenti,A.J. and Holman,R.T., *Biochim. Biophys. Acta,* **618**, 1-12 (1980).
71. Mahfouz,M.M., Johnson,S. and Holman,R.T., *Lipids,* **15**, 101-106 (1980).
72. Mahfouz,M.M., Johnson,S. and Holman,R.T., *Biochim. Biophys. Acta,* **663**, 58-68 (1981).
73. Mahfouz,M.M., Smith,T.L. and Kummerow,F.A., *Lipids,* **19**, 214-222 (1984).
74. Mazuzawa,Y., Prasad,M.R. and Lands,W.E.M., *Biochim. Biophys. Acta,* **919**, 297-306, (1987).
75. Menon,N.K. and Dhopeshwarkar,G.A., *Biochim. Biophys. Acta,* **751**, 14-20 (1983).
76. Moore,C.E., Alfin,-Slater,R.B. and Aftergood,L., *Am. J. Clin. Nutr.*, **33**, 2318-2323 (1980).
77. Normann,P.T., Thomassen,M.S., Christiansen,E,N. and Flatmark,T., *Biochim. Biophys. Acta,* **664**, 416-427 (1981).
78. Norseth,J., *Biochim. Biophys. Acta,* **575**, 1-9 (1979).
79. Ohlrogge,J.B., Emken,E.A. and Gulley,R.M., *J. Lipid Res.,* **22**, 955-960 (1981).
80. Ohlrogge,J.B., Gulley,R.M. and Emken.E.A., *Lipids* **17**, 551-57 (1982).
81. Ohlrogge,J.B. in *Dietary Fats and Health,* pp. 359-374 (1983) (edited by E.G. Perkins and W.J. Visek, Am. Oil Chem. Soc. Champaign, Ill).
82. Okuyama,H., Lands,W.E.M., Gunstone,F.D. and Barve,J.A., *Biochemistry,* **11**, 4392-4398 (1972).
83. Ono,K. and Fredrickson,D.S., *J. Biol. Chem.,* **239**, 2482-2488 (1964).
84. Osmundsen,H., *Ann. NY. Acad. Sci.,* **386**, 13-29 (1982).
85. Pettersen,J. and Opstvedt,J., *Lipids,* **23**, 720-26 (1988).
86. Pettersen,J. and Opstvedt,J., *Lipids,* **27**, 761-69 (1992).
87. Pollard,M.R., Gunstone,F.D., James,A.T. and Morris,L.J., *Lipids,* **15**, 306-314 (1980).
88. Pollard,M.R., Gunstone,F.D., Morris,L.J. and James,A.T., *Lipids,* **15**, 690-693 (1980).
89. Reichwald-Hacker,I., Grosse-Oetringhaus,S., Kiewitt,I. and Muhkerjee,K.D., *Biochim. Biophys. Acta,* **575**, 327-334 (1979).
90. Reichwald-Hacker,I., Ilsemann,K. and Muhkerjee,K.D., *J. Nutr.*, **109**, 1051-1056 (1979).
91. Reitz,R.C., El-Sheikh,M., Lands,W.E.M., Ismail,J.A. and Gunstone,F.D., *Biochim. Biophys. Acta,* **176**, 480-490 (1969).
92. Report of the British Nutrition Foundation (BNF) Task Force July 1995: *Trans Fatty Acids,* BNF, London.
93. Rocquelin,G., Guenot,L., Astorg,P.O. and David,M., *Lipids,* **24**, 775-780 (1989).
94. Rosenthal,M.D. and Doloresco,M.A., *Lipids,* **19**, 869-74 (1984).
95. Royce,S.M., Holmes,R.P., Takagi,T. and Kummerow,F.A., *Am. J. Clin. Nutr.*, **39**, 215-222 (1984).
96. Sprecher,H. in *Essential Fatty Acids and Eicosanoids,* pp. 18-22 (1992) (edited by A. Sinclair and R. Gibson, Am. Oil Chem. Soc., Champaign Ill.)
97. Svensson,L., *Lipids,* **18**, 171-78 (1983).
98. Sugano,M. and Ikeda,I., *Curr. Opinion Lipidology,* **7**, 38-42 (1996).
99. Thomassen,M.S., Christiansen,E.N. and Norum,K.R., *Biochem. J.,* **206**, 195-202 (1982).
99A. Thomson,A.B., Garg,M., Keelan,M. and Doring,K., *Digestion,* **55**, 405-409 (1994).
100. Van Houvelingen,A.C. and Hornstra,G., *World. Rev. Nutr. Diet.*, **75**, 175-178 (1994).
101. Verkamp,J.H. and Zevenbergen,J.L., *Biochim. Biophys. Acta,* **878**, 102-109 (1986).
102. Wahle,K.W.J. and Peacock,L.L., *Biochim. Biophys. Acta,* **1301**, 141-149 (1996).
103. Watanabe,M., Sugano,M., Murakami,H. and Omura,H., *J. Clin. Biochem. Nutr.,* **2**, 41-53 (1987).
104. Wolff,R.L., Combe,N. and Etressangles,B., *Rev. Franc. Corps Gras,* **31**, 161-70 (1984).

105. Wood,R. and Chumbler,F., *Lipids,* **13**, 75-84 (1978).
106. Wood,R., in *Geometrical and Positional Fatty Acids Isomers,* pp. 213-281 (1979) (edited by E.A. Emken and H.J. Dutton, Am. Oil Chem. Soc., Champaign, Ill).
107. Wood,R., *Lipids,* **14**, 975-982 (1979).
108. Zevenbergen,J.L., Houtsmüller,U.M.T. and Gottenbos,J.J., *Lipids,* **23**, 178-186 (1988).
109. Zevenbergen,J.L. and Haddeman,E., *Lipids,* **24**, 555-563 (1989).

CHAPTER 6

BIOCHEMISTRY OF TRANS POLYUNSATURATED FATTY ACIDS

Jean-Louis Sébédio and Jean-Michel Chardigny
INRA, Unité de Nutrition Lipidique, 17 rue Sully, 21034 Dijon Cedex, France

A. Introduction
B. Effect of *Trans* Polyunsaturated Fatty Acids on the Metabolism of Linoleic Acid *in Vivo.*
C. Metabolism of *Trans* Polyunsaturated Fatty acids *in Vivo.*
D. Desaturation and Elongation of *Trans* C18 Polyunsaturated Fatty Acids *in Vitro.*
E. Catabolism of *Trans* Polyunsaturated Fatty Acids.
F. Effects of *Trans* Polyunsaturated Fatty Acids on Eicosanoid Synthesis.
G. Human Studies.
H. Other Studies
I. Conclusion

A. INTRODUCTION

While *trans* isomers of monounsaturated fatty acids are generally produced by catalytic hydrogenation or biohydrogenation, *trans* isomers of polyunsaturated fatty acids are formed as a result of heat or frying treatments and deodorization of oils (see Chapter 2) [1-5]. Furthermore, minor quantities of positional and geometrical isomers of linoleic acid may be found in partially hydrogenated oils [6]. Among the *trans* polyunsaturated fatty acids found in food products, the mono-*trans* isomers of linoleic ($18:2(n$-$6)$) and of α-linolenic ($18:3(n$-$3)$) acids are the most common while only minor quantities of di-*trans* isomers are detected generally [7].

[1]Footnote

Phone: +33.(0)3.80.63.31.10
Fax: +33.(0)3.80.63.32.23
e-mail: sebedio@dijon.inra.fr

Former studies on *trans* polyunsaturated fatty acids (*trans* PUFA) focused mainly on the di-*trans* isomer of linoleic acid, 9-*trans*,12-*trans*-18:2, which was commercially available even if detected as a minor isomer in food items. *In vitro* studies on the other mono-*trans* isomers of linoleic and linolenic acids only started a few years ago when the different isomers were synthesised [8-15].

B. EFFECT OF *TRANS* POLYUNSATURATED FATTY ACIDS ON THE METABOLISM OF LINOLEIC ACID *IN VIVO*

While data on the absorption of *trans* monoenes have been published (see Chapter 5), only limited data dealing with the absorption of *trans* polyunsaturated fatty acids are available [16-18]. However, studies carried out with the all *trans*-18:2 and the 18:3 isomers did not shown any differences in their absorption compared to the *cis* isomers.

Selinger and Holman showed that the di-*trans* 18:2 isomer when given to rats that have been fed a fat-free diet together with linoleic and linolenic acids decreased the levels of the polyunsaturated fatty acids formed from these precursors [19]. Privett *et al.* [20] reported that dietary mono-*trans* 18:2 did not impair the conversion of linoleic into arachidonic acid using animals previously fed a fat-free diet. Feeding di-*trans* linoleate impaired the conversion of oleic acid into 20:3(n-9) and of linoleic into arachidonic acid [20]. Anderson *et al.* [21] also showed that the presence of the mono-*trans* 18:2 isomers in the diet did not affect the level of arachidonic acid in the liver and that the di-*trans* 18:2 isomer exerted an inhibition of arachidonic acid synthesis only at 18:2 di-*trans*/linoleic acid ratios which would be much higher that those observed in the human diet [21]. Studies of Berdeaux *et al.* [22-23] also confirmed that when fed at low levels in the diet, the mono-*trans* isomers of linoleic acid did not induce any modification in the arachidonic acid content in different organs. However, in the presence of linoleic acid at 1.1% of the calories, dietary di-*trans* 18:2 may affect essential fatty acid metabolism by reducing the Δ6 desaturase activity [24].

The di-*trans* 18:2 was found to be incorporated like the saturated fatty acids preferentially in the 1-acyl position of phosphatidylcholine rather than into the 2-acyl position [19]. Similar results were obtained by Privett *et al.* [25]. Furthermore, the di-*trans* 18:2 isomer, palmitic and stearic acids were incorporated mainly at the α positions of the triacylglycerol while the 9-*cis*,12-*trans*-18:2, like linoleic acid was incorporated in the β position.

C. METABOLISM OF *TRANS* POLYUNSATURATED FATTY ACIDS *IN VIVO*

The *trans*,*trans*-18:2 isomer was reported initially by Knipprath and Mead [26] to be converted into a di-*trans* isomer of eicosatetraenoic acid, but the work of Privett *et al.* [20] demonstrated that no such conversion occurs.

Studies carried out by Karney and Dhopeshwarkar [27] on the developing rat brain also demonstrated a lack of direct conversion of the *trans,trans*-18:2 isomer. However, analysis of the brain polyunsaturated fatty acids after intracranial injection showed that 9-*trans*,12-*trans*-18:2 was desaturated into 6-*cis*,9-*trans*,12-*trans*-18:3 followed by an elongation to form 8-*cis*,11-*trans*,14-*trans*-20:3. These authors concluded that the conversion of 9-*trans*,12-*trans*-18:2 was blocked at the Δ5 desaturation step rather than at the Δ6 desaturation one.

The metabolism of the mono-*trans* isomers of linoleic acid was studied as early as 1963 by Blank and Privett [28] where the utilization of silver nitrate-thin layer chromatography and of infrared spectroscopy permitted them to show that methyl 9-*cis*,12-*trans*-linoleate had been converted to a *trans* eicosatetraenoic acid which was not fully identified. Further work in the field allowed characterization of the different *trans* polyunsaturated fatty acids formed by elongation and desaturation of the mono-*trans* isomers of linoleic acid [20,22,23,29]. All the studies demonstrated that at least one *trans* isomer of linoleic acid, 9-*cis*,12-*trans*-18:2, could be desaturated and elongated into a *trans* isomer of arachidonic acid, 5-*cis*,8-*cis*,11-*cis*,14-*trans*-20:4. However, all these studies do not agree as to whether the other mono-*trans* isomer, 9-*trans*,12-*cis*-18:2, could be converted into a *trans* isomer of arachidonic acid. Moreover, Beyers and Emken [30] and Berdeaux *et al.* [23] showed that only the 9-*trans*,12-*cis*-18:2 was elongated into the 11-*trans*,14-*cis*-20:2, while Ratnayake *et al.* [29] suggested that both the mono-*trans* isomers of 9,12-18:2 could be elongated into *trans* isomers of 11,14-20:2.

Studies on mice using isotopically labelled 18:2 isomers (9-*cis*,12-*cis*; 9-*cis*,12-*trans* and 9-*trans*,12-*cis*) showed that the metabolites are incorporated into liver, plasma, heart and brain lipids (Figure 6.1) [30]. Analysis of the metabolites showed a similar pattern in plasma and in liver lipids. For the brain, the levels of the all *cis* metabolites were about four times higher than the 9-*cis*,12-*trans*-18:2 metabolites especially for the 20:4. However, the amount of 5-*cis*,8-*cis*,11-*cis*,14-*trans*-20:4, 5-*cis*,8-*cis*,11*trans*,14-*cis*-20:4 and 5-*cis*,8-*cis*,11-*cis*,14-*cis*-20:4 in the brain were 31, 15 and 10 times lower than their concentrations in the liver lipids. In contrast to the brain and liver lipid incorporation, the data for the heart were somewhat different, the concentration of the 20:4 metabolite of the 9-*cis*,12-*trans*-18:2 being 2.3 times that of the 9-*cis*,12-*cis*-18:2 metabolites.

Furthermore, as shown by Berdeaux *et al.* [23] for the rat, the long-chain *trans* 20:4 can be incorporated into different liver lipid classes (Table 6.1). However no modifications of the fatty acid composition of the phospholipid classes were induced by small quantities of isomers of linoleic acid in the diet (0.6% of the energy). The 14-*trans*-20:4 isomer was only detected in three phospholipid classes, phosphatidylethanolamine, phosphatidylinositol and phosphatidylcholine, the highest quantity (0.8%) being found in phosphatidylinositol. It should be noted that this phospholipid class also contained the highest amount of 20:3(*n*-9) (2.8%). The lowest level of 14-

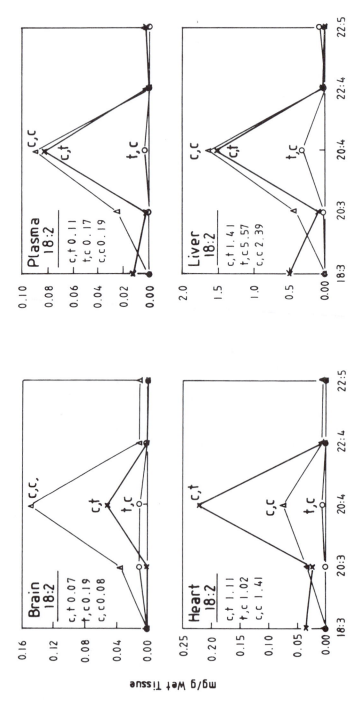

Fig. 6.1. Desaturated and elongated metabolites of 9-*trans*,12-*cis*-18:2-d4 (TC), 9-*cis*,12-*trans*-18:2-d0 (CT) and 9-*cis*,12-*cis*-18:2-d2 (CC) in total lipids from mouse liver, plasma, heart and brain after feeding the experimental diet for 4 days. The concentrations of the dietary 18:2 fatty acids incorporated are listed in the upper left corner of each panel. Reproduced by kind permission of the authors and *Biochimica et Biophysica Acta* [30].

TABLE 6.1

Fatty acid composition (wt%) in liver phosphatidylethanolamine (PE), phosphatidylinositol (PI), phosphatidylcholine (PC), triacylglycerols (TG) and cholesterol esters (CE) of rats fed with a mixture of linoleic acid isomers (9-*cis*,12-*cis*-18:2: 80, 9-*cis*,12-*trans*-18:2 :9, 9-*trans*,12-*cis*-18:2: 11, 9-*trans*,12-*trans*-18:2: traces) [24]

	Phospholipids			Non-phosphorus lipids	
	PE	PI	PC	TG	CE
16:0	12.5±2.68	5.2±0.65	15.1±1.39	25.7±2.22	12.5±2.95
18:0	22.7±0.96	42.6±2.92	20.6±1.98	2.5±0.29	2.2±0.36
Σ16:1	0.4±0.22	0.8±0.11	1.3±0.24	7.5±2.31	9.0±3.2
Σ18:1	5.0±0.76	3.5±0.26	9.6±1.35	49.4±1.79	44.7±4.02
18:2(*n*-6)	3.5±0.74	1.3±0.15	10.8±1.67	7.2±2.69	7.5±2.15
9-c,12-t-18:2	0.07±0.02	traces[a]	0.09±0.03	0.37±0.04	0.21±0.07
9-t,12-c-18:2	0.35±0.11	traces	0.76±0.11	0.98±0.22	0.39±0.07
18.3 (*n*-3)	ND[b]	ND	0.07±0.02	0.6±0.26	0.6±0.26
20:2(*n*-6)	ND	ND	0.11±0.08	0.05±0.03	traces
11t,12c-20:2	ND	ND	ND	ND	ND
20:3(*n*-6)	0.5±0.07	202±0.46	1.5±0.09	0.1±0.05	0.3±0.07
20:3(*n*-9)	0.4±0.12	2.8±0.64	0.7±0.17	0.15±0.02	0.3±0.07
20:4(*n*-6)	263±1.44	33.6±3.41	23.2±1.46	1.2±0.47	8.7±2.32
14t-20:4	0.15±0.03	0.83±0.10	0.30±0.07	0.03±0.01	0.06±0.02
20:5(*n*-3)	0.1±0.02	ND	0.2±0.10	0.0±0.11	0.5±0.13
22:5(*n*-6)	0.7±0.40	0.2±0.04	0.4±0.14	traces	traces
22:6(*n*-3)	18.8±1.98	2.4±0.38	7.6±0.95	0.4±0.20	traces
others	8.5	4.6	7.2	3.2	13.0.

[a] <0.05;
[b] <0.01

trans-20:4 (0.2%) was observed in phosphatidylethanolamine, the phospholipid which contained the lowest quantity of 20:3(*n*-9). The phosphatidylcholine showed an intermediate situation. It therefore appears that 14-*trans* 20:4 behaves like its structural analogue, the *trans* Δ14 bond being recognised as a single bond by enzymatic systems as already noted for the 18:3 isomers [31].

Ratnayake *et al.* also showed [29] that the 9-*cis*,12-*trans*-18:2 and 9-*trans*,12-*cis*-18:2 were not the only *trans* 18:2 isomers which were desaturated and elongated into *trans* isomers of arachidonic acid. Similarly, 9-*cis*,13-*trans*-18:2, which is the major *trans* polyunsaturated fatty acid in partially hydrogenated vegetable oils was shown to be desaturated and elongated to 5-*cis*,8-*cis*,11-*cis*,15-*trans*-20:4. This fatty acid was detected further in different tissues of rats fed a partially hydrogenated canola oil.

Some *trans* 18:3 isomers can be desaturated and elongated into *trans* isomers of eicosapentaenoic (EPA) and docosahexaenoic acids. *In vivo*, among the different isomers formed during heat treatment of oils (9-*trans*,12-*cis*,15-*cis*-18:3, 9-*cis*,12-*cis*,15-*trans*-18:3 and to a lesser extent 9-*trans*,12-*cis*,15-*trans*-18:3), the 9-*cis*,12-*cis*,15-*trans*-18:3 seems to be the preferential substrate for the conversion into 5-*cis*,8-*cis*,11-*cis*,14-*cis*,17-*trans*-20:5, 7-*cis*,10-*cis*,13-*cis*,16-*cis*,19-*trans*-22:5 and into 4-*cis*,7-*cis*,10-*cis*,13-*cis*,16-*cis*,19-*trans*-22:6 [32,33]. First evidence of this conversion was reported by Grandgirard *et al.*

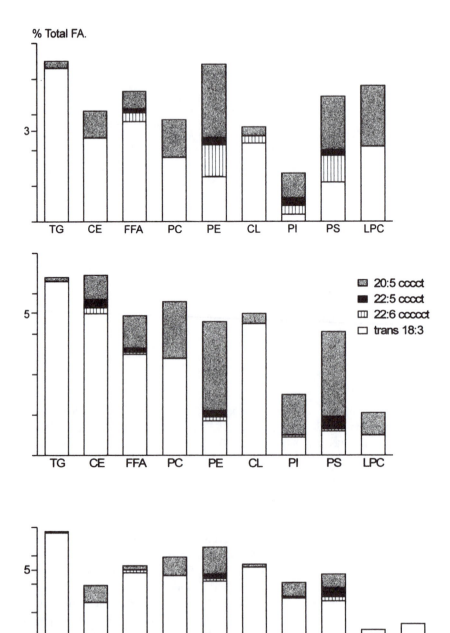

Fig. 6.2. *Trans* (*n*-3) fatty acids in liver (upper), kidney (medium) and heart (lower) lipid classes from rats fed a diet containing 10% of a heated linseed oil for 3 weeks [7,33]. Abbreviations: TG, triacylglycerols; CE, cholesteryl esters; FFA, free fatty acids; PC, phosphatidylcholine; PE, phosphatidylethanolamine; DPG, diphosphatidylglycerol; PI, phosphatidylinositol; PS, phosphatidylserine; LPC, lysophosphatidylcholine.

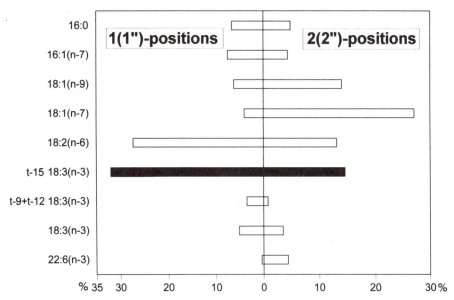

Fig. 6.3. Distribution of the main fatty acids esterified to positions 1(1") and 2(2") in cardiolipins from rat mitochondria of rats fed a concentrate of linolenic acid isomers following a fat-free diet. Results are expressed as percentages relative to total fatty acids esterified to a given pair of positions. Reproduced by kind permission from the authors and *Biochimica et Biophysica Acta* [31].

by feeding rats with a heated linseed oil which contained high levels of geometrical isomers of linolenic acid [32]. The mono-*trans* geometrical isomers of 20:5(n-3) and 22:6(n-3) were identified in liver lipids after isolation of these compounds by a combination of preparative high-performance liquid chromatography followed by silver nitrate thin-layer chromatography. The structures were elucidated using partial hydrazine reduction, oxidative ozonolysis and Fourier-transform-infrared spectroscopy, techniques often used in structural analysis.

These *trans* (n-3) long-chain polyunsaturated fatty acids were found to be incorporated into almost all lipid classes, including phospholipids of different tissues, such as liver, heart, kidneys, adrenals and testes (Figure 6.2). The incorporation of the different *trans* long-chain PUFA is quite different among the different tissues that have been considered. However, while the 17-*trans* isomer of EPA is incorporated in all the neutral lipid classes and in most of the phospholipid classes except sphingomyelin, the *trans* isomers of 22:5(n-3) and 22:6(n-3) were never detected in either the triacylglycerols or the phosphatidylcholine (Figure 6.2).

In contrast, a preferential incorporation of 9-*cis*,12-*cis*,15-*trans*-18:3 into rat mitochondria cardiolipin was demonstrated by Wolff *et al.* [31]. 9-*cis*,12-*cis*,15-*trans*-18:3 accumulates in cardiolipin at a higher level than in other phospholipids (11 times in liver and 5-7 times in heart), representing 22-24% of the total fatty acids in cardiolipin. Furthermore, the 15-*trans*-18:3 is esterified to both the 1 (1") and 2 (2") positions of liver mitochondria

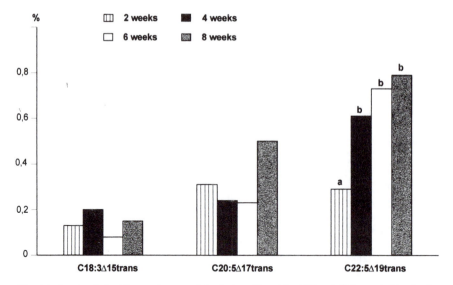

Fig. 6.4. Levels of the 15-*trans* isomer of linolenic acid and its C20 and C22 metabolites in the rat retina as a function of the duration of the experimental diet. Values having a different superscript are statistically different.

cardiolipin, with a marked selectivity for positions 1 ($1''$) (Figure 6.3). The selectivity ratio ($1(1'')/2(2'')$) is similar to what was found for linoleic acid, its structural analogue. It therefore appears that the *trans*-15 ethylenic bond would be perceived as a single bond by the enzymatic systems that ensure acylation of cardiolipin. Interestingly, Waku and Lands [34] showed that the acylation of 9-*cis*,12-*trans*-18:2 in phosphatidylcholine (position 2) was close to that of 18:1. On the other hand, the acylation rate of 9-*trans*,12-*cis*-18:2 was similar to that of linoleic acid. These data enhance the hypothesis that a *trans* double bond located closer to the methyl end of the carbon chain may be recognised as a saturated bond by acylation enzymes. That is not the case when the *trans* double bond is closer to the carboxyl end.

Recently, Chardigny *et al.* [35] showed that other 18:3 geometrical isomers could be converted into long-chain *trans* polyunsaturated fatty acids. This was for example the case for the 9-*trans*,12-*cis*,15-*cis*-18:3 and of 9-*trans*,12-*cis*,15-*trans*-18:3, which were converted into 5-*cis*,8-*cis*,11-*trans*,14-*cis*,17-*cis*-20:5 and 5-*cis*,8-*cis*,11-*trans*,14-*cis*,17-*trans*-20:5, respectively. Other *trans* 20:5 isomers were detected. This indicates that the other 18:3 isomers (9-*cis*,12-*trans*,15-*trans* and 9-*cis*,12-*trans*,15-*cis*) may be converted to a lesser extent into long-chain metabolites [36]. In any case, the conversion was better if the geometry of the Δ9 ethylenic bond was *cis* [37]. This confirms the previous data found for the (*n*-6) family [20].

The *trans* long-chain (*n*-3) polyunsaturated fatty acids may also be incorporated into brain structures and into retina [38,39]. Feeding the animals for 8 weeks with a heated oil containing a mixture of 18:3 geometrical isomers resulted in the incorporation of *trans* 18:3 but also of *trans* 20:5 and 22:6

TABLE 6.2

Contents of 22:5(n-6), 19-*trans*-22:6 and 22:6(n-3) in synaptosomes, brain capillaries, myelin and sciatic nerve of rats fed isomers of linolenic acid (A and B) or a control diet (C) [38]

Fatty acids	Synaptosomes			Brain capillaries and microvessels			Myelin			Sciatic nerve		
	A	B	C	A	B	C	A	B	C	A	B	C
22:5(n-6)	1.92	10.89	2.27	2.01	10.19	1.53	0.61	2.95	0.81	0.31	1.71	0.36
19t-22:6	0.49	1.13	–	0.51	0.95	–	0.12	0.22	–	0.08	0.28	–
22:6(n-3)	15.65	4.20	15.32	14.18	4.26	13.73	3.44	1.00	4.07	2.07	0.86	2.17

TABLE 6.3

Relative rates of *in vitro* desaturation of 18:2(*n*-6) and itsgeometrical
isomers in rat liver microsomes and brain homogenates [22,45].

Substrates	Liver microsomes	Brain homogenates
9-*cis*,12-*cis*-18:2	1.00	1.00
9-*cis*,12-*trans*-18:2	0.74	0.27
9-*trans*,12-*cis*-18:2	0.26	-
9-*trans*,12-*trans*-18:2	-	0.18

isomers [39]. The incorporation of *trans* 22:6 was found to be time-dependent, whereas the *trans* 20:5 only increased strongly after 8 weeks of diet, after being almost constant between 2 and 6 weeks (Figure 6.4). The incorporation of *trans* 22:6 reached 0.8% of the total fatty acids after 8 weeks of feeding which represents almost 4% of the total C22:6 fatty acids. This incorporation is comparable to what was observed for the liver (4.5% of the total C22:6. Along with the retina, synaptosomes and brain microvessels were shown to contain the highest levels of *trans* 22:6 [38]. This fatty acid was also observed in myelin and sciatic nerve but to a lesser extent. However, the ratios of *trans* 22:6 to *cis* 22:6 were similar in all the tissues studied (Table 6.2). When the diet was deficient in linolenic acid (group B in Table 6.2), the incorporation of *trans* 22:6 was doubled. However, the decrease in the amount of 22:6 in the tissue was not balanced by the increase in 19-*trans*-22:6 (comparison of groups A and B). Instead, a large increase in 22:5(*n*-6) was observed, as has previously been described in (*n*-3) fatty acid deficiency [40].

While the *trans* isomers of linoleic acid can inhibit the conversion of 18:2(*n*-6) to 20:4(*n*-6), especially at the Δ6 desaturation step as described earlier, the 18:3 isomers when fed as heated linseed oil (HLO) resulted in a decrease in 20:4(*n*-6) and in the 20:4(*n*-6)/18:2(*n*-6) ratio in liver phospholipid compared to what was observed for the animals fed the fresh linseed oil (LO) [41]. *In vitro* assays using the liver microsomes of the rats fed the HLO and the LO diets showed that the Δ6 desaturase activity (18:2 → 18:3(*n*-6)) was not significantly altered. On the contrary, the Δ5 desaturase activity (20:3 → 20:4(*n*-6)) was higher in the HLO group compared to the LO group. However, the 18:3 geometrical isomers fed to animals as a purified fraction were shown to increase the Δ6 desaturase activity of linolenic acid when rats were fed a (*n*-3) deficient diet [42]. This may indicate that the presence of *trans* 18:3 in the diet induces increased rates of 18:3(*n*-3) desaturation, related to a requirement of the liver for *cis* (*n*-3) fatty acids.

D. DESATURATION AND ELONGATION OF *TRANS* C18 POLYUNSATURATED FATTY ACIDS *IN VITRO*

The main conversion pathway of 18 carbon polyunsaturated fatty acids involves a desaturation of the carbon chain at the Δ6 position. This

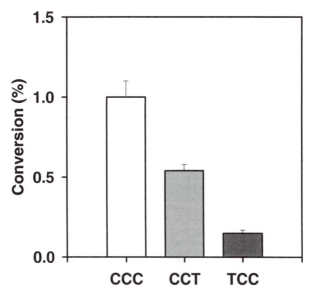

Fig. 6.5. Relative conversion of 18:3 isomers by rat liver microsomes under desaturation conditions. 5 mg of microsomal protein were incubated for 15 minutes with 15 nmoles of one of the radiolabelled substrates [43].

desaturation is known to be the regulatory step in the biosynthesis of 20-, and 22-carbon PUFA. Moreover, a competition between the different substrates occurs at this step. For example, the Δ6 desaturase affinity is higher for α-linolenic acid than for linoleic acid. This conversion step of *trans* PUFA has been studied using rat liver microsomes [22,43,44] and brain homogenates [45].

The *in vitro* Δ6 desaturation of 18:2 isomers was reported by Cook [45] and Berdeaux [22]. The conversion of 9-*cis*,12-*trans*-18:2 was described to be lower than the desaturation of linoleic acid in the liver, but the difference between both substrates is more important in brain homogenates (Table 6.3). The conversion of 9-*trans*,12-*cis*-18:2 and 9-*trans*,12-*trans*-18:2 were lower than those of the 9-*cis* isomers (Table 6.3). Together, these results confirm the hypothesis of Brenner [37], suggesting that a 9-*cis* double bond is required for the Δ6 desaturation of fatty acids, but it is not mandatory as 9-*trans* PUFA are slightly desaturated.

Similarly, recent studies on the Δ6 desaturation of *trans* isomers of α-linolenic acid showed that the 9-*trans* isomer was less converted by microsomes under desaturation conditions than the 15-*trans* isomer [43], which has its Δ9 double bond with a *cis* configuration. The relative conversions were 1.00, 0.54 and 0.15 for 9-*cis*,12-*cis*,15-*cis*-18:3, 9-*cis*,12-*cis*,15-*trans*-18:3 and 9-*trans*,12-*cis*,15-*cis*-18:3, respectively (Figure 6.5). These different levels of conversion could explain the differences observed in the 20:5 *trans* isomer contents of liver lipids from rats fed *trans* 18:3 isomers (see above). The conversion of 15-*trans*-18:3 was more extensively studied, as this isomer is the precursor of *trans* isomers of EPA and DHA [32]. Using liver microsomes

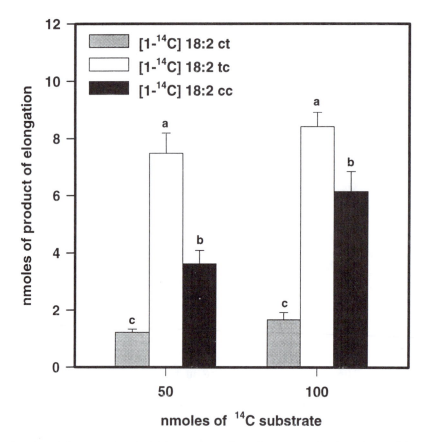

Fig. 6.6. Relative conversion of 18:2 isomers by rat liver microsomes under elongation conditions. 5 mg of microsomal protein were incubated for 30 minutes with 50 of 100 nmoles of one of the radiolabelled substrates [22].

from rats fed a fat-free diet, it was shown that the Δ6 desaturase was saturated when the 18:3(*n*-3) substrate reached 60 nmoles per 5 mg of microsomal protein. On the other hand, it appeared that the saturating substrate level was not reached when the 15-*trans* isomer of α-linolenic acid was the substrate. These data suggested that the affinity of the desaturation system was lower for the *trans* isomer than for linolenic acid [44].

Aside from Δ6 desaturation, the C18 polyunsaturated fatty acids might be chain-elongated into 20:2(*n*-6) and 20:3(*n*-3) isomers. To our knowledge, only elongation studies on *trans* isomers of linoleic acid have been performed. Berdeaux [22] showed that 9-*trans*,12-*cis*-18:2 was a good substrate for the elongation, as the production of 11-*trans*,14-*cis*-20:2 was about twice that observed with linoleic acid when the substrates were present at 50 nmoles. On the other hand, 9-*cis*,12-*trans*-18:2 was poorly elongated (Figure 6.6).

The data on elongation together with those obtained on the Δ6 desaturation can be well correlated to what is observed *in vivo* (see above) [20,22,30]. The

TABLE 6.4.

Rat liver mitochondrial oxidation of the geometric isomers of uniformly ^{14}C-labelled 9,12-octadecadienoic acid [47].

	9-*cis*,12-*cis*-18:2	18:2 mono-*trans*	9-*trans*,12-*trans*-18:2
Total oxidation	3.97 ± 0.21	6.09 ± 0.70	5.71 ± 0.24
[14]CO$_2$	1.40 ± 0.10	1.57 ± 0.08	1.65 ± 0.13
% oxidised	27 ± 1	40 ± 5	39 ± 2

TABLE 6.5.

Effects of dietary 9-*trans*,12-*trans*-18:2 and 9-*cis*,12-*cis*-18:2 (5 % in the diet) on arachidonic acid and PGF$_{2\alpha}$ contents in rat serum [48]

Dietary fatty acids	20:4(*n*-6) (%)	PGF$_{2\alpha}$ (ng/mL)
9-*cis*,12-*cis*-18:2	30.0	27.4
9-*trans*,12-*trans*-18:2	0.8	0.3

TABLE 6.6.

Effects of dietary 9-*cis*,12-*trans*-18:2 and 9-*cis*,12-*cis*-18:2 on platelet 20:3(*n*-6) and 20:4(*n*-6) contents and on prostaglandin concentration in the rat [49])

Dietary fatty acids	20:3(*n*-6) (%)	PGE$_1$ (ng/mL)	20:4(*n*-6) (%)	PGE$_2$ (ng/mL)
9-*cis*,12-*cis*-18:2	0.3	5.69 ± 0.69	17.6	24.89 ± 4.35
9-*trans*,12-*trans*-18:2	tr.	0.22 ± 0.02	1.4	0.15 ± 0.03
9-*cis*,12-*cis*-18:2 + 9-*trans*,12-*trans*-18:2	0.1	3.51 ± 0.58	14.1	11.64 ± 2.63

Rats were fed 5% of purified 18:2 fatty acids as methyl esters
Values of the percentages of fatty acids were obtained on platelet lipids pooled from 3 rats. Values for prostaglandins are means ± SEM of 8 rats.

9-*trans* fatty acids are preferentially converted into higher metabolites through the chain-elongation pathway, whereas the 9-*cis* ones are desaturated.

E. CATABOLISM OF *TRANS* POLYUNSATURATED FATTY ACIDS

The β-oxidation of geometrical isomers of linoleic acid was assessed some years ago using uniformly labelled substrates in a soybean oil vehicle [46]. The [14]CO2 production was measured for 51 h after feeding labelled linoleic acid, linolelaidic acid or a mixture of the two mono-*trans* isomers of linoleic acid. The [14]CO2 production of animals fed mono-*trans* 18:2 (70%) and 9-*trans*,12-*trans*-18:2 (68%) was slightly higher than the production of rats fed linoleic acid (64%, P < 0.10). This difference was correlated with a preferential esterification of *cis,cis*-18:2 (linoleic acid) into liver phospholipids. One year later, Anderson [47] reported that the mitochondrial oxidation of the *trans*

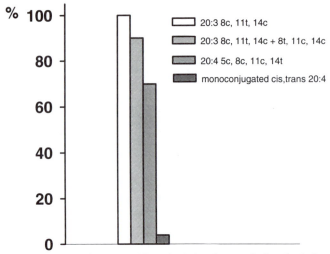

Fig. 6.7. Relative inhibition of the PGE$_1$ biosynthesis by sheep vesicular glands from 20:3(*n*-6) (30 μg) in the presence of 30 μg of mono-*trans* isomers of 20:3(*n*-6) or 20:4(*n*-6) [56].

isomers of 18:2 (mono-*trans* and di-*trans*) was greater than the oxidation of linoleic acid (Table 6.4). The same study also demonstrated that the metabolic pathway was β-oxidation, as illustrated by the acetoacetic acid production. As both mono-*trans* isomers of 18:2 are now available in radiolabelled form [14], it would be of great interest to carry out metabolic studies with each of these.

F. EFFECTS OF *TRANS* POLYUNSATURATED FATTY ACIDS ON EICOSANOID SYNTHESIS

Several years ago, some studies were carried out on the effects of 9-*trans*,12-*trans*-18:2 on eicosanoid production. At high dietary levels (5 % by weight), it was suggested that 9-*trans*,12-*trans*-18:2 decreased the production of prostanoids (Table 6.5) and altered platelet aggregation correlatively [48]. It was then suggested that the decrease in rat platelet production of eicosanoids was greater than the decrease in their precursors, *i.e.* 20:3(n-6) and 20:4(n-6) (Table 6.6) [49], indicating that dietary 9-*trans*,12-*trans*-18:2 or metabolic derivatives may inhibit some enzymes involved in prostaglandin production [50]. Using lower levels of 9-*trans*,12-*trans*-18:2 in the diet (up to 2.5 % of the diet), Hwang *et al.* [51] showed that 9-*trans*,12-*trans*-18:2 can alter the production of arachidonic acid metabolites in rat platelets (TxB$_2$, PGF$_{2α}$,12-HETE) only when the dietary 9-*trans*,12-*trans*-18:2 level is similar or greater than the level of linoleic acid.

Together, all the studies indicate that when fed at high levels and alone, 9-*trans*,12-*trans*-18:2 alters the prostanoid production by decreasing the substrate availability, because an effect of di-*trans*-18:2 on the phospholipase activity was not supported by the data from Kinsella's group [52,53]. However, these data are unrealistic nutritionally, considering that the quantities of the di-*trans*

TABLE 6.7.

Effects of 14-*trans*-20:4 on 20:4(*n*-6) (7.5μM) metabolism by rat platelets [58].

	Control	20:4 Δ14 t (7.5 μM)	20:4 Δ14 t (15 μM)	20:4 Δ14 t (22.5 μM)
TxB$_2$	339 ± 32[a]	255 ± 37[b]	172 ± 37[c]	126 ± 27[d]
HHT	331 ± 49[a]	257 ± 22[b]	154 ± 19[c]	136 ± 19[c]
12-HETE	483 ± 72[a]	644 ± 59[b]	739 ± 74[c]	749 ± 89[c]

Results (pmol/10^8 platelets) are means ± SD of 12 determinations.
Values in rows having different superscripts are significantly different (P < 0.05).

TABLE 6.8.

Inhibition (%) of aggregation of rat platelets stimulated by 20:4(*n*-6) (7.5 μM) in the presence of increasing concentrations of 20:4(*n*-6) and 14-*trans*-20:4. [58].

Fatty acids	0.0 μM	7.5 μM	15.0 μM	22.5 μM
20:4(*n*-6)	0	a	a	a
14-*trans*-20:4	0	b	c	d

Values having a different superscript are significantly different (P < 0.05).

TABLE 6.9.

Metabolites (pmol/10^8 platelets) formed from [1-^{14}C]-14-*trans*-20:4 (7.5 μM) in the presence of 20:4(*n*-6) [58].

20:4(*n*-6) (μM)	0	7.5	15
M1	nd	73.7 ± 22.1	74.2 ± 21.0
M2	traces	136.7 ± 25.4	155.1 ± 33.9

nd: not detected. Results are means ± SD of 6 determinations.

TABLE 6.10.

Effects of 20:5 and 22:6 isomer and C22 fatty acids on the metabolism of 20:4(*n*-6) (2.5 μM) by human platelets [60].

	20:4(*n*-6) (control)	20:4(*n*-6) + 20:5(*n*-3)	20:4(*n*-6) + *trans*-17-20:5
T × B$_2$	42.0 ± 4.4[a]	32.8 ± 3.5[b]	30.6 ± 3.1[b]
HHT	91.9 ± 17.2[a]	68.3 ± 16.0[a,b]	65.8 ± 12.7[b]
12-HETE	49.4 ± 9.7[a]	50.6 ± 10.5[a]	108.8 ± 20.1[b]
	20:4(*n*-6) (control)	20:4(*n*-6) + 22:6(*n*-3)	20:4(*n*-6) + 19-*trans*-22:6
T × B$_2$	42.6 ± 10.0[a]	27.6 ± 7.6[b]	22.7 ± 6.2[c]
HHT	111.4 ± 18.6[a]	65.1 ± 14.7[b]	48.1 ± 9.8[c]
12-HETE	83.6 ± 18.5[a]	102.6 ± 37.4[a]	90.2 ± 37.2[a]

Results (means ± SD, pmol/10^8 platelets) were calculated for concentrations inducing 50% of inhibition of aggregation. Values in rows having different superscripts are significantly different (P < 0.05).
HHT = Hydroxyheptadecatrienoic acid; 12-HETE = 12 hydroxyeicosatetraenoic acid

isomer are much smaller than the quantities of the mono-*trans* isomers present in the human diet [7,54]. It has been reported recently that dietary 9-*cis*,12-*trans*-18:2 and 9-*trans*,12-*cis*-18:2 did not affect the PGI$_2$ production in the rat aorta and that they slightly suppressed rat platelet aggregation in response to collagen [55]. However, the detailed results are not yet published.

Nugteren [56] studied the effects of several 20:3 isomers and of 14-*trans*-20:4 on the biosynthesis of PGE$_1$ from 20:3(*n*-6). The results are summarised in Figure 6.7. The different mono-*trans* isomers of 20:3 and 20:4 poorly inhibited the PGE$_1$ production when the double bonds were located at positions 8, 11, 14 and 5, 8, 11, 14, respectively. On the other hand, the occurrence of a conjugated *cis-trans* double bond system in the same molecules (20:3 and 20:4) produced a strong inhibition of this metabolic pathway.

Mono-*trans* isomers of 18:2(*n*-6) and 18:3(*n*-3) are converted into higher metabolites (see above), including C20 and C22 PUFA containing one (or two) *trans* double bonds. The effects of geometrical isomers of 20-carbon PUFA have also been assessed, generally using fully synthetic molecules [57-59], but the first study on the effects of 17-*trans*-20:5 and 19-*trans*-22:6 was carried out using fatty acids purified from rat liver lipids fed the *trans* precursor (*i.e.* 18:3 *trans* isomers) [60].

The effects of 14-*trans*-20:4 on the metabolism of arachidonic acid in eicosanoids was studied recently by Berdeaux *et al.* [58] using rat platelets. It was shown that this structural analogue of 20:3(*n*-9) induced an inhibition of the conversion of 20:4(*n*-6) into thromboxane B2 or HHT (cycloxygenase pathway) and increased the production of 12-hydroxyeicosatetraenoic acid (12-HETE) through the 12-lipoxygenase pathway (Table 6.7). These data were well correlated to platelet aggregation (Table 6.8) and were different from those obtained with 20:3(*n*-9). Mead acid only induced a slight decrease in the 12-HETE production. These data confirm a specific effect of the *trans* double bond located at the Δ14 position. Moreover, using radiolabelled 14-*trans*-20:4, it was shown that it was metabolized by platelets into two metabolites, called M1 and M2 (Table 6.9). One of them is probably a product of the 12-lipoxygenase pathway, as its production is lowered when platelets are preincubated with baicalein, a selective 12-lipoxygenase inhibitor [58]. The production of this unknown metabolite is greatly enhanced when arachidonic acid or 12-HPETE are present in the incubation medium. The data suggest that a sufficient "peroxide tone" is needed to produce this unknown metabolite [61]. They enhance the hypothesis about the origin of the metabolite, which might be a *trans* isomer of 12-HETE. However, its structure needs to be fully elucidated. Berdeaux *et al.* [58] demonstrated that 14-*trans*-20:4 is also metabolized into another product, but to a lesser extent. This second unknown metabolite might be produced by the cycloxygenase pathway, but its exact origin remains still unclear.

The effects of *trans* isomers of long-chain (*n*-3) PUFA were studied earlier, as it was possible to purify 17-*trans*-20:5 and 19-*trans*-22:6 from rat liver lipids fed heated linseed oil. The first study on this topic was carried out by O'Keefe

TABLE 6.11.

(*N*-3) fatty acid concentrations required to produce a 50 % inhibition (IC50) of platelet aggregation induced by 20:4(*n*-6) (2.5 μM) [60].

Samples	IC50 (μM)			
	EPA	17-*trans*-20:5	DHA	19-*trans*-22:6
A	1.8	29.9	9.2	10.6
B	27.4	50.2	0.7	1.1
C	10.8	66.5	0.9	1.2
D	5.3	14.3	9.4	5.8
E	4.2	16.0	19.5	12.2
F	11.2	40.5	4.0	3.6
G	3.4	22.2	0.6	0.01
H	3.3	14.7	0.7	0.1
I	1.0	8.9		
Means ± SD	7.6 ± 8.26[a]	29.2 ± 19.46[b]	5.6 ± 6.75[a]	4.3 ± 4.77 [a]

Human platelet samples were different for C20 and C22 fatty acids. Values having a different superscript are significantly different (P < 0.05).

et al. [60], using washed human platelets. In their experimental conditions, 19-*trans*-22:6 appeared to be an inhibitor of the cycloxygenase pathway, as assessed by the TxB$_2$ and HHT production (Table 6.10). On the other hand, 17-*trans*-20:5 inhibited the 12-lipoxygenase pathway (Table 6.10). At the same time, 17-*trans*-20:5 was less antiaggregant than EPA, whereas 19-*trans*-22:6 and DHA had a similar antiaggregant effect (Table 6.11). These specific effects of 19-*trans*-22:6 on the cycloxygenase pathway and of 17-*trans*-20:5 on the lipoxygenase pathway were not clearly understood, as they were not well correlated to the platelet aggregation response. More recently, the effects of these isomers were studied in conditions which represent better the physiological situation, after incorporation into platelet lipids [57]. Thrombin and collagen stimulation of platelets enriched in 20:5 or 22:6 PUFA showed that the occurrence of a *trans* double bond at the (*n*-3) position decreased the antiaggregant effects of both 20:5 and 22:6 fatty acids. Similarly, the stimulation of these platelets with U46619, a stable analogue of thromboxane A$_2$, showed that platelets enriched in *trans* PUFA were more sensitive than when enriched with *cis* PUFA. These data suggest that the incorporation of *trans* PUFA may modify the sensitivity of the TxA$_2$ receptor of the platelet membranes. This hypothesis is enhanced by the lack of effect on the production of eicosanoids when platelets were triggered by collagen [57].

As several 9-*trans*-18:3 isomers are converted *in vivo* into 20:5 isomers (see above), we have recently carried out studies on the effects of 11-*trans*-20:5 and 11-*trans*,17-*trans*-20:5 on platelet aggregation. However, this study was performed using washed rat platelets as human platelets are not devoid of long-chain *trans* PUFA [62]. Washed rat platelets were stimulated by arachidonic acid in the presence of increasing quantities of EPA or one of its *trans* isomers [59]. It appeared that 20:5 isomers with a *trans* double bond at

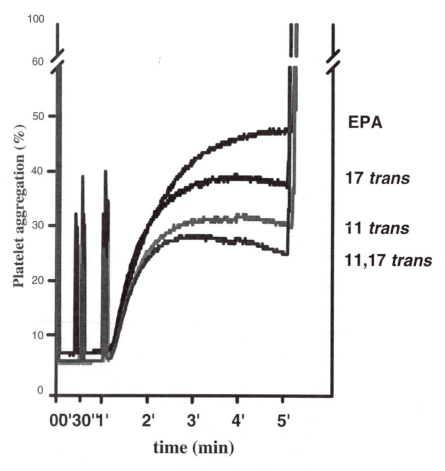

Fig. 6.8. Typical aggregation response of rat platelets triggered with arachidonic acid (5 μM) and EPA or one of its *trans* isomers (5 μM) [59].

the $\Delta 11$ position were inhibitors of the cycloxygenase activity [59] and were more antiaggregant that EPA or 17-*trans*-20:5 (Figure 6.8).

In order to measure these antiaggregant effects of 11-*trans*-20:5 fatty acids, it would be of interest to study the metabolism of these fatty acid isomers. However, labelled molecules are needed for this purpose.

Additionally, studies with other experimental models would allow a better understanding of the role of *trans* PUFA on the physiological functions, particularly in the cardiovascular system.

G. HUMAN STUDIES

While numerous data are available on the *trans* monoethylenic isomers (see Chapter 5), much less information is available for the *trans* polyenes. Geometrical linoleic acid isomers were reported in human tissues, including

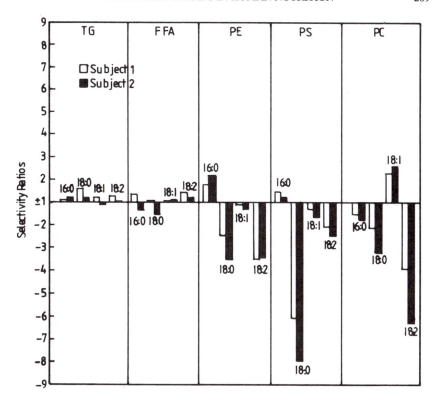

Fig. 6.9. Selectivity ratios for comparison of the incorporation of deuterium labelled 12-*cis*,15-*trans*-18:2 relative to 16:0, 18:0, 9-*cis*-18:1 and 9-*cis*,12-*trans*-18:2 into plasma triacylglycerols (TG), free fatty acids (FFA), phosphatidylethanolamine (PE), phosphatidylserine (PS) and phosphatidylcholine (PC). reproduced by kind permission of the authors and *Lipids* [69].

milk [63-66], adipose tissue, kidney, heart and liver [67]. Adlof and Emken [67] reported that adipose tissue had the highest concentration of *trans* 18:2 (9-*cis*,12-*trans* and 9-*trans*,12-*cis*) while the amount present in the brain was too small to be quantified. Only traces of the di-*trans* isomers were detected. While 18:3 geometrical isomers were reported in human milk [64], none of the 18:3 isomers were detected by Adlof and Emken in the other human tissues [67]. Longer-chain (*n*-3) polyunsaturated fatty acids, especially the metabolites of 9-*cis*,12-*cis*,15-*trans*-18:3, were also detected in human platelets [62]. The effects of the long-chain *trans* polyunsaturated fatty acids on platelet aggregation and eicosanoid production were described earlier in this chapter.

The effects of *trans* fatty acids, including *trans* mono- and di-enoic isomers on the biosynthesis of long-chain polyunsaturated fatty acids in premature infants were studied recently by Koletzko [68]. *Trans* octadecenoic acid and total *trans* isomers, including 9-*cis*,12-*trans*-18:2, 9-*trans*,12-*cis*-18:2 and 9-*trans*,12-*trans*-18:2 in plasma lipid fractions were not related to either linoleic or linolenic acids but these were correlated inversely to the long-

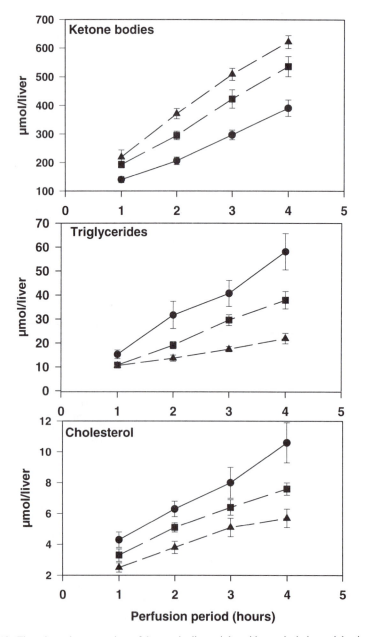

Fig. 6.10. Time-dependent secretion of ketone bodies, triglycerides and cholesterol by isolated perfused rat livers in the presence of 18:2 di-*trans* (Δ), 18:2 mono-*trans* (■) or linoleic acid (●) [70].

chain (*n*-3) and (*n*-6) polyunsaturated acids. These data may indicate a potential impairment of essential fatty acid metabolism.

The metabolism in humans of the 12-*cis*,15-*trans*-18:2, which can be formed from linolenic acid by catalytic hydrogenation, was studied by Emken *et al.* using triacylglycerols containing deuterium-labelled fatty acids [69]. Analyses performed on two human subjects indicated that the turnover for triacylglycerols, cholesterol esters, phosphatidylethanolamine and phosphatidylcholine of all the deuterated fatty acids tested (16:0, 18:0, 18:1, 9-*cis*,12-*cis*-18:2 and 9-*cis*,15-*trans*-18:2) were similar. Furthermore, the selectivity ratios (Figure 6.9) for the deuterium-labelled isomers incorporated into the triacylglycerol and free fatty acid classes were small compared to what was observed for the phospholipids. For example, the incorporation of 9-*cis*,15-*trans*-18:2 into phosphatidylethanolamine, phosphatidylserine and phosphatidylcholine is limited compared to 9-*cis*,12-*cis*-18:2 while the mono-*trans* 18:2 isomer is preferentially incorporated into phosphatidylethanolamine and phosphatidylserine relative to 16:0. It appears that the 9-*cis*,15-*trans*-18:2 must have been utilized mainly for energy, considering that most of the selectivity ratios for the phospholipids are negative and that no large positive ratios for the neutral lipids were observed.

H. OTHER STUDIES

The oxidative metabolism of geometrical isomers of linoleic acid was studied some years ago by Fukuda *et al.* [70], using an isolated perfused rat-liver model. The ketone body production was increased by *trans* isomers of 18:2 (Figure 6.10). As 9-*trans*-12-*trans*-18:2 was the most potent stimulus of ketone body production, it was suggested that the number of *trans* double bonds was an important parameter for such an effect. The hepatic secretion of triacylglycerols and cholesterol was also reported (Figure 6.10). These were inversely related to the ketone body production. These data are of particular interest, as they suggest that *trans* isomers of 18:2 might be hypolipidemic, by depressing the efflux of cholesterol and triacylglycerols in the blood stream. Similar results were also obtained when rats were fed a cholesterol-enriched diet [71]. Further studies on these aspects would be mandatory to determine the hypolipemic effects of *trans* 18:2 versus linoleic acid. The same study with *trans* 18:3 isomers would also be of interest.

Recently, a lack of impact of *trans* fatty acids on cancer was reported by Ip & Marschall [72]. However, it was shown that 9-*trans*,12-*trans*-18:2 was less inhibitory of the thymidine incorporation into DNA in 7800NJ rat hepatoma and T47D human breast cancer cells [73]. However, in this study, the authors showed also that linoleic acid was more inhibitory than conjugated linoleic acid, which is generally considered to be an inhibitor of cell proliferation [74-78] (see Chapter 9). Additional studies on these potent effects of *trans* PUFA are needed.

Besides the *trans* polyunsaturated fatty acids resulting from technological processes of fats and oils, some unusual *trans* polyunsaturated fatty acids are present in certain seeds from which it is possible to extract oil. One of these fatty acids is columbinic acid (5-*trans*,9-*cis*,12-*cis*-18:3), which is linoleic acid

TABLE 6.12.

Fatty acid composition (%) of serum or rats fed a fat-free diet (ffd) and the same diet supplemented with columbinic acid for 24 and 48 hours [84].

Fatty acids	ffd	ffd + columbinic 24 h	ffd + columbinic 48 h
14:0	0.9 ± 0.05[a]	0.8 ± 0.01	0.7 ± 0.1
16:0	29.8 ± 0.6	22.5 ± 0.6*	22.2 ± 0.5*
16:1	6.6 ± 0.5	4.8 ± 0.5	4.3 ± 0.4*
18:0	12.4 ± 0.5	12.4 ± 0.5	13.0 ± 0.2
18:1	21.9 ± 1.0	17.5 ± 0.5*	15.5 ± 0.6*
18:2(*n*-6)	7.0 ± 0.3	3.0 ± 0.3	3.2 ± 0.3*
18:3 columbinic	-	13.4 ± 0.3	15.6 ± 0.6*
20:3(*n*-6) +			
7-*trans*,11-*cis*,14-*cis*-20:3	1.3 ± 0.2	2.6 ± 0.1*	2.7 ± 0.2*
20:4(n-6)	12.3 ± 0.4	20.0 ± 0.5	19.3 ± 0.4

[a]: Means ± SEM of 6 determinations. *: P < 0.01 compared to ffd

TABLE 6.13.

Fatty acid composition (%) of liver or rats fed a fat-free diet (ffd) and the same diet supplemented with columbinic acid for 24 and 48 hours [84].

Fatty acids	ffd	ffd + columbinic 24 h	ffd + columbinic 48 h
14:0	0.4 ± 0.1	0.3 ± 0.1	0.3 ± 0.1
16:0	21.4 ± 0.5	19.5 ± 0.2*	19.4 ± 0.6
16:1	5.7 ± 0.5	4.0 ± 0.3	3.7 ± 0.2*
18:0	13.9 ± 1.0	15.1 ± 0.5	15.2 ± 0.9
18:1	21.0 ± 1.1	17.4 ± 0.9	16.5 ± 0.9*
18:2(*n*-6)	7.8 ± 0.4	4.0 ± 0.1	3.5 ± 0.2*
18:3 columbinic	-	9.4 ± 0.4	10.1 ± 0.5
20:3(*n*-6) +			
7-*trans*,11-*cis*,14-*cis*-20:3	1.4 ± 0.1	1.4 ± 0.1*	2.0 ± 0.2
20:4(*n*-6)	14.9 ± 0.9	15.8 ± 0.5	15.1 ± 0.6
9-*trans*,13-*cis*,16-*cis*,22:3	-	1.6 ± 0.1	2.2 ± 0.2
22:4(*n*-6)	1.2 ± 0.2	1.4 ± 0.1	2.0 ± 0.2
22:5(*n*-6)	2.9 ± 0.1	3.0 ± 0.1	4.0 ± 0.4
22:5(*n*-3)	0.6 ± 0.04	0.6 ± 0.1	0.9 ± 0.2
22:6(*n*-3)	3.0 ± 0.2	3.4 ± 0.1	3.6 ± 0.5

[a]: Means ± SEM of 6 determinations. *: P < 0.01 compared to ffd

with addition of a *trans* Δ5 bond. It has been introduced by Houtsmüller [79] to study some effects of essential fatty acids. When supplemented to rats fed an essential fatty acids-deficient diet, columbinic acid improves some biochemical and/or physiological aspects of the deficiency. For example, the skin permeability is restored [80]. Moreover, the content of some hepatic enzymes is reduced, *i.e.* fatty acid synthase [81], acetyl-CoA carboxylase [82] and microsomal fatty acid desaturase [83].

Columbinic acid is incorporated into the tissue lipids of rats [84]. This incorporation seems to be at the expense of linoleic acid. The decrease in the linoleic acid content of serum lipids allows a recovery of the arachidonic acid content [84]. By this way, the 20:3(n-9) content is decreased in EFA-deficient rats fed columbinic acid. On the other hand, columbinic acid is chain elongated, but not desaturated. The chain elongation induces the incorporation of 7-*trans*,11-*cis*,14-*cis*-20:3, but also of 9-*trans*,13-*cis*,16-*cis*-22:3 in serum (Table 6.12), liver (Table 6.13), lung, spleen and kidney lipids.

Columbinic acid, also a substrate for cycloxygenase and lipoxygenase enzymes [85], is converted into hydroxy-derivatives, but not into cyclized structures. Washed human platelets and to a greater extent ram seminal vesicle microsomes are able to convert columbinic acid into 9-hydroxy-(5-*trans*,10-*trans*,12-*cis*)-octadecatrienoic acid. On the other hand, the lipoxygenase product from washed human platelets, soybean lipoxygenase and neonatal rat epidermal homogenates was identified as being 13-hydroxy-(5-*trans*,9-*cis*,11-*trans*)-octadecatrienoic acid. This latter compound may explain the effects of columbinic acid on skin dermatitis, as its application to paws of essential fatty acid deficient rats resolved the dermatitis as did the application of columbinic acid itself. The cycloxygenase product was not active.

The effects of columbinic acid were compared to those of linoleic acid on tumor growth in mice. Abraham & Hillyard [86] reported that columbinic acid induced a tumor growth rate similar to oleic acid (control), whereas linoleic acid induced heavier adenocarcinomas after 7 weeks of dietary supplementation. (1.58 ± 0.67 *versus* 1.02 ± 0.62 g, P < 0.05). Columbinic acid is also known as an inhibitor of fatty acid synthesis in rat hepatocytes [87].

I. CONCLUSION

Most of the studies on *trans* polyunsaturated fatty acids have been carried out so far *in vitro* and *in vivo* using rats or mice as models. Only a few data on humans have been reported, while human studies using *trans* monounsaturated fatty acids, mainly found in partially hydrogenated oils have been the subject of numerous papers (see Chapter 5). Even if the quantities of *trans* polyunsaturated fatty acids ingested mainly through refined vegetable oils are small compared to the amount of *trans* monoenes, it would be interesting to carry out human studies using either stable isotopes or refined oils enriched in *trans* polyenes. This would permit evaluation of the nutritional impact of these *trans* C18 polyunsaturates.

LIST OF ABBREVIATIONS

DHA, docosahexaenoic acid; DNA, deoxyribonucleic acid; EPA, eicosapentaenoic acid; HETE, hydroxy eicosatetraenoic acid; HHT, hydroxy heptadecatrienoic acid; HLO, heated linseed oil; HPETE, hydroperoxy eicosatetraenoic acid; LO, linseed oil; PC, phosphatidylcholine; PE, phosphatidylethanolamine; PG, prostaglandin; PI, phosphatidylinositol; PUFA, polyunsaturated fatty acids; Tx, thromboxane.

REFERENCES

1. Ackman,R.G., Hooper,S.N. and Hooper,D.L., *J. Am. Oil Chem. Soc.*, **51**, 42-49 (1974).
2. Grandgirard,A., Sébédio,J.L. and Fleury,J., *J. Am. Oil Chem. Soc.*, **61**, 1561-1568 (1984).
3. Sébédio,J.L., Grandgirard,A., Septier,C. and Prévost,J., *Rev. Fr. Corps Gras*, **34**, 15-18 (1987).
4. Sébédio,J.L., Grandgirard,A. and Prévost,J., *J. Am. Oil Chem. Soc.*, **65**, 362-366 (1988).
5. Wolff,R.L. and Sébédio,J.L., *J. Am. Oil Chem. Soc.*, **71**, 117-126 (1994).
6. Ratnayake,W.M.N., Hollywood,R., O'Grady,E. and Beare-Rogers,J.L., *J. Am. Oil Chem. Soc.*, **67**, 804-810 (1990).
7. Chardigny,J.M., Sébédio,J.L. and Berdeaux,O., in *Advances in Applied Lipid Research*, pp. 1-33 (1996) (edited by F.D. Padley, Jai Press Inc, London).
8. Rakoff,H. and Emken,E.A., *Chem. Phys. Lipids*, **31**, 215-225 (1982).
9. Rakoff,H. and Emken,E.A., *J. Labelled Compds. Radiopharm.*, **19**, 19-33 (1982).
10. Rakoff,H., *Chem. Phys. Lipids*, **35**, 117-125 (1984).
11. Vatèle,J.M., Dong Doan,H., Chardigny,J.M., Sébédio,J.L. and Grandgirard,A., *Chem. Phys. Lipids*, **74**, 185-193 (1994).
12. Vatèle,J.M., Dong Doan,H., Fenet,B., Chardigny,J.M. and Sébédio,J.L., *Chem. Phys. Lipids*, **78**, 65-70 (1995).
13. Eynard,T., Vatèle,J.M., Poullain,D., Noël,J.P., Chardigny,J.M. and Sébédio,J.L., *Chem. Phys. Lipids*, **74**, 175-184 (1994).
14. Berdeaux,O., Vatèle,J.M., Eynard,T., Nour,M., Poullain,D., Noël,J.P. and Sébédio,J.L., *Chem. Phys. Lipids*, **78**, 71-80 (1995).
15. Adlof,R.O., Rakoff,H. and Emken,E.A., *J. Am. Oil Chem. Soc.*, **68**, 303-306 (1991).
16. Emken,E.A., *J. Am. Oil Chem. Soc.*, **60**, 995-1004 (1983).
17. Ono,K. and Fredrickson,D.S., *J. Biol. Chem.*, **239**, 2482-2488 (1964).
18. Trus,M., Grandgirard,A. and Sébédio,J.L., *Reprod. Nutr. Develop.*, **31**, 294 (1991).
19. Selinger,Z.V.I. and Holman,R.T., *Biochim. Biophys. Acta*, **106**, 56-62 (1965).
20. Privett,O.S., Stearns,E.M. and Nickell,E.C., *J. Nutr.*, **92**, 303-310 (1967).
21. Anderson,R.L., Fullmer,C.S. and Hollenbach,E.J., *J. Nutr.*, **105**, 393-400 (1975).
22. Berdeaux O. (1996). PhD thesis, Université de Bordeaux I.
23. Berdeaux,O., Sébédio,J.L., Chardigny,J.M., Blond,J.P., Mairot,T., Vatèle,J.M., Poullain,D. and Noël,J.P., *Grasas Aceit.*, **47**, 86-99 (1996).
24. Schimp,J.L., Bruckner,G. and Kinsella,J.E.,, *J. Nutr.*, **112**, 722-735 (1982).
25. Privett,O.S., Nutter,L.J. and Lightly,F.S., *J. Nutr.*, **89**, 257-264 (1966).
26. Knipprath,W.G. and Mead,J.F., *J. Am. Oil Chem. Soc.*, **41**, 437-440 (1964).
27. Karney,R.I. and Dhopeshwarkar,G.A., *Biochim. Biophys. Acta*, **531**, 9-15 (1978).
28. Blank,M.L. and Privett,O.S., *J. Lipid Res.* **4**, 470-476 (1963).
29. Ratnayake,W.M.N., Chen,Z.Y., Pelletier,G. and Weber,D., *Lipids*, **29**, 707-714 (1994).
30. Beyers,E.C. and Emken,E.A., *Biochim. Biophys. Acta*, **1082**, 275-284 (1991).
31. Wolff,R.L., Combe,N.A., Entressangles,B., Sébédio,J.L. and Grandgirard,A., *Biochim. Biophys. Acta*, **1168**, 285-291 (1993).
32. Grandgirard,A., Piconneaux,A., Sébédio,J.L., O'Keefe,S.F., Sémon,E. and Le Quéré,J.L., *Lipids*, **24**, 799-804 (1989).
33. Grandgirard,A., *les Cahiers de l'ENSBANA*, **8**, 49-67 (1992).
34. Waku, K. and Lands, W.E.M., *J. Lipid Res.*, **9**, 12-18 (1968).
35. Chardigny,J.M., Sébédio,J.L., Grandgirard,A., Martine,L., Berdeaux,O. and Vatèle,J.M., *Lipids*, **31**, 165-168 (1996).
36. Piconneaux,A. *PhD thesis*, University of Dijon (1987).
37. Brenner,R.R., *Lipids*, **6**, 567-571 (1971).
38. Grandgirard,A., Bourre,J.M., Julliard,F., Homayoun,P., Dumont,O., Piciotti,M. and Sébédio,J.L., *Lipids*, **29**, 251-258 (1994).
39. Chardigny,J.M., Bron,A., Sébédio,J.L., Juaneda,P. and Grandgirard,A., *Nutr. Res.*, **14**, 909-917 (1994).
40. Bourre,J.M., Pascal,G., Durand,G, Masson,M, Dumont,O. and Piciotti,M., *J. Neurochem.*, **43**, 342-348 (1984).
41. Blond,J.P., Henchiri,C., Précigou,P., Grandgirard,A. and Sébédio,J.L., *Nutr. Res.*, **10**, 69-79 (1990).
42. Blond,J.P., Chardigny,J.M., Sébédio,J.L. and Grandgirard,A., *J. Food Lipids*, **2**, 99-106 (1995).
43. Chardigny,J.M., Blond,J.P., Saget,L., Sébédio,J.L., Eynard,T., Poullain,D., Vatèle,J.M. and Noël,J.P., in *Proceedings of the 21st Congress of the International Society for Fat Research (ISF)* 1995, pp. 219-221 (PJ Barnes & Associates).

44. Chardigny,J.M., Blond,J.P., Bretillon,L., Mager,E., Poullain,D., Martine,L., Vatèle,J.M., Noël,J.P. and Sébédio,J.L., *Lipids, 32*, 731-735 (1997) .
45. Cook,H.W., *Lipids, 16*, 920-926 (1981).
46. Anderson,R.L. and Coots,R.H., *Biochim. Biophys. Acta, 144*, 525-531 (1967).
47. Anderson,R.L., *Biochim. Biophys. Acta, 152*, 531-538 (1968).
48. Hwang,D.H. and Kinsella,J.E., *Prost. Med., 1*, 121-130 (1978).
49. Kinsella,J.E., Hwang,D.H., Yu,P., Mai,J. and Shimp,J., *Biochem. J., 184*, 701-704 (1979).
50. Kinsella,J.E., Bruckner,G., Mai,J. and Shimp,J. *Am. J. Clin. Nutr., 34*, 2307-2318 (1981).
51. Hwang,D.H., Chanmugan,P. and Anding,R., *Lipids, 17*, 307-313 (1982).
52. Chern,J.C., Bruckner,G. and Kinsella,J.E., *Nutr. Res., 3*, 571-581(1983).
53. Chern,J.C., Bruckner,G. and Kinsella,J.E., *Nutr. Res., 4*, 79-82 (1984).
54. Entressangles,B., *Cah. Nutr. Diet., 21*, 309-321 (1986).
55. Sugano,M. and Ikeda,I., *Curr. Opin. Lipid., 7*, 38-42 (1996).
56. Nugteren,D.H., *Biochim. Biophys. Acta, 210*, 171-176 (1970).
57. Chardigny,J.M., Sébédio,J.L., Juanéda,P., Vatèle,J.M. and Grandgirard,A., *Nutr. Res., 15*, 1463-1471 (1995).
58. Berdeaux,O., Chardigny,J.M., Sébédio,J.L., Mairot,T., Poullain,D., Vatèle,J.M. and Noël,J.P., *J. Lipid Res., 37* , 2244-2250 (1996).
59. Loï,C., Chardigny,J.M., Berdeaux,O., Vatèle,J.M., Poullain,D., Noël,J.P. and Sébédio, J.L., *Thromb. Haemost.,* in press (1998).
60. O'Keefe,S.F., Lagarde,M., Grandgirard,A. and Sébédio,J.L. *J. Lipid Res., 31*, 1241-1246 (1990).
61. Croset,M. and Lagarde,M., *Lipids, 20*, 743-750 (1985).
62. Chardigny,J.M., Sébédio,J.L., Juanéda,P., Vatèle,J.M. and Grandgirard,A., *Nutr. Res., 13*, 1105-1111 (1993).
63. Boatella,J., Rafecas,M., Codony,R., Gibert,A., Rivero,M., Turmo,R., Infante,D. and Valverde,F.S., *J. Pediatr. Gastroenterol. Nutr., 16*, 432-434 (1993).
64. Chardigny,J.M., Wolff,R.L., Mager,E., Sébédio,J.L., Martine,L. and Juanéda,P., *Eur. J. Clin. Nutr., 49*, 523-531 (1995).
65. Chen,Z.Y., Pelletier,G., Hollywood,R. and Ratnayake,W.M.N., *Lipids, 30*, 15-21 (1995).
66. Koletzko,B., Mrotzek,M., Eug,B. and Bremer,H.J., *Am. J. Clin. Nutr., 47*, 954-959 (1988).
67. Adlof,R.O. and Emken,E.A., *Lipids, 21*, 543-547 (1986).
68. Koletzko,B., *Acta Paediatr., 81*, 302-306 (1992).
69. Emken,E.A., Rohwedder,W.K., Adlof,R.O., Rakoff,H. and Gulley,R.M., *Lipids, 22*, 495-504 (1987).
70. Fukuda,N., Etoh,T., Wada,K., Hidaka,T., Yamamoto,K., Ikeda,I. and Sugano,M., *Ann. Nutr. Metab., 39*, 185-192 (1995).
71. Fukuda,N., Igari,N., Etoh,T., Hidaka,T., Ikeda,I. and Sugano,M.., *Nutr. Res., 13*, 779-786 (1993).
72. Ip,C. and Marshall,J.R., *Nutr. Rev., 54*, 138-145 (1996).
73. Desbordes,C. and Lea,M.A., *Anticancer Res,. 15*, 2017-2022 (1995).
74. Ha,Y.L., Grimm,N.K. and Pariza,M.W., *Carcinogenesis, 8*, 1881-1887 (1987).
75. Schultz,T.D., Chew,B.P., Seaman,W.R. and Luedecke,L.O., *Cancer Lett., 63*, 125-133 (1992).
76. Schultz,T.D., Chew,B.P. and Seaman,W.R., *Anticancer Res., 12*, 2143-2146 (1992).
77. Ip,C., Chin,S.F., Scimeca,J.A. and Pariza,M.W., *Cancer Res., 51*, 6118-6124 (1991).
78. Ip,C., Singh,M., Thompson,H.J. and Scimeca,J.A., *Cancer Res., 54*, 1212-1215 (1994).
79. Houtsmüller,U.M.T., *Prog. Lipid Res., 20*, 889-896 (1981).
80. Houtsmüller,U.M.T. and Van Der Beek,A., *Prog. Lipid Res., 20*, 219-228 (1981).
81. Schwartz,R.S. and Abraham,S., *Biochim. Biophys. Acta, 711*, 316-326 (1982).
82. Abraham,S., Hillyard,L.A., Lin,C.Y. and Schwartz,R.S., *Lipids, 18*, 820-829 (1983).
83. De Alaniz,M.J.T., De Gomez Dumm,I.N.T. and Brenner,R.R., *Lipids, 21*, 425-429 (1986).
84. Mandon,E.C., De Gomez Dumm,I.N.T. and Brenner,R.R., *Arch. Latinoam. Nutr., 36*, 401-414 (1986).
85. Elliot,W.J., Morrison,A.R., Sprecher,H.W. and Needleman,P., *J. Biol. Chem., 260*, 987-992 (1985).
86. Abraham,S. and Hillyard,L.A., *J. Natl. Cancer Inst., 71*, 601-605 (1983).
87. Mikkelsen,L., Hansen,H.S., Grunnet,N. and Dich,J., *Biochim. Biophys. Acta, 1166*, 99-104 (1993).

CHAPTER 7

LIPOPROTEIN METABOLISM AND TRANS *FATTY ACIDS*

Ronald P. Mensink[a] **and Peter L. Zock**[b]

[a]*Department of Human Biology, Maastricht University, 6200 MD Maastricht, the Netherlands*
[b]*Division of Human Nutrition & Epidemiology, Agricultural University, 6700 EV Wageningen, the Netherlands*

A. INTRODUCTION

Lipoproteins transport cholesterol through the blood vessels. In man, about 60 to 70 percent of the total serum cholesterol level is transported by the low

Footnote

Address all correspondence to:
Ronald P. Mensink, Department of Human Biology, Maastricht University, Universiteitssingel 50/ P.O. Box 616, 6200 MD Maastricht, the Netherlands

Telephone:+31-43-3881308
FAX: +31-43-3670976
E-mail: r.mensink@hb.unimaas.nl

density lipoproteins (LDL), 20 to 30 percent by the high-density lipoproteins (HDL), while the remainder is carried mainly by the very-low density lipoproteins (VLDL). In addition to the absolute level of serum total cholesterol, the distribution of cholesterol over the various lipoproteins is a major risk factor for coronary heart disease. Epidemiological studies have shown that the risk for coronary heart disease is positively related to LDL-cholesterol levels, but negatively to that of HDL-cholesterol [10].

Lipoprotein cholesterol levels can be changed by modifying dietary fat intake and diet is therefore an important tool for lowering the risk of coronary heart disease. Relative to an iso-energetic amount of carbohydrates, saturated fatty acids increase serum LDL-cholesterol levels. This effect is mainly due to lauric acid (C12:0), myristic acid (C14:0) and palmitic acid (C16:0). Effects of stearic acid (C18:0) are comparable to those of carbohydrates. Oleic acid (*cis*-C18:1(n-9)), the most abundant monounsaturated fatty acid in the diet, may slightly decrease LDL-cholesterol concentrations. Linoleic acid (*cis*, *cis*-C18:2(n-6)) reduces LDL-cholesterol levels and increases those of HDL-cholesterol, but a review of recent studies [21] has suggested that this effect is less than was predicted by studies performed in the early sixties [14]. All these fatty acids increase HDL-cholesterol relative to carbohydrates, but the effect diminishes as the number of double bonds in the fatty acid molecule increases [21].

The effects in humans of dietary *trans* fatty acids on serum lipoprotein cholesterol levels have been studied also, though less extensively. The purpose of the present chapter, therefore, is to give an overview on the effects of *trans* fatty acids on serum lipid and lipoprotein concentrations and metabolism. Also, effects of *trans* fatty acids on other risk indicators for coronary heart disease will be discussed. The emphasis will be on *trans* monounsaturated fatty acids with eighteen carbon atoms and one double bond (*trans*-C18:1), because these are the most common in the diet [30].

B. *TRANS* FATTY ACIDS AND SERUM LIPIDS AND LIPOPROTEINS

1. Earlier Studies

Results of many earlier studies on the cholesterolemic effects of dietary *trans* fatty acids are difficult to interpret, because fatty acid intakes between the different dietary regimens were not very well balanced. For example, increased intakes of *trans* fatty acids were in general accompanied by higher intakes of saturated fatty acids and lower intakes of polyunsaturated fatty acids. This makes it impossible to fully attribute the observed effects to an increased intake of *trans* isomers.

Anderson and coworkers made a comparison between the effects on serum cholesterol and triglyceride concentrations of partially hydrogenated corn oil (PHCO), and those of a mixture of corn oil and olive oil [4]. *Trans* isomers in the PHCO-diet provided 14.7% of daily energy intake, 11% from *trans*-C18:1

TABLE 7.1.

Results of earlier studies on the effects of *trans* fatty acids on serum total cholesterol levels.

Ref. no.	First author	Hydrogenated oil in the diet	Conclusion
4	Anderson	Corn oil	Hypercholesterolemic
8	De Iongh	Soybean oil Whale oil	Very small effect
9	Erickson	Soybean oil	No effect
15	Laine	Soybean oil	Slight negative effect
17	Mattson	Soybean oil	No effect
18	McOsker	Soybean oil Cottonseed oil	No effect
32	Vergroesen	Olive oil	Moderate effect

and 3.7% from *trans* isomers of linoleic acid. When corrected for small differences in the intake of polyunsaturated fatty acids, the results demonstrated that replacement of each 1% of energy from carbohydrates in the diet by *trans* fatty acids increased serum cholesterol levels by about 0.031 mmol/L (1.2 mg/dL). This effect is about half that of the saturated fatty acids [14]. The diet high in *trans* fatty acids also increased serum triglyceride concentrations.

Later studies yielded contradictory results, however. Mattson and coworkers fed a group of 33 men for 21 days formula diets high in *cis* unsaturated fatty acids [17]. For the next 28 days, one group of 17 men continued on the same diet, while for the other men a part of the oleic acid in the diet was replaced by *trans* fatty acids. These *trans* isomers were provided by a transesterified, hydrogenated, soybean oil and provided 18% of daily energy intake, of which 14% was supplied by *trans*-C18:1. Cholesterol intake was 500 mg/day. At the end of the study, no differences in plasma total cholesterol or triglyceride concentrations between the two groups were observed.

Vergroesen and Gottenbos reported two studies in which the effects of *trans*-C18:1 on serum cholesterol levels were compared with those of oleic acid [32]. Liquid formula diets were fed for 20 or 28 days to healthy volunteers. Hydrogenated olive oil was used as a source for *trans*-C18:1, which provided 14% of daily energy intake. The results demonstrated that *trans*-C18:1 was only hypercholesterolemic when the cholesterol intake was 250 mg/day. When the diets were cholesterol-free, no effects on serum total cholesterol levels were observed. From this study it can be estimated that, at a cholesterol intake of 250 mg/day, each 1% of energy from *trans*-C18:1 increases the serum total cholesterol level with about 0.033 mmol/L (1.3 mg/dL). This estimate is very close to that obtained from the study of Anderson and coworkers [4].

Results from other studies were also not conclusive (Table 7.1). However, when the data from all these studies are pooled [4,8,9,17,18,32], it appears that dietary *trans* fatty acids increase the serum total cholesterol level. Table 7.2 shows the results of 4 different regression equations relating dietary *trans*-fatty

TABLE 7.2.

Estimated mean changes in serum total cholesterol concentrations when 1% of energy in the diet from carbohydrates diet is replaced isocalorically by *trans* fatty acids (carb $\Rightarrow$ *trans*), saturated fatty acids (carb $\Rightarrow$ sat), *cis*-monounsaturated fatty acids (carb $\Rightarrow$ mono) or *cis*-polyunsaturated fatty acids (carb $\Rightarrow$ poly). Results from the earlier studies (see Table 7.1).*

Equation	Change per percent of energy replaced
I	mmol/L $0.028 \times$ (carb $\Rightarrow$ *trans*) mg/dL $\quad 1.10 \times$ (carb $\Rightarrow$ *trans*) $\quad\quad\quad\quad P < 0.01$
II	mmol/L $0.029 \times$ (carb $\Rightarrow$ *trans*) $+ 0.029 \times$ (carb $\Rightarrow$ sat) mg/dL $\quad 1.11 \times$ (carb $\Rightarrow$ *trans*) $+ 1.12 \times$ (carb $\Rightarrow$ sat) $\quad\quad\quad\quad\quad P = 0.05 \quad\quad\quad\quad\quad\quad P < 0.01$
III	mmol/L $0.032 \times$ (carb $\Rightarrow$ *trans*) $+ 0.030 \times$ (carb $\Rightarrow$ sat) $+ 0.007 \times$ (carb $\Rightarrow$ mono) mg/dL $\quad 1.26 \times$ (carb $\Rightarrow$ *trans*) $+ 1.17 \times$ (carb $\Rightarrow$ sat) $\;\; + 0.29 \times$ (carb $\Rightarrow$ mono) $\quad\quad\quad\quad\quad P = 0.04 \quad\quad\quad\quad\quad\quad P < 0.01 \quad\quad\quad\quad\quad\quad P = 0.51$
IV	mmol/L $0.025 \times$ (carb $\Rightarrow$ *trans*) $+ 0.022 \times$ (carb $\Rightarrow$ sat) $- 0.008 \times$ (carb $\Rightarrow$ poly) mg/dL $\quad 0.96 \times$ (carb $\Rightarrow$ *trans*) $+ 0.87 \times$ (carb $\Rightarrow$ sat) $\;\; - 0.31 \times$ (carb $\Rightarrow$ poly) $\quad\quad\quad\quad\quad P = 0.01 \quad\quad\quad\quad\quad\quad P = 0.19 \quad\quad\quad\quad\quad\quad P = 0.49$

* The regression coefficient predicts the mean change in serum total cholesterol concentration for a group of subjects, when 1% of daily dietary energy intake as carbohydrates is replaced isocalorically by a particular fatty acid. For an "average" group of adult men or women with a daily energy intake of 10 MJ (2400 kcal), 1% of energy is equivalent to about 6 gram of carbohydrates or 2.7 gram of fatty acid.

acid intake to changes in serum total cholesterol levels. For equation 1, only *trans* fatty acids were put into the regression equation. The estimated change in the serum total cholesterol level, when 1% of dietary carbohydrates is replaced isocalorically by *trans* fatty acids, was 0.028 mmol/L (1.10 mg/dL). As it is possible that this effect is due to other fatty acids in the diet - for example, when high intakes of *trans* fatty acids are related to high intakes of saturated fatty acids, regression models that included other fatty acids were also evaluated. In this way, differences were corrected for intakes of other fatty acids. The estimate changed slightly depending on the other fatty acids in the model. However, the coefficient for *trans* fatty acids was statistically significant for each equation and close to 0.028 mmol/L (1.10 mg/dL), very similar to the estimate of Anderson and co-workers [4] and of Vergroesen and Gottenbos [32]. For an "average" group of adult men or women with a daily energy intake of 10 MJ (2400 kcal), 1% of energy is equivalent to about 6 gram of carbohydrates or 2.7 gram of fatty acid.

2. *Recent Studies*

The discussion about the effects of *trans* fatty acids on serum lipoproteins started again after the study of Mensink and Katan [20]. In that study, 25 healthy men and 34 healthy women consumed three mixed natural diets. The nutrient composition of the diets was identical, except that 10% of daily energy intake was provided by either a mixture of the cholesterol-raising saturated fatty acids (lauric, myristic, and palmitic acids), oleic acid, or *trans* isomers of oleic acid. These three diets were consumed by each subject in random order for three weeks. Compared with the oleic-acid diet, serum LDL-cholesterol concentrations increased by 0.37 mmol/L (14 g/dL) on the *trans*-fatty-acid diet and by 0.47 mmol/L (18 mg/dL) on the saturated-fat diet. Differences in LDL between the three diets were significantly different. Serum HDL-cholesterol levels decreased significantly by 0.17 mmol/L (6.6 mg/dL) on the *trans*-fatty-acid diet, but were the same on the saturated-fat and the oleic-acid diets. Thus, these results suggest that *trans*-C18:1 has an unfavorable effect on the serum lipoprotein profile. This study was criticized for several reasons, however. First of all, the amount of *trans* fatty acids in the diet exceeded those in common diets. Therefore, it is possible that *trans*-C18:1 does not have affect serum lipoprotein concentrations at lower intakes. Secondly, it can be argued that the *trans* fatty acids were not obtained, as in practice, by hydrogenation, but mainly by isomerization, which may have confounded the results. Indeed, a high-oleic-acid sunflower oil was treated in such a way that half of the oleic acid was converted into *trans*-C18:1, while during industrial hydrogenation most of the *trans*-C18:1 is formed from linoleic acid.

Zock and Katan therefore decided to examine the effects of *trans*-C18:1 at lower intakes [35]. In their study, three different diets were given to 26 men and 30 women, all normolipidemic and apparently healthy. The first diet provided 7.7% of daily energy intake from *trans*-C18:1, which was replaced by either stearic acid or linoleic acid. These three diets were given in random order for three weeks to each subject. Mean serum LDL-cholesterol concentrations were 3.07 mmol/L (119 mg/dL) on the *trans*-diet, and decreased to 3.00 mmol/L (116 mg/dL) on the stearate-diet, and to 2.83 mmol/L (109 mg/dL) on the linoleate-diet. HDL-cholesterol concentrations were 1.37 mmol/L (53 mg/dL) on the *trans*-diet, and increased to 1.41 mmol/L (55 mg/dL) on the stearate-diet, and to 1.47 mmol/L (57 mg/dL) on the linoleate-diet. Concentrations of LDL-cholesterol and HDL-cholesterol were significantly different on the linoleate-diet as compared with the other two diets. LDL-cholesterol levels on the *trans*-diet and the stearate-diet were similar, while the difference of 0.03 mmol/L (1.2 mg/dL) in HDL-cholesterol just did not reach statistical significance. Thus, this study showed that *trans*-C18:1 has an unfavorable effect on serum lipoproteins at lower intakes also. However, as in the first study [20], a high-oleic acid sunflower oil was used to make a fat high in *trans*-C18:1. Therefore, this study did not answer the question whether or not the source, and way of production of *trans*-C18:1 is important for its effects on serum lipoproteins.

Judd and colleagues examined the effects of *trans* monounsaturated fatty acids from hydrogenated vegetable oils [13]. Twenty-nine healthy women and 29 healthy men ate four different diets, each for six weeks in random order.

The first diet (oleic-acid diet) provided 16.7% of energy from oleic acid. For the other three diets, oleic acid was replaced by 3.1% or 6.0% of energy from *trans* fatty acids (Mod-*trans* and high-*trans* diets, respectively), or by 5.6% of energy from the cholesterol-raising saturated fatty acids (Sat-diet). *Trans*-C18:1 provided more than 97% of all *trans* fatty acids. The intake of other nutrients was kept constant. Compared with the oleic-acid diet, LDL-cholesterol concentrations increased by 0.20 mmol/L (7.7 mg/dL) on the mod-*trans* diet, by 0.26 mmol/L (10 mg/dL) on the high-*trans* diet, and by 0.30 mmol/L (12 mg/dL) on the sat-diet. HDL-cholesterol levels were slightly decreased by 0.02 mmol/L (0.8 mg/dL) and 0.04 mmol/L (1.5 mg/dL) on the mod-*trans* diet and the high-*trans* diet, and increased by 0.05 mmol/L (1.9 mg/dL) on the sat-diet. Thus, this study clearly demonstrated that *trans* isomers formed by hydrogenation of high-linoleic acid vegetable oils have similar effects on the serum lipoprotein profile as those formed by isomerization of high-oleic acid vegetable oils.

The effects of *trans*-C18:1 from a mixture of hydrogenated rapeseed oil plus palmolein was studied by Nestel and coworkers [24]. Twenty-seven mildly hypercholesterolemic men first received for a 2-week control period a diet, which resembled their habitual fat intake. Subsequently, a diet high in oleic acid was provided for 3 weeks. During the next two 3-week dietary periods, the men received in random order an elaidic acid-enriched diet or a palmitic acid-enriched diet. The fatty acids under study differed about 7% of daily energy intake between the diets. As compared with the oleic-acid diet, plasma LDL-cholesterol concentrations were 0.36 mmol/L (14 mg/dL) higher on the elaidic-acid diet and 0.26 mmol/L (10 mg/dL) on the palmitic-acid diet. Thus, this study suggests that the LDL-cholesterol increasing effect of *trans*-C18:1 is similar to that of palmitic acid. Both the oleic-acid and the elaidic-acid diets lowered HDL-cholesterol by 0.10 mmol/L (4 mg/dL) relative to the palmitic-acid diet.

Lichtenstein *et al.* investigated in 14 men and women with moderate hypercholesterolemia the effects on plasma lipoproteins of replacing corn oil with a margarine containing hydrogenated corn oil [16]. Energy intake from *trans*-C18:1 was increased by 3.7%, mainly at the expense of oleic and linoleic acids. LDL-cholesterol concentrations were, nearly significantly, increased by 0.26 mmol/L (10 mg/dL) on the margarine-diet, while no effects were observed on HDL-cholesterol concentrations.

The effects of partially hydrogenated fish oil (PHFO) on serum lipid and lipoprotein concentrations were examined by Almendingen and coworkers [2]. Hydrogenated fish oil contains a complex mixture of *trans*-monoenes and *trans*-polyenes with 18, 20 and 22 carbon atoms. Thirty-one healthy young men consumed three test diets, each for three weeks in random order. For the first diet, a margarine was produced from butterfat, while the two other fats were made from either partially hydrogenated fish oil or from partially hydrogenated soybean oil (PHSO). *Trans*-monoenes contributed respectively 0.9, 8.4 and 6.6% to total energy intake. Consumption of *trans*-polyenes was

TABLE 7.3.

Sources of *trans* fatty acids in the recent studies.

Ref. no.	First author	Source of *trans* fatty acids in the diet	% of energy from *trans*
13	Judd	Hydrogenated vegetable oils	3.7% and 6.4%
16	Lichtenstein	Hydrogenated corn oil	4.2%
20	Mensink	Hydrogenated and isomerized high-oleic acid sunflower oil	10.9%
24	Nestel	Hydrogenated rapeseed/palmolein oils	7.0%
35	Zock	Hydrogenated and isomerized high-oleic acid sunflower oil	7.7%

1.5% of total energy intake on the PHFO-diet, and negligible on the other two diets. Thus, as expected, the PHFO-diet did not only contain more *trans*-polyenes, but also more *trans*-monoenes with longer chain-lengths as compared with the PHSO-diet. The serum LDL-cholesterol concentration was significantly elevated with 0.36 mmol/L (14 mg/dL) and 0.23 mmol/L (9 mg/dL) on the PHFO-diet as compared with respectively the PHSO-diet and the butter-fat diet. The difference in LDL-cholesterol concentrations of 0.13 mmol/L (5 mg/dL) between the latter two diets was not significant. HDL-cholesterol levels were significantly decreased by 0.07 mmol/L (2.7 mg/dL) on the PHFO-diet. These results suggest that *trans* fatty acids from hydrogenated vegetable and fish oils may be different. It was concluded, however, that both hydrogenated oils are not advisable alternatives for the cholesterol-raising butter-fat diet.

3. Meta-Analysis

In general, recent studies indicate that *trans*-C18:1 increases LDL-cholesterol concentration as compared with oleic and/or linoleic acids, two *cis* unsaturated fatty acids. Effects on HDL-cholesterol are less consistent: some studies found no effects, while other studies showed a decrease, especially at higher intakes of *trans*-C18:1. To calculate the quantitative effects of *trans*-C18:1, the data of these studies were combined by multiple regression analysis [37]. As in previous meta-analyses [14,21], carbohydrates were chosen as point of reference. This means that the effects of carbohydrates on serum lipids and lipoproteins was set at zero. The study of Almendingen *et al.* [2], although very well-controlled, was not included, because the diets differed to a considerable extent in myristic-acid intake. Some studies [36], but not all [29], have suggested that this saturated fatty acid may have a very pronounced effect on serum LDL- and HDL-cholesterol concentrations. Thus, as long as the quantitative effects of myristic acid on serum lipoproteins are still uncertain, it is not possible to adjust differences between the diets in serum lipoprotein concentrations for differences in myristic-acid intake.

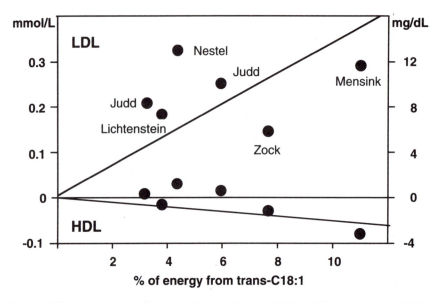

Fig. 7.1. Effects of exchanging dietary carbohydrates for *trans*-C18:1 on LDL-cholesterol and HDL-cholesterol concentrations. Data are derived from five recent studies [13,16,20,24,35]. The regression coefficient is 0.034 mmol/L (1.3 mg/dL) for LDL-cholesterol, which depicts the predicted change in serum LDL-cholesterol for a group of subjects when 1% of daily energy intake from carbohydrates is replaced by *trans*-C18:1. The regression coefficient for HDL-cholesterol was -0.004 mmol/L (-0.1 mg/dL).

Some characteristics of the studies are presented in Table 7.3. Daily energy intake from *trans*-C18:1 ranged between 3.7 to 10.9%. The source and way of production of *trans*-C18:1 was very different between the studies, which makes it probable that each fat had his own characteristic distribution of positional *trans*-C18:1 isomers. Despite these differences, when all the data were combined into a linear regression model, a clear relationship was found between the energy intake from *trans*-C18:1 and LDL-cholesterol concentrations (Figure 7.1). As can be seen, none of the data points was extremely deviant, which suggests that the position of the double bond and the way of production is not very important for the hypercholesterolemic effect of *trans*-C18:1. It was found that each increment of 1% of dietary energy as *trans*-C18:1 at the expense of carbohydrates results in an increase in LDL cholesterol concentrations of 0.034 mmol/L (1.3 mg/dL). The regression coefficient for HDL was very close to zero (-0.004 mmol/L or -0.1 mg/dL), which suggests that the effects of *trans*-C18:1 on HDL are very similar to those of carbohydrates.

The results were subsequently combined with those of a previous meta-analysis that described the effects of a mixture of saturated fatty acid, oleic acid and linoleic acid on serum lipid and lipoprotein concentrations [21]. Figure 7.2 shows, that the effects of *trans*-C18:1 on LDL cholesterol concentrations are very similar to that of a mixture of saturated fatty acids. Both *cis* unsaturated fatty acids have an LDL cholesterol-lowering effect

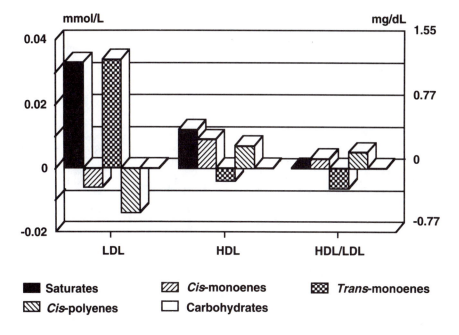

Fig 7.2. Bar graphs showing predicted changes in serum LDL-cholesterol and HDL-cholesterol concentrations and on the HDL to LDL cholesterol ratio when 1% of energy as carbohydrates is replaced by fatty acids of a particular class.

relative to carbohydrates, the effect of linoleic acid being the strongest. The results for HDL show that *trans*-C18:1 is the only fatty acid that will not elevate HDL cholesterol when it replaces carbohydrates in the diet. As some studies suggest that risk for coronary heart disease is negatively related to the HDL to LDL cholesterol ratio [33], effects of the various dietary fatty acids on this ratio were also calculated. The HDL to LDL cholesterol ratio did not change if saturates were replaced by carbohydrates, but it decreased if carbohydrates were replaced by *trans*-C18:1. Replacement of carbohydrates by *cis* unsaturated fatty acids increased the HDL to LDL cholesterol ratio. Thus, these results suggest that the effects of *trans*-C18:1 on the serum lipoprotein profile are at least as bad as that of a mixture of saturated fatty acids.

4. Lipoprotein[a]

Lipoprotein[a] (Lp[a]) is a macromolecular complex that closely resembles the composition of LDL, but has an extra glycoprotein, the so-called apoprotein[a]. The Lp[a] concentration in the blood is largely under genetic control and does not change much with age [7,25]. Most subjects have Lp[a] levels below 150 mg/L, but levels in some may well exceed 400 mg/L [7,25]. Such subjects have a markedly increased risk for coronary heart disease, a relationship that is not confounded by serum LDL- or HDL-cholesterol concentration [25,26]. Despite the structural similarity between

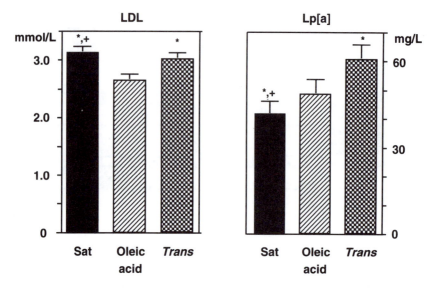

Fig. 7.3. Effects of dietary *trans*-C18:1 on serum LDL-cholesterol and Lp[a] concentrations [20,22]. Subjects received for 3-week periods, diets enriched in either 11% of energy from the cholesterol-raising saturated fatty acids (lauric, myristic, and palmitic acids), or from oleic acid, or from *trans*-C18:1. * P < 0.020 as compared with the oleic-acid diet; + P < 0.020 as compared with the *trans*-diet.

Lp[a] and LDL, Lp[a] concentrations are, as opposed to LDL-cholesterol levels, hardly affected by dietary changes. However, several recent studies have indicated that dietary *trans* fatty acids increase Lp[a] concentrations.

Hornstra *et al.* have shown that in healthy normocholesterolemic men replacement of the habitual fat in the Dutch diet with palm oil resulted in a significant decrease in serum Lp[a] concentrations [12]. Although the fatty-acid composition of the diets differed in many aspects, it was speculated that the more than 50% reduction in the consumption of *trans* fatty acids on the palm-oil enriched diet may have been responsible for the decrease in Lp[a] levels. It was also reported that the change in Lp[a] was dependent on the initial Lp[a] levels.

Nestel and coworkers found that after consumption of 7% of energy from elaidic acid, Lp[a] levels were 296 units/L and decreased significantly to 249 units/L on a palmitic acid-rich diet, and to 236 units/L on an oleic acid-rich diet [24]. The difference between the *trans* and oleic acid-diet failed to reach statistical significance.

Mensink *et al.* measured Lp[a] concentrations in serum samples from healthy normocholesterolemic volunteers after consumption for three weeks of diets enriched in either 11% of energy from the cholesterol-raising saturated fatty acids (lauric, myristic, and palmitic acids), oleic acid or *trans*-C18:1 [22]. The median level of Lp[a] was the lowest on the saturated fatty acid diet, increased on the oleic acid diet and was the highest on the *trans*-C18:1 diet. The difference in Lp[a] between the three diets was highly significant. As shown in Figure 7.3, the diets affected mean Lp[a] and LDL-cholesterol concentrations differently. This demonstrates that dietary effects on these two lipoprotein particles are controlled by different mechanisms. In a second study, subjects received 8% of energy from the saturated fatty acid stearic acid, or from linoleic acid, or from trans-C18:1 for three weeks each. Median Lp[a] levels were 69 mg/L on both the stearate-diet and linoleate-diet, and increased significantly to 85 mg/L on the trans-C18:1 diet. In both studies, changes in Lp[a] were positively related to initial levels, which suggests that subjects with high Lp[a] levels would benefit most from a reduced intake of trans-C18:1.

In a study carried out by Lichtenstein and co-workers, effects on Lp[a] concentrations were examined when corn oil in the diet was replaced by a corn-oil margarine [16]. Although the margarine diet provided 3.7% of energy more from trans-C18:1, plasma Lp[a] concentrations were not different between the diets. However, the margarine-diet also contained some saturated fatty acids, which might have partly counteracted the Lp[a]-increasing effect of trans-C18:1 [22]. As discussed, changes are most pronounced in subjects with relatively high Lp[a] levels, and only subjects with relatively low Lp[a] concentrations were involved in the study [16]. Therefore, the statistical power of this study may have been too low to demonstrate any effect of diet on Lp[a]. Results are therefore not necessarily in disagreement with those of other studies.

Finally, Almendingen et al. demonstrated that the adverse effects of trans fatty acids on Lp[a] are not limited to those from hydrogenated vegetable oils: trans fatty acids produced by hydrogenation of fish oil also increased Lp[a] relative to saturated fatty acids [2].

C. MECHANISMS

Although most studies have shown that, relative to oleic acid, trans fatty acids increase LDL-cholesterol concentrations and lower those of HDL-cholesterol, the mechanisms of these effects are - as for other dietary fatty acids - still obscure.

1. LDL-Receptor Activity

The steady-state concentration of LDL-cholesterol is determined by four variables: the LDL-production rate, its receptor-dependent and receptor-independent uptakes, and the affinity of the LDL-particle for its receptor.

Dietschy and colleagues have now hypothesized [27], after a series of studies in the hamster, that dietary fatty acids affect the concentration of LDL mainly by changing the rate of the receptor-dependent uptake of this particle by the liver (Figure 7.4). As a secondary consequence, the rate of LDL-production will be influenced, because the LDL-receptor is also involved in the uptake of intermediate-density lipoproteins (IDL), a lipoprotein that can be converted into LDL and is formed from VLDL. Thus, when the LDL-receptor is down-regulated, the uptake of IDL is also decreased. The production rate of

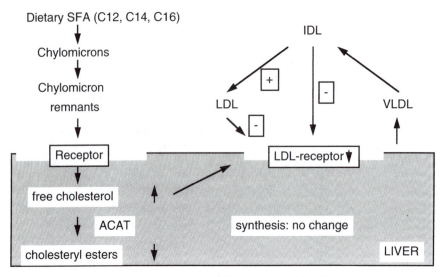

Fig. 7.4. Simplified scheme for the effects of the LDL-cholesterol-raising dietary saturated fatty acids (SFA) on LDL-metabolism.

After absorption, dietary saturated fatty acids are transported through the blood by chylomicrons, which are converted into chylomicron remnants. These particles are taken up by the liver and the fatty acids become a substrate for the enzyme acyl-CoA:cholesterol acyltransferase (ACAT), which catalyses the formation of cholesteryl esters from free cholesterol plus a fatty acid. As this enzyme has a low affinity for the cholesterol-raising saturated fatty acids, the concentration of free cholesterol in the liver increases. This will result in a decreased LDL-receptor activity, increased conversion of IDL into LDL, and into higher LDL-cholesterol concentrations. VLDL-output will not change.

This model has been postulated by Dietschy and colleagues [27] after a series of studies in the hamster. It is not know to what extent findings from this animal model may be extrapolated to man.

LDL will then increase, as increased amounts of IDL will then be converted into LDL. The receptor-independent uptake of LDL and the affinity of the LDL-particle for its receptor are not sensitive to changes in dietary fatty-acid intake.

The regulation of the hepatic LDL-receptor activity is controlled by a putative regulatory pool of free cholesterol. At a constant intake of cholesterol, the size of this pool is mainly determined by the fatty-acid composition of the diet. Fatty acids from the diet that reach the liver can be esterified to free cholesterol by the enzyme acyl-CoA:cholesterol acyltransferase (ACAT). If the affinity of the enzyme for a specific fatty acid is low, then the concentration of free cholesterol in the liver increases. This will result in a decreased LDL-receptor activity, increased conversion of IDL into LDL, and thus into higher LDL-cholesterol levels. This hypothesis is supported by animal studies.

In one of the studies by the group of Dietschy, hamsters were fed for four weeks diets enriched in oleic acid, *trans*-C18:1, myristic acid (C14:0) or octanoic acid (C8:0) [34]. The group of animals that were fed the C8:0-diet were considered as the control group, because octanoic acid is very rapidly oxidized and hence does not alter the fatty-acid profile of the liver. Myristic acid, a very potent cholesterol-raising fatty acid, was used as a positive

TABLE 7.4.

Effects of feeding four different dietary fatty acids on plasma LDL-cholesterol concentration, relative hepatic LDL-receptor activity, and relative LDL-cholesterol production rate in the hamster [34].*

	Unit	C8:0	C14:0	Oleic acid	*Trans*-C18:1
Plasma LDL-cholesterol	mmol/L	2.09	5.02	1.03	1.94
	mg/dL	80	194	40	75
LDL-receptor activity	%	100	63	126	100
LDL-cholesterol production	%	100	118	75	100

* Hepatic LDL-receptor activity and LDL-cholesterol production rate are expressed relative to levels on the C8:0-diet.

control. Some of the results of this study are summarized in Table 7.4. As hypothesized, changes in plasma LDL-cholesterol were negatively related to changes in LDL-receptor activity and positively to the LDL-cholesterol production rate. Cholesterol synthesis, dietary cholesterol absorption, and cholesterol delivery to the liver were not affected by any of the four diets. Compared with the oleic-acid diet, the *trans*-C18:1-diet decreased the concentration of cholesteryl esters in the liver. These findings suggest that the concentration of the putative regulatory free cholesterol pool in the liver was increased on the *trans*-C18:1 diet. This would have caused the decreased LDL-receptor activity, increased LDL-production, and finally the increased plasma LDL-cholesterol concentrations in the animals fed the *trans*-C18:1 diet as compared with those on the oleic-acid diet. However, it is still unknown to what extent findings from this animal model may be extrapolated to man.

2. Endogenous Cholesterol Synthesis

Cuchel *et al.* have examined in eleven subjects the effects of replacing corn oil in the diet for a corn oil margarine in stick form on endogenous cholesterol synthesis [6]. Consumption of the margarine resulted in an increased intake of almost 4% of energy from *trans*-C18:1, mainly at the expense of *cis* unsaturated fatty acids. During the final week of each dietary phase, which lasted 32 days, cholesterol fractional and absolute synthetic rates (C-FSR and C-ASR, respectively) were determined. Total cholesterol concentrations increased from 5.01 mmol/L on the corn-oil diet to 5.30 mmol/L on the margarine-diet. On the corn-oil diet, the C-FSR was on average 0.0668 pools/day and decreased to 0.0466 pool/day on the margarine-diet. This difference, however, failed to reach statistical significance. The C-ASR was on average 1.7101 g/day on the oil diet and 1.1761 g/day on the margarine-diet. Again, this difference was not statistically significant. It was therefore concluded that the increase in cholesterol concentration after margarine consumption was not due to increased endogenous cholesterol synthesis, but at least partly, by a decreased catabolic rate of cholesterol. This variable was not measured, however, while the results do to some extent suggest that dietary

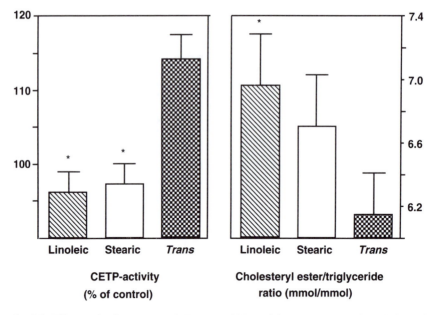

Fig. 7.5. Effects of dietary *trans*-C18:1 on CETP-activity and the molar cholesteryl ester/triglyceride ratio in HDL [31]. Subjects received for 3-week periods, diets enriched in either 8% of energy 'from the saturated fatty acid stearic acid, or from linoleic acid, or from *trans*-C18:1. * P < 0.020 as compared with the *trans*-diet.

fatty acids may have affected endogenous cholesterol synthesis. If so, this would not be in agreement with the model from the hamster studies proposed by Dietschy and coworkers [27]. Further studies in this area are therefore required to delineate by which mechanisms *trans* and other dietary fatty acids change LDL-metabolism in man.

3. Cholesteryl Ester Transfer Protein

Cholesteryl ester transfer protein (CETP) catalyses the transfer of plasma cholesteryl esters from HDL to apolipoprotein B-containing lipoproteins such as VLDL and LDL. During this process, the cholesteryl esters are exchanged for triglycerides. As dietary *trans* fatty acids simultaneously increase LDL-cholesterol and reduce HDL-cholesterol concentrations, some investigators have hypothesized that CETP might be involved. Indeed, Abbey and Nestel have demonstrated that the activity of CETP in plasma was on average 24% higher when 6% of energy from oleic acid in the diet was replaced by *trans*-C18:1 [1]. In addition, changes in plasma CETP-activity were positively related to the proportion of *trans*-C18:1 in plasma. Also, plasma CETP-activity was inversely correlated with HDL-cholesterol concentrations after the elaidic acid-rich diet.

Van Tol *et al.* have also found that mean CETP-activity was about 18% higher when the intake of *trans*-C18:1 was increased to 8% of energy at the expense of either linoleic or stearic acids [31]. Analysis of the lipid composition in HDL showed that the increased CETP-activity on the *trans*-diet was accompanied by a lower molar ratio of cholesteryl esters to triglycerides in HDL (Figure 7.5). This indicates that the high plasma CETP-activity on the *trans*-diet indeed affected the exchange of lipid transfer reactions between lipoproteins *in vivo*.

D. OTHER RISK INDICATORS

Most studies have focused on the effects of *trans* fatty acids on serum lipids and lipoproteins. A few studies have also examined the effects of hydrogenated fats on other risk indicators for coronary heart disease.

1. LDL Oxidation

As *trans* fatty acids are not synthesised by the human body, a higher dietary intake will be reflected by a higher proportion of these fatty acids in the LDL-particle. Nestel *et al.* have examined if this made the LDL-particle more susceptible to oxidation [24], because oxidatively modified LDL is more atherogenic than native LDL [28]. However, it was reported that *trans*-C18:1 did not affect the susceptibility of LDL to oxidation relative to oleic or palmitic acids. Cuchel *et al.* found that consumption of hydrogenated corn oil did not change the susceptibility of LDL to oxidation either [6]. Therefore, it seems that *trans*-C18:1 does not have an important impact on LDL-oxidizability.

2. Hemostasis

Arterial thrombus formation is the resultant of the balance between several processes: platelet aggregation, coagulation, and fibrinolysis. A stable thrombus is formed by aggregation and coagulation, while fibrinolysis is an important determinant for dissolution of the thrombus. These processes are difficult to measure in man, because measurements are generally made *in vitro* and in venous blood samples. Also, many different methods are being used, which makes it difficult to compare the results between studies. Finally, many assays measure the total amount of circulating coagulation or fibrinolytic factors, which are mainly present in the plasma in a non-activated form. However, despite these problems to quantify arterial thrombosis tendency, the importance of platelet aggregation, coagulation and fibrinolysis is stressed by the results of many epidemiological studies. For example, there are strong indications that increased concentrations of the coagulation factors fibrinogen and factor-VII, and of plasminogen activator inhibitor type 1 (PAI-1), a fibrinolytic factor, are associated with increased risk for coronary heart disease [11,19]. Although it is known that hemostatic variables can be modified by dietary lipids, only a few investigators have examined the effects of isomeric fatty acids on platelet function.

Mutanen and colleagues have compared the effects on *in vitro* aggregation of platelet-rich plasma from healthy subjects after consumption for 5 weeks of a diet enriched in *trans*-C18:1 or stearic acid (Mutanen M., personal communication). It was demonstrated that collagen-induced platelet aggregation was significantly decreased after the *trans*-diet. A similar pattern was observed when ADP was used as an agonist to induce platelet aggregation, although the difference between the two diets failed to reach statistical significance. Thus, this study suggests that *trans*-C18:1 has a favorable effect on *in vitro* platelet aggregation as compared with stearic acid. However, the relevance of these findings for the in vivo situation, as well as the effects on platelet aggregation of *trans* fatty acids relative to those of other fatty acids, remains to be determined.

The effects of *trans* isomers from *n*-3 polyunsaturated fatty acid have been examined by Chardigny and coworkers [5]. Human platelets were enriched with eicosapentaenoic or docosahexaenoic acids (EPA and DHA, respectively) or their respective $\Delta 17$ or $\Delta 19$ *trans*-isomers. Platelets were then stimulated with collagen or with U-46619, a stable thromboxane A_2 analogue. Results indicated that collagen-induced platelet aggregation was increased by *trans n*-3 polyunsaturated fatty acids. No effects were observed when U46619 was used to induce aggregation.

Almendingen *et al.* have found that a diet enriched in PHSO increased the concentration of PAI-1 as compared with a diet high in PHFO or butter fat [3]. Fibrinogen levels were slightly increased on the butter diet as compared with the PHFO-diet. No effects on the levels of factor VII were observed. The PHSO-diet had an adverse effect on PAI-1, however. Thus, this study suggests that the butter-diet may be pro-coagulant relative to the PHFO-diet, and that the PHSO-diet may have unfavorable effects on the fibrinolytic system. Mutanen and coworkers found no differences in coagulation and fibrinolytic factors after consumption of a diet high in stearic acid or *trans*-C18:1 [23], however. Therefore, more controlled studies are needed before any definite conclusions can be drawn about the effects of dietary *trans* fatty acids on hemostatic variables.

E. CONCLUSIONS

The effects of dietary *trans* fatty acids on serum cholesterol and lipoprotein concentrations have been investigated for many decades. Results of the earlier studies that only examined effects on serum total cholesterol levels were not conclusive. However, when the data from all these studies are combined, the estimated rise in the serum total cholesterol level is about 0.028 mmol/L (1.10 mg/dL) for each 1% of dietary carbohydrates that is replaced isocalorically by *trans* fatty acids.

More recent studies show that *trans*-C18:1 increases concentrations of LDL-cholesterol particularly. It is estimated that replacing 1% of energy from carbohydrates by *trans*-C18:1 raises the LDL-cholesterol level by 0.034 mmol/L (1.3 mg/dL). *Trans*-C18:1 does not have a pronounced effect on HDL-cholesterol as compared with carbohydrates. However, as carbohydrates lower HDL-cholesterol relative to oleic acid [21], it can be deduced that *trans*-C18:1

lowers HDL-cholesterol concentrations as compared with their *cis*-isomer oleic acid (Figure 7.2). The mechanisms of these effects are still unknown, as for other dietary fatty acids, although part of the effects of *trans*-C18:1 on the serum lipoprotein profile might be explained by an increased CETP-activity. Also, *trans*-C18:1 has an adverse effect on the atherogenic Lp[a].

Some studies have addressed whether *trans* fatty acids modify other risk indicators for coronary heart disease. It was found that *trans*-C18:1 does not affect the oxidizability of LDL and has no beneficial effects on hemostatic variables. Therefore, diets for the prevention of coronary heart disease should not only be low in cholesterol and saturated fatty acids, but also in *trans* fatty acids, as will be discussed at length in the next chapter.

ABBREVIATIONS

C-ASR, Cholesterol absolute synthetic rate; C-FSR, Cholesterol fractional synthetic rate; CETP, Cholesteryl ester transfer protein; DHA, Docosahexaenoic acid; EPA, Eicosapentaenoic acid; HDL, High-density lipoproteins; IDL, Intermediate-density lipoproteins; LDL, Low-density lipoproteins; Lp[a], Lipoprotein[a]; PAI-1, Plasminogen activator inhibitor type 1; PHCO, Partially hydrogenated corn oil; PHFO, Partially hydrogenated fish oil; PHSO, Partially hydrogenated soybean oil; VLDL, Very-low density lipoproteins.

REFERENCES

1. Abbey,M. and Nestel,P.J., *Atherosclerosis*, **106**, 99-107 (1994).
2. Almendingen,K., Jordal,O., Kierulf,P., Sandstad,B. and Pedersen,J.I., *J. Lipid Res.*, **36**, 1370-1384 (1995).
3. Almendingen,K., Seljeflot,I., Sandstad,B. and Pedersen,J.I., *Arterioscler. Thromb. Vasc. Biol.*, **16**, 375-380 (1996).
4. Anderson,J.T., Grande,F. and Keys,A., *J. Nutr.*, **75**, 388-394 (1961).
5. Chardigny,J.M., Sébédio,J.L., Juanéda,P., Vatèle,J.M. and Grandgirard,A., *Nutr. Res.*, **15**, 1463-1471 (1995).
6. Cuchel,M., Schwab,U.S., Jones,P.J.H., Vogel,S., Lammi-Keefe,C., Li,Z., Ordovas,J., McNamara,J.R., Schaefer,E.J. and Lichtenstein,A.H., *Metabolism*, **45**, 241-247 (1996).
7. Dahlén,G.H., in *Lipoprotein[a]*, pp. 151-173 (1990) (edited by A.M. Scanu, Academic Press Inc., San Diego, CA).
8. De Iongh,H., Beerthuis,R.K., den Hartog,C., Dalderup,L.M. and van der Spek,P.A.F., *Biblthca. Nutr. Dieta.*, **7**, 137-152 (1965).
9. Erickson,B.A., Coots,R.H., Mattson,F.H. and Kligman,A.M., *J. Clin. Invest.*, **43**, 2017-2025 (1964).
10. Gordon,D.J. and Rifkind,B.M., *New Engl. J. Med.*, **321**, 1311-1316 (1989).
11. Hamsten,A., De Faire,U., Walldius,G., Dahlén,G., Szamosi,A., Landou,C., Blombäck,M. and Wiman,B., *Lancet*, **2**, 3-9 (1987).
12. Hornstra,G., van Houwelingen,A.C., Kester,A.D.M. and Sundram,K., *Atherosclerosis*, **90**, 91-93 (1990).
13. Judd,J.T., Clevidence,B.A., Muesing,R.A., Wittes,J., Sunkin,M.E. and Podczasy,J.J., *Am. J. Clin. Nutr.*, **59**, 861-868 (1994).
14. Keys,A., Anderson,J.T. and Grande,F., *Metabolism*, **14**, 776-784 (1965).
15. Laine,D.C., Snodgrass,C.M., Dawson,E.A., Ener,M.A., Kuba,K. and Frantz,I.D., *Am. J. Clin. Nutr.*, **35**, 683-690 (1982).

16. Lichtenstein,A.H., Ausman,L.M., Carrasco,W., Jenner,J.L., Ordovas,J.M. and Schaefer,E.J., *Arterioscler. Thromb.*, **13**, 154-161 (1993).
17. Mattson,F.H., Hollenbach,E.J. and Kligman,A.M., *Am. J. Clin. Nutr.*, **28**, 726-731 (1975).
18. McOsker,D.E., Mattson,F.H., Sweringen,B.H. and Kligman,A.M., *JAMA*, **180**, 380-385 (1962).
19. Meade,T.W., Ruddock,V., Chakrabarti,R. and Miller,G.J., *Lancet*, **342**, 1076-1079 (1993).
20. Mensink,R.P. and Katan,M.B., *New Engl. J. Med.*, **323**, 439-445 (1990).
21. Mensink,R.P. and Katan,M.B., *Arterioscler. Thromb.*, **12**, 911-919 (1992).
22. Mensink,R.P., Zock,P.L., Katan,M.B. and Hornstra,G., *J. Lipid Res.*, **33**, 1493-1501 (1992).
23. Mutanen,M. and Aro,A., *Thromb. Haemost.*, **77**, 99-104 (1997).
24. Nestel,P., Noakes,M., Belling,B., McArthur,R., Clifton,P., Janus,E. and Abbey,M., *J. Lipid Res.*, **33**, 1029-1036 (1992).
25. Sandkamp,M., Funke,H., Schulte,H., Köhler,E. and Assmann,G., *Clin. Chem.*, **36**, 20-23 (1990).
26. Seed,M., Hoppichler,F., Reaveley,D., McCarthy,S., Thompson,G.R., Boerwinkle,E., and Utermann,G., *New Engl. J. Med.*, **332**, 1494-1499 (1990).
27. Spady,D.K., Woollett,L.A. and Dietschy,J.M., *Annu. Rev. Nutr.*, **13**, 355-381 (1993).
28. Steinberg,D., Parthasarathy,S., Carew,T.E., Khoo,J.C. and Witzum,J.L., *New Engl. J. Med.*, **320**, 915-924 (1989).
29. Tholstrup,T., Marckmann,P., Jespersen,J., Vessby,B., Jart,A. and Sandström,B., *Am. J. Clin. Nutr.*, **60**, 919-925 (1994).
30. Van den Reek,M.M., Craig-Schmidt,M.C., Weete,J.D. and Clark,A.J., *Am. J. Clin. Nutr.*, **43**, 530-537 (1986).
31. Van Tol,A., Zock,P.L., van Gent,T., Scheek,L.M. and Katan,M.B., *Atherosclerosis*, **115**, 129-134 (1995).
32. Vergroesen,A.J. and Gottenbos,J.J., in *The role of fats in human nutrition*, pp. 1-41 (1975) (edited by A.J. Vergroesen, Academic Press, New York).
33. Watts,G.F., Lewis,B., Brunt,J.N., Lewis,E.S., Coltart,D.J., Smith,L.D., Mann,J.I. and Swan,A.V., *Lancet*, **339**, 563-569 (1992).
34. Woollett,L.A., Daumerie,C.M. and Dietschy,J.M., *J. Lipid Res.*, **35**, 1661-1673 (1994).
35. Zock,P.L. and Katan,M.B., *J. Lipid Res.*, **33**, 399-410 (1992).
36. Zock,P.L., de Vries,J.H.M. and Katan,M.B., *Arterioscler. Thromb.*, **14**, 567-575 (1994).
37. Zock,P.L., Mensink,R.P. and Katan,M.B., *Am. J. Clin. Nutr.*, **61**, 617 (1995).

CHAPTER 8

EPIDEMIOLOGICAL STUDIES OF TRANS FATTY ACIDS AND CORONARY HEART DISEASE

Antti Aro, MD

Department of Nutrition, National Public Health Institute, FIN-00300 Helsinki, Finland[1]

A. Introduction
B. Review of Individual Studies
    1. Case-control studies of fatal coronary heart disease
    2. Case-control studies of non-fatal coronary heart disease
    3. Studies of coronary artery disease diagnosed by angiography
    4. Cohort studies
    5. Other studies
C. Assessment of *Trans* Fatty acid Intake
    1. Biochemical methods
    2. Nutritional methods
D. Association between *Trans* Fatty Acids and Coronary Heart Disease
    1. Case-control studies
    2. Cohort studies
    3. Effect of different *trans* isomers
E. Summary and Conclusions

A. INTRODUCTION

Dietary isomeric *trans* fatty acids from hydrogenated vegetable and fish oils affect serum lipoproteins in a way which is supposed to increase the risk of coronary heart disease (CHD) [2,3,26,29]. Observational studies on the associations between dietary *trans* fatty acids and CHD have not produced convincing evidence on the putative harmful effects of these fatty acid isomers, however, and recent reports of British and US working groups concluded that data are equivocal and additional research is needed to resolve questions

[1]Fax: +358 9 47 44 85 91, E-mail: Antti.Aro@ktl.fi
Supported by a grant from the Research Council for Health, Academy of Finland

about the independent effects and mechanism of action of *trans* fatty acids [9,13].

Particular difficulties are caused in the studies of dietary *trans* fatty acids by the fragmentary knowledge of the amounts of *trans* fatty acids in foods, the tendency of subjects with CHD to change their dietary habits, either unintentionally or by prescription, and the recent activity of the food industry to change the fat composition of margarines in Europe [42]. The epidemiology of *trans* fatty acids and CHD has been recently reviewed by Allison [1], particularly focusing on statistical aspects. In the following, the epidemiological studies on associations between measures of *trans* fatty acid intake and risk of CHD are reviewed with particular emphasis on the effects of study design and the assessment of *trans* fatty acid intake on the results and conclusions.

B. REVIEW OF INDIVIDUAL STUDIES

1. Case-Control Studies of Fatal Coronary Heart Disease

Thomas *et al.* (1983) studied postmortem adipose tissue samples from 136 subjects who had died of ischaemic heart disease and 95 control subjects who had died of other causes, from different areas of England and Wales (see Table 8.1) [38]. Fatty acids were analysed by GLC using a 40-ft packed column. Cases had somewhat higher proportions of C16-18 *trans* fatty acids but the case-control difference was statistically significant for C16:1 *trans* fatty acids only. No case-control differences in C20-22 fatty acids were observed. The results were not adjusted for putative confounding factors. It was discussed that the use of margarines containing hydrogenated fats differed between different areas in the UK, but it was not reported how cases and controls were distributed within the areas.

In a later study in Scotland with a similar analytical method, postmortem adipose tissue samples from 27 subjects who had died of ischaemic heart disease and 27 subjects who had died of unrelated causes were compared [37]. Again cases showed higher proportions of C16:1 *trans* fatty acids compared with controls but the differences in longer-chain fatty acids were non-significant.

Roberts *et al.* (1995) analysed postmortem adipose tissue samples from 66 cases of sudden cardiac death from Southampton in southern England and compared them with samples from 286 healthy control subjects, matched for age and gender, and living in the same area [30]. Only 18-carbon fatty acids were analysed. Cases had lower C18 *trans* fatty acids compared with control subjects, due to a lower proportion of C18:1 *trans* fatty acids. By univariate analysis there was a significant inverse association between risk of sudden cardiac death and C18:1 *trans* fatty acids divided into quintiles but not between risk of sudden cardiac death and C18:2 *trans* fatty acids. By multivariate analysis, including age, cigarette smoking, treated hypertension, diabetes, oleic

TABLE 8.1.

Studies on associations between *trans* fatty acids and coronary heart disease.

Study	Subjects	Measure of *trans* fatty acid intake	Main results
1. Case-control studies of fatal coronary heart disease			
Thomas *et al.* 1983 [38]	Cases: 136 British men who died of CHD; Controls: 95 British men who died of other causes	Analysis of fatty acids in post-mortem adipose tissue samples	C16:1 *trans* fatty acids higher in cases
Thomas & Winter 1987 [37]	Cases: 27 British men who died of CHD; Controls: 27 British men who dies of other causes	Analysis of fatty acids in post-mortem adipose	C16:1 *trans* fatty acids higher in cases
Roberts *et al.* 1995 [30]	Cases: 66 British men who died of CHD; Controls: 286 healthy living British men	Analysis of C18 fatty acids in adipose tissue samples	C18:1 *trans* fatty acids lower in cases. No difference in C18:2 *trans*
2. Case-control studies of non-fatal coronary heart disease			
Thomas *et al.* 1995 [36]	Cases: 59 British men with ischaemic changes in electrocardiogram; Controls: 61 British men without ECG changes	Analysis of fatty acids in adipose tissue	No significant case-control differences
Ascherio *et al.* 1994 [6]	Cases: 239 US subjects with first myocardial infarction; Controls: 282 age-matched subjects from the same area	Food frequency questionnaire	High *trans* fatty acid intake positively associated with CHD risk; inverse association with moderate intake
Aro *et al.* 1995 [5]	Cases: 671 men with first myocardial infarction; Controls: 717 men without myocardial infarction from 9 European countries	Analysis of fatty acids in adipose tissue	No overall association between C18:1 *trans* fatty acids and risk of CHD. Significant between-countries differences
3. Studies of coronary artery disease (CAD) diagnosed by angiography			
Siguel & Lerman 1993 [32]	Cases: 47 US subjects with CAD; Controls: 56 US subjects without angiography, studied earlier	Analysis of fatty acids in total plasma	C16:1 *trans* directly and C18:2 *cis* inversely related with CAD
Hodgson *et al.* 1993, 1996 [14,15]	191 non-diabetic subjects with angiography for chest pain	Analysis of fatty acids in platelets	No association between total *trans* fatty acids and severity of CAD. Direct association between C18:1 9*t* and 10*t* and C18:2 *cis* and CAD score

TABLE 8.1 (continued).

Study	Subjects	Measure of *trans* fatty acid intake	Main results
van de Vijver *et al.* 1996 [41]	Cases: 83 Dutch subjects with severe CAD; Controls: 78 subjects with no or mild CAD	Analysis of fatty acids in plasma phospholipids	No case-control difference in total *trans* and C16:1, C18:1 or C18:2 *trans* fatty acids

4. Cohort studies

Willett *et al.* 1993 [44]	85 065 female US nurses without diagnosed CHD at baseline. 8-year follow-up	Semiquantitative food frequency questionnaire	*Trans* fatty acids in highest quintile associated with increased risk of CHD
Ascherio *et al.* 1996 [7]	43 757 male US health professionals free of diagnosed CHD at baseline. 6-year follow-up	Semiquantitative food frequency questionnaire	Increased RR of CHD in upper quintiles of *trans* fatty acid intake, attenuated by adjustment for confounders including fibre
Kromhout *et al.* 1995 [23]	12 763 men aged 40-59 years at baseline from 16 cohorts in 7 countries. 25-year follow-up	Analyzed average diets of cohorts based on weighed diet records of 498 men	Significant direct correlation between intakes of saturated and *trans* fatty acids and risk of CHD death
Pietinen *et al.* 1997 [27]	21 930 male smokers from Finland, aged 50-69 years and free of diagnosed CHD at baseline. Average follow-up 6.1 years	Modified self-administered dietary history	Increased RR of major CHD event and CHD in the highest quintile of *trans* fatty acid intake

and linoleic acid in addition to *trans* fatty acids, only cigarette smoking remained significantly associated with increased risk of sudden cardiac death.

2. Case-Control Studies of Non-Fatal Coronary Heart Disease

In a third study by Thomas *et al.*, 59 male subjects with electrocardiographic (ECG) evidence of ischaemia but without symptomatic CHD were compared with 61 controls without ECG changes and matched for body mass index and smoking [36]. No statistically significant case-control differences in C16:1, C18:1 and C18:2 *trans* fatty acids were found.

In Boston, USA, Ascherio *et al.* (1994) studied patients with a first acute myocardial infarction in 1982-1983 and compared them with age-matched control subjects derived from local population registers [6]. Originally, 340 case-control pairs were enrolled but, after exclusion of subjects with a history of diabetes or hypercholesterolaemia, 238 patients and 282 control subjects

TABLE 8.2.

Relative risk (RR) of myocardial infarction in quintiles of energy-adjusted *trans* fatty acid intake in a case-control study of US men. Data from Ascherio *et al.* (1994) [6].

	Quintile 1	Quintile 2	Quintile 3	Quintile 4	Quintile 5
Median intake (g/d)	3.05	3.72	4.36	5.01	6.47
No of cases/controls	37/53	42/61	27/58	43/56	90/54
RR[a] (95% CI)	1.00	1.00 (0.56-1.79)	0.67 (0.36-1.24)	1.12 (0.63-2.00)	2.44 (1.42-4.19)
Multivariate RR[b] (95% CI)	1.00	0.89 (0.48-1.65)	0.52 (0.26-1.02)	0.93 (0.50-1.75)	2.28 (1.27-4.10)
Multivariate RR[c] (95% CI)	1.00	0.81 (0.42-1.57)	0.40 (0.19-0.83)	0.72 (0.36-1.48)	2.03 (0.98-4.22)

[a] Adjusted for age and gender;
[b] Adjusted for age, gender, smoking, history of hypertension, BMI, alcohol intake, family history of CHD and physical activity;
[c] Additionally adjusted for intake of saturated fat, monounsaturated fat, linoleic acid and cholesterol.

TABLE 8.3.

Risk of acute myocardial infarction in quartiles of adipose tissue C18:1 *trans* fatty acid distribution of control subjects in the multi-centre case-control EURAMIC study. Data from Aro *et al.* (1995) [5].

	Quartile 1	Quartile 2	Quartile 3	Quartile 4
All 10 centres				
Median of C18:1 *trans* (%)	0.45	1.29	1.80	2.51
No of cases/controls	182/181	125/178	182/179	182/179
Crude Odds Ratio	1.00	0.70	1.01	1.01
Multivariate Odds Ratio[a] (95% CI)	1.00	0.68 (0.41-1.13)	1.05 (0.63-1.75)	0.97 (0.56-1.67)
In 8 centres (Spanish centres excluded)				
Median of C18:1 *trans* (%)	1.12	1.55	1.98	2.63
No of cases/controls	108/143	113/143	158/140	147/143
Crude Odds Ratio (95% CI)	1.00	1.05 (0.74-1.49)	1.49 (1.07-2.10)	1.36 (0.97-1.91)
Multivariate Odds Ratioa (95% CI)	1.00	1.16 (0.79-1.71)	1.53 (1.02-2.28)	1.44 (0.94-2.20)

[a] adjusted for age, centre, smoking and BMI.

remained for analysis. Dietary intake was assessed with a validated, semi-quantitative food frequency questionnaire. Cases had a higher mean intake of *trans* fatty acids than controls but they also had higher intakes of total fat and saturated and *cis*-unsaturated fatty acids. The relative risk of myocardial infarction was increased in the highest quintile of energy-adjusted *trans* fatty acids as compared with the lowest quintile but not in the other quintiles. In multivariate analysis with adjustment for numerous confounding factors including dietary saturated and monounsaturated fatty acids, linoleic acid and cholesterol, the increased relative risk in the highest quintile remained borderline significant, whereas the relative risk of myocardial infarction was significantly reduced in the third quintile of *trans* fatty acid intake as compared with the lowest quintile (Table 8.2). Hydrogenated vegetable fats, mainly margarines, contributed to 74 % of total *trans* fatty acid intake, and the overall association between *trans* fatty acids and risk of myocardial infarction was almost entirely accounted for by *trans* fatty acids from hydrogenated vegetable fats.

In the European multicentre, case-control EURAMIC study, male patients with a first acute myocardial infarction and control subjects without a history of myocardial infarction were studied during 1991-1992 in eight European countries and Israel [5]. Adipose tissue samples were available from 671 cases and 717 control subjects, aged 70 years or less, for the analysis of fatty acid composition. The method was chosen for the analysis of the whole fatty acid composition in a large number of samples and not primarily for the separation of *trans* isomers. The C18:1 *trans* isomers could be measured in the majority of samples but the C16 and C20-22 isomers were not included in the analysis, because in many of the participating countries these were below the detection limit in a majority of the samples.

There was no overall difference of adipose tissue C18:1 *trans* fatty acids between cases and control subjects, and the risk of myocardial infarction did not differ significantly over quartiles of adipose tissue C18:1 *trans* fatty acids (Table 8.3). The mean proportions of C18:1 *trans* fatty acids varied up to six-fold between the countries reflecting considerable differences in the intake of *trans* fatty acids, lowest in Spain and highest in Norway and the Netherlands (Figure 8.1). After exclusion of the two participating centres from the south of Spain, showing outlying low values for C18:1 *trans* fatty acids and high levels of oleic acid, the multivariate odds ratio for myocardial infarction increased to 1.53 and 1.44 in the two highest quartiles vs. the lowest quartile (Table 8.3). Within the countries the associations between *trans* fatty acids and risk of myocardial infarction differed considerably. In Norway and Finland, C18:1 *trans* fatty acids showed a significant positive association with risk of myocardial infarction whereas an inverse association was evident in Spain and Russia (Figure 8.1) [5].

In a further attempt to study the reasons for the association between adipose tissue *trans* fatty acids and risk of CHD, the gas chromatograms of the Norwegian and Finnish samples of the EURAMIC study were reintegrated and

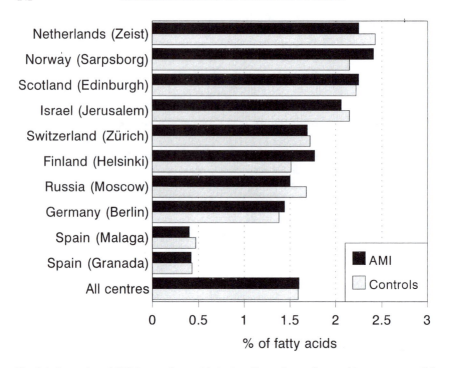

Fig. 8.1. Proportion of C18:1 *trans* fatty acids in the adipose tissue of men with acute myocardial infarction (AMI) and control men in the multi-centre, case-control EURAMIC study. Data from Aro *et al*. [5].

the C16:1, C20:1 and C22:1 *trans* isomers were measured in addition to the C18:1 *trans* isomers [I. Salminen, personal communication]. All *trans* fatty acid classes were significantly higher in the samples from Norway compared with those from Finland. In Norway mean C18:1 (2.46 %) and total *trans* fatty acid levels (4.30 %) were higher in cases with myocardial infarction than in controls (2.11 % and 3.85 %, respectively) but no significant case-control difference was found for the other *trans* fatty acids. No difference was found between cases and controls in C16:1 *trans* fatty acids, neither in Norway (0.72 % in cases and 0.71 % in controls) nor in Finland (0.31 % and 0.36 %, respectively).

3. *Studies of Coronary Artery Disease Diagnosed by Angiography*

Siguel and Lerman in the USA studied 47 subjects with a substantive obstruction of at least one coronary artery at angiography compared with apparently healthy subjects from a previous study published six years earlier. The coronary arteries of the controls had not been examined. The fatty acid composition of whole plasma was determined by GLC. Patients with coronary artery disease (CAD) had higher proportions of total and C16:1 *trans* fatty acids than controls but no significant difference was found in C18:1 *trans* fatty acids [32]. A major difference was found in plasma linoleic acid which contributed by 41.05 and 46.34 % to total fatty acids of cases and controls, respectively.

In Australia Hodgson *et al.* studied 191 non-diabetic subjects who had undergone coronary angiography for the investigation of chest pain, thought to be due to CHD in 97 % of the cases [14]. The fatty acid composition of blood platelets was analysed by GLC and the degree of coronary artery disease was quantified by calculating an angiographic score reflecting the proportion of coronary endothelial surface covered by atheroma. Platelet total *trans* fatty acids, C16:1 *trans* and C18:1 11*t* (vaccenic acid) isomers did not show significant associations with the CAD score in multiple regression analysis adjusting for age, gender, duration of smoking, hypertension and serum total cholesterol concentration. However, platelet C18:1 9*t* (elaidic acid) and 10*t* isomers that were significantly intercorrelated (r=0.84) were significantly associated with the CAD score. The proportion of linoleic acid of platelet fatty acids also showed a significant positive association with the extent of coronary artery disease, as reported in an earlier article [15].

Results of a third study based on coronary angiography were recently published from the Netherlands by van de Vijver *et al.* [41]. Subjects with severe coronary artery disease (n=83) showing a more than 80 % stenosis in one coronary vessel were compared with a control group of 78 subjects with less than 50% stenosis in the coronary arteries. The two groups were comparable with respect to age, gender, smoking status, body mass index, blood pressure and total fat intake. No statistically significant differences were found between cases with severe CAD and controls with no or mild CAD in total plasma phospholipid *trans* fatty acids or the C16:1, C18:1 or C18:2 *trans* fatty acid classes. Even the adjusted odds ratios for tertiles of total *trans* fatty acids failed to show differences between the second and third tertile compared with the lowest one.

4. Cohort Studies

i) The Nurses' Health Study. In 1976, when the study began, 121,700 US registered nurses completed questionnaires on medical history including information on previous and parental myocardial infarction and various CHD risk factors. Every two years, follow-up questionnaires were sent and in 1980 data on usual dietary intake were collected in a group of 85,095 women without diagnosed CHD, stroke, diabetes or hypercholesterolaemia by means of a semi-quantitative food frequency questionnaire [44].

During 8 years of subsequent follow-up, there were 431 cases of new CHD including both non-fatal myocardial infarction and CHD death. The diagnosis of CHD was based on self-reporting of myocardial infarction on a follow-up questionnaire or fatal CHD reported by family members and the National Death Index and subsequently documented by medical and death records. After adjustment for age and total energy intake, the relative risk of CHD was

TABLE 8.4.

Relative risk (RR) of CHD according to energy-adjusted intake of *trans* fatty acids in the in the 8-year follow-up of the Nurses' Health Study cohort. Data from Willett *et al.* (1993) [44].

	Quintile 1	Quintile 2	Quintile 3	Quintile 4	Quintile 5
No of cases	80	89	70	86	106
Person-years at follow-up	130 345	133 898	132 898	133 439	132 260
Age-adjusted RR (95% CI)	1.00	1.15 (0.85-1,56)	1.03 (0.74-1.42)	1.16 (0.85-1.59)	1.50 (1.12-2.00)
Multivariate RR[a] (85% CI)	1.00	1.12 (0.82-1.52)	0.97 (0.71-1.6)	1.12 (0.82-1.54)	1.35 (1.00-1.82)
Multivariate RR[b] (95% CI)	1.00	1.12 (0.81-1.55)	0.99 (0.6-1.43)	1.16 (0.80-1.70)	1.47 (0.98-2.20)

[a] Adjusted for age, smoking, BMI, hypertension, alcohol intake, menopausal status, postmenopausal estrogen use, energy intake, family history of myocardial infarction before age 60;
[b] Additionally adjusted for intake of saturated fat, monounsaturated fat and linoleic acid, and use of multivitamin supplements.

significantly increased in the highest quintile of *trans* fatty acid intake compared with the lowest (Table 8.4). Additional adjustments for known CHD risk factors, intakes of saturated and monounsaturated fatty acids and linoleic acid, and the use of multivitamin supplements, did not change the increased relative risk in the highest quintile substantially. In a subgroup of nurses who had not changed their margarine consumption between 1970 and 1980 (*n* = 69,181), the multiple-adjusted relative risk of CHD in the highest quintile was 1.67 (95 % CI 1.05-2.66) compared with the lowest quintile. Within this group, an increased relative risk of CHD was found in subjects who had consumed *trans* isomers mainly from vegetable fats but not in those who had consumed isomers from animal fats [44].

ii) The Health Professionals Follow-up Study. In 1986 a cohort follow-up study of 51,529 male subjects (dentists, veterinarians, pharmacists, optometrists, osteopathic physicians and podiatrists) aged 40 to 75 years was started in the USA. Adequate food frequency questionnaires were obtained from 43,757 men free of diagnosed cardiovascular disease and diabetes [7]. Follow-up questionnaires were sent at two-year intervals and information on reported CHD was collected by a method similar to that adopted in the Nurses' Health Study.

During a six-year follow-up, 505 non-fatal myocardial infarctions and 229 CHD deaths were documented. The relative risks of myocardial infarction and CHD death were significantly higher in the two or three upper quintiles of energy- and age-adjusted *trans* fatty acid intake but the risk ratios were reduced to statistically non-significant levels after adjustment for multiple confounders including dietary fibre (Table 8.5). In this study high intakes of saturated fatty acids were associated with an increased risk of CHD of similar magnitude to that associated with *trans* fatty acids and, after adjustment for confounders including dietary fibre, the relative risk remained significantly increased in the highest quintile of saturated fat intake for fatal CHD only. Intake of *trans* fatty acids was between 11 and 12 % of saturated fatty acid intake in each quintile, with twice higher mean intake of *trans* fatty acids in the highest quintile of saturated fatty acids compared with the lowest quintile [7].

iii) The Seven Countries Study. In the baseline surveys of the Seven Countries Study between 1958 and 1964, 12,763 men aged 40 to 59 years forming 16 cohorts in seven countries (USA, Finland, the Netherlands, Italy, Croatia and Serbia in former Yugoslavia, Greece and Japan) were studied for CHD risk factors [21]. The vital status of all participants has been verified at regular intervals during 25 years of follow-up. Dietary information was collected by weighed diet records from random samples of 8 to 45 men in 14 cohorts between 1959 and 1964 and in two cohorts around 1970 [22]. In 1985-1986, the average intakes of foods consumed in the 16 cohorts were calculated and in 1987 equivalent food composites representing the average food intakes at baseline were collected locally in each country. The fatty acid composition of the food composites was analysed centrally in the Netherlands [23].

TABLE 8.5.

Relative risk (RR) of CHD according to energy-adjusted intake of *trans* fatty acids in the 6-year follow up of the Health Professionals Follow-up Study cohort. Data from Ascherio *et al.* (1996) [7].

	Quintile 1	Quintile 2	Quintile 3	Quintile 4	Quintile 5
Median intake (g/d)	1.5	2.2	2.7	3.3	4.3
Person-years at follow-up	44 764	47 378	48 173	48 158	48 310
Total CHD[a]					
No of cases	112	140	147	154	181
Age-adjusted RR (95% CI)	1.00	1.24 (0.97-1.59)	1.33 (1.04-1.70)	1.40 (1.10-1.78)	1.57 (1.24-1.98)
Multivariate RR[b] (95% CI)	1.00	1.12 (0.86-1.44)	1.12 (0.87-1.48)	1.12 (0.86-1.46)	1.21 (0.93-1.58)
Fatal CHD					
No of cases	27	51	39	56	56
Age-adjusted RR (95% CI)	1.00	1.88 (1.19-2.98)	1.50 (0.92-2.42)	2.10 (1.34-3.29)	1.99 (1.27-3.12)
Multivariate RR[b] (95% CI)	1.00	1.63 (1.01-2.62)	1.18 (0.71-1.96)	1.59 (0.98-2.60)	1.41 (0.86-2.32)

[a] Includes non-fatal myocardial infarction and fatal CHD;
[b] Adjusted for age, BMI, smoking habits, alcohol consumption, physical activity, history of hypertension or high blood cholesterol, history of myocardial infarction before age 60, profession, and energy-adjusted fibre intake.

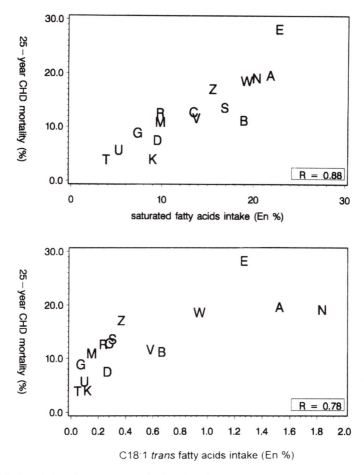

Fig. 8.2. Associations between average intake of total saturated fatty acids and average intake of C18:1 *trans* fatty acids and 25-year mortality rates from coronary heart disease in the Seven Countries Study. Observe the different scales on the x-axis. Symbols: A, US Railroad; B, Belgrade, Serbia; C, Crevalcore, Italy; D, Dalmatia, Croatia; E, East Finland; G, Corfu, Greece; K, Crete, Greece; M, Montegiorgio, Italy; N, Zutphen, Netherlands; R, Rome, Italy; S, Slavonia, Croatia; T, Tanushimaru, Japan; U, Ushibuka, Japan; V, Velika Krsna, Serbia; W, West Finland; Z, Zrenjanin, Serbia. Adapted with permission of the authors and of Academic Press, Inc. from Kromhout *et al.* [23].

The average population intakes of C18:1 *trans* fatty acids showed a significant positive association with the respective 25-year CHD mortality rates (Figure 8.2). In accordance with the previous findings in the Seven Countries Study [21], the intake of saturated fatty acids was strongly related with the 25-year death rates from CHD. Even the intakes of individual saturated fatty acids including lauric, myristic, palmitic and stearic acid showed significant positive associations with CHD mortality [23].

iv) The Finnish α-Tocopherol, β-Carotene Cancer Prevention Study. During 1985-1988, 29,133 male smokers were recruited in southern Finland in a randomized, double-blind, placebo-controlled primary prevention trial

EPIDEMIOLOGICAL OF *TRANS* FATTY ACIDS

TABLE 8.6.

Relative risk (RR) of major coronary event and coronary death according to energy-adjusted intake of *trans* fatty acids in the six-year follow-up of the Finnish ATBC study cohort. Data from Pietinen *et al.* (1997) [27].

	Quintile 1	Quintile 2	Quintile 3	Quintile 4	Quintile 5
Major coronary event[a]					
Median intake (g/d)	1.3	1.7	2.0	2.7	6.2
No of cases	244	288	268	297	302
Person-years at follow-up	24 675	25 241	25 675	25 675	25 703
Age-adjusted RR (95% CI)	1.00	1.13 (0.95-1.34)	1.02 (0.86-1.22)	1.14 (0.96-1.34)	1.19 (1.00-1.41)
Multivariate RR[b] (95% CI)	1.00	1.10 (0.93-1.31)	0.97 (0.82-1.16)	1.07 (0.90-1.28)	1.14 (0.95-1.35)
Coronary death					
Median intake (g/d)	1.3	1.7	2.0	2.7	5.6
No of cases	109	122	136	111	157
Person-years at follow-up	25 070	25 756	26 112	26 265	26 186
Age-adjusted RR (95% CI)	1.00	1.05 (0.81-1.36)	1.14 (0.86-1.47)	0.92 (0.71-1.20)	1.38 (1.08-1.76)
Multivariate RR[b] (95% CI)	1.00	1.05 (0.81-1.35)	1.12 (0.87-1.45)	0.90 (0.69-1.18)	1.39 (1.09-1.78)

[a] nonfatal myocardial infarction or CHD death;
[b] adjusted for age, supplementation group, smoking, BMI, blood pressure, intakes of energy, alcohol and fibre, years of education and physical activity

undertaken to determine, whether supplementation with α-tocopherol, β-carotene, or both would reduce the incidence of lung cancer. The main results indicated no reduction in the incidence of lung cancer after 5-8 years of supplementation [35]. The baseline diet of the men was assessed using a self-administered, modified diet history method which was adequately completed by 21,930 men without diagnosed CHD, stroke or diabetes and without exercise-related chest pain. Data on non-fatal myocardial infarction were obtained from the National Hospital Discharge Register and CHD deaths were identified through the Central Population Register by review of the death certificates.

During a median follow-up time of 6.1 years, 1,399 major coronary events including 635 CHD deaths were documented [27]. Men in the highest quintile of *trans* fatty acid intake had a moderately increased relative risk of CHD compared with those in the lowest quintile. The risk was attenuated slightly by multivariate adjustment for age, supplementation group, smoking, BMI, blood pressure, intakes of energy, alcohol and fibre, years of education and physical activity (Table 8.6). The relative risk in the highest quintile was somewhat greater for CHD death and remained statistically significant even after adjustment for confounders. The *trans* fatty acids contributing to the intake in the highest quintile were mainly derived from margarines and less from butter. The median intake of *trans* fatty acids of the whole group was 2 g/d but it was considerably higher, 6.2 g/d in the highest quintile of *trans* fatty acids. There was no association between intakes of saturated and *cis* monounsaturated fatty acids, linoleic and α-linolenic acids and the risk of CHD [27].

5. Other Studies

In a case-control study conducted in Athens, Greece in 1990-1991, 329 subjects with a first myocardial infarction or positive coronary angiogram were compared with 570 hospital patients with other conditions. CHD was not associated with the intakes of saturated, monounsaturated and polyunsaturated fatty acids but cooking with margarine was associated with a relative risk of 1.87, but with wide 95 % confidence intervals (0.82-4.28). No information was given on *trans* fatty acids in margarines or the frequency of margarine consumption [40].

Another study conducted in Italy between 1983 and 1992 suggested an association between margarine intake and risk of CHD also. In 429 women with a diagnosis of acute myocardial infarction who were compared with 866 female hospital controls, medium to high margarine consumption was associated with an increased risk of CHD, with a multivariate odds ratio of 1.5 (95% CI 1.0-2.2). The report did not include information on *trans* fatty acids. It was reported that the margarine consumption of Italian women is very low but no exact figures were provided [34].

The relationship between the intake of *trans* fatty acids as assessed by a food frequency questionnaire and serum lipid levels in a group of 748 men was studied by Troisi *et al.* (1992). A statistically significant association was found between *trans* fatty acid intake and serum LDL- and HDL-cholesterol concentrations. However, the correlations were of low order, $r=0.07$ for LDL-cholesterol, $r=0.09$ for HDL-cholesterol and $r=0.12$ for the ratio of LDL- to HDL-cholesterol [40]. Hudgins *et al.* (1991) studied adipose tissue *trans* fatty acids and CHD risk factors in 76 men and could not find significant associations between total adipose *trans* fatty acids and serum lipid concentrations [16]. In the case-control study by Siguel and Lerman (1993) the proportion of C16:1 *trans* fatty acids in plasma showed significant positive correlations with serum total and LDL-cholesterol and negative correlations with HDL-cholesterol and HDL/total cholesterol [32]. On the other hand, in the Dutch case-control study by van de Vijver *et al.* (1996), no significant correlations were found between *trans* fatty acids in plasma phospholipids and LDL- and HDL-cholesterol levels [41].

C. ASSESSMENT OF *TRANS* FATTY ACID INTAKE

1. Biochemical Methods

In many of the studies the proportions of *trans* fatty acids in plasma [32,41] or platelets [14] or adipose tissue [5,30,36-38] fatty acids were used as biomarkers of dietary intake. Plasma or serum is easily obtainable but its *trans* fatty acids are indicators of short-term intake [25]. Measurement of *trans* fatty acids in plasma or blood cell membranes may be useful for the quality control of dietary interventions but it is less suitable for the assessment of usual intakes, particularly in subjects who may have changed their dietary habits recently. Therefore it is a particular problem that all three studies on subjects with coronary artery disease diagnosed at angiography used short-term biomarkers of *trans* fatty acid intake [14,32,41].

The turnover of fatty acids in the adipose tissue is slow and its fatty acid composition reflects dietary intake over a period of 1-2 years. For this reason, the fatty acid composition of adipose tissue is to be preferred for the assessment of long-term intake of *trans* fatty acids. The proportions of those fatty acids that are not synthesised in the organism, *i.e.* essential polyunsaturated fatty acids and *trans* fatty acids, are significantly associated with dietary intake but the intakes of saturated and *cis*-monounsaturated fatty acids are poorly reflected in the fatty acid composition of tissues [17,24,43]. Therefore, when *trans* fatty acids in the adipose tissue are used to assess dietary intake of *trans* fatty acids they may also act as surrogate markers of saturated fatty acids from foods that contain both saturated fatty acids and *trans* fatty acids, such as butter and margarines.

Both in plasma and adipose tissue the fatty acid composition is determined as relative proportions of total fatty acids and not as absolute concentrations. Changes in the major fatty acids will cause reciprocal changes in the proportions of other fatty acids. In the study of Siguel and Lerman, the most marked case-control difference was a significantly lower proportion of linoleic acid in the plasma of cases [32]. A more than 10% difference in linoleic acid that contributed to almost one-half of all plasma fatty acids affected the proportions of all other fatty acids. Thus the higher proportions of *trans* fatty acids of the cases were largely due to passive replacement of linoleic acid and did not definitely indicate higher dietary intake. In the Australian study that was also based on coronary angiography [14,15], the severity of coronary artery disease was positively associated with the proportion of linoleic acid in platelets but this difference did not affect other fatty acids, because linoleic acid comprised only 5-6% of total fatty acids in platelets.

In the adipose tissue, oleic acid is the major fatty acid with a 40-50% proportion. In the EURAMIC study, the Spanish participants were characterized by very low proportions of *trans* fatty acids combined with high levels of oleic acid. The possible confounding by oleic acid was one of the reasons why the results of the study were analysed also after exclusion of the Spanish groups [5].

2. Nutritional Methods

The dietary history interview is generally considered the reference method for assessing dietary intake. This method is laborious and time-consuming and, therefore, in epidemiological studies with large numbers of participants it is usually replaced by methods and modifications that are based on self-administered questionnaires.

The food frequency questionnaire that was used in the US studies was developed for the purposes of the Nurses' Health Study. Originally it comprised 61 food items [44]. The data for the *trans* fatty acid composition of foods were derived from two studies of selected foods [12,33] and included all C18 *trans* fatty acids [44]. The method was subsequently updated and modified to comprise 131 foods [7]. The methods were validated in random groups of women and men by comparison with diet records and adipose tissue fatty acid composition [17,24]. In women, dietary intake of *trans* fatty acids and adipose tissue *trans* fatty acids showed a Spearman correlation of 0.51. Total *trans* fatty acid intake by the food frequency questionnaire was estimated to be 3.4 g/d or 5.8% of fat intake. *Trans* fatty acids comprised 4.3% of adipose tissue fatty acid methyl esters. In men, the Spearman correlation between dietary and adipose tissue *trans* fatty acids was 0.29, the total *trans* fatty acid intake was 3 g/d or 4.7% of fat intake, and *trans* fatty acids contributed by 4.2% to adipose tissue total fatty acids [17,24].

Animal studies suggest that the proportion of *trans* fatty acids in adipose tissue is roughly one-half of their contribution to dietary fat intake. This is supported by the EURAMIC study results from Norway where a mean adipose tissue total *trans* fatty acid value of 3.9% was in agreement with the reported mean consumption of 7-8 g/d or 8-9% of dietary fats in the Norwegian population [19]. The adipose tissue fatty acid composition of the US validations studies suggested mean *trans* fatty acid intakes in the order of 7-8% of dietary fats. Hunter and Applewhite estimated that the availability of *trans* fatty acids to the US population was about 7.6 g/d both in 1970 and 1984, 80% derived from vegetable fats [18]. As hydrogenated vegetable fats are very low in C16:1 *trans* fatty acids it can be estimated that C18 *trans* fatty acids, that are the basis for the calculations in the epidemiological studies that used the food frequency questionnaires, comprise 90% or more of the total *trans* fatty acid intake in the USA. The proportions of *trans* fatty acids in ruminant animal fats is relatively constant and not subject to large calculation errors. In the food composition studies, margarines have been well represented [12,33] but less information is available on the composition of fats used by the food industry and catering services.

In the Finnish ATBC study, a very low mean *trans* fatty acid intake of 2 g/d or about 2% of total fat intake was found [27]. This is in agreement with the finding of a lower than average proportion of *trans* fatty acids in adipose tissue in the Finnish subjects of the EURAMIC study [5]. However, the adipose tissue fatty acid composition would suggest a somewhat higher dietary intake of *trans* fatty acids for the Finnish men, too. Thus it seems probable that the nutritional methods have underestimated dietary *trans* fatty acid intake and the underestimation is particularly due to poor knowledge of the composition of industrial fats. More reliable data will be soon available for Europe from the TRANSFAIR study based on the analysis of representative food samples from 14 European countries [42].

In the Seven Countries Study, nutritional and biochemical methods were combined [23]. The diets were analysed by weighed seven-day diet records in a sub-sample of 498 men at baseline. In 1987-1988 equivalent food composites representing the average diet of each cohort at baseline were collected in each country and analysed centrally. This allowed a more exact analysis of the fatty acid composition of the average diets, provided that the composition of the foods had stayed similar during the 20-30 year period between the dietary survey and the collection and analysis of foods. However, the results could not be adjusted for individual factors nor for the effects of other fatty acids.

D. ASSOCIATION BETWEEN *TRANS* FATTY ACIDS AND CORONARY HEART DISEASE

1. Case-Control Studies

In the postmortem studies of Thomas *et al.*, higher C16:1 *trans* fatty acids in victims of CHD death was the main finding [37,38]. The authors suggested that this was the result of consumption of margarines containing hydrogenated fish oils but they could not explain why there was no difference in C20-22 fatty acids that are abundant in partially hydrogenated fish oils. Methodological problems may be responsible for some of the findings. Separation of C20-22 *trans* isomers is difficult even with more developed capillary GLC methods using polar stationary phases and it is unclear how well the method that was used was able to separate the *cis* and *trans* isomers of C16:1. Even in the case-control study of Siguel and Lerman the main finding concerning *trans* fatty acids was an increased proportion of C16:1 *trans* fatty acids in the plasma of cases with angiographically documented coronary artery disease [32]. The origin of the C16:1 *trans* fatty acids in human tissues is poorly known. They can be formed by retroconversion of C18:1 *trans* fatty acids [11], and it is not known to what extent dietary intake of C16:1 *trans* contributes to the amounts found in the organism. If it is assumed that C16:1 *trans* fatty acids in plasma reflect dietary intake of C16:1 *trans* fatty acids, the finding of the US case-control study, higher C16:1 *trans* together with significantly lower linoleic acid in the cases [32], would be compatible with an increased intake of dairy fat and other ruminant fats in the cases with coronary artery disease, since hydrogenated fish oils were not consumed by the population that was studied.

Studies on subjects with coronary artery disease diagnosed at angiography share the common problem of possible recent dietary changes caused by awareness of the disease or dietary prescription as part of therapy. People who undergo angiography are often considered candidates for coronary by-pass surgery and it is reasonable to assume that they have been subjected to both pharmacological and non-pharmacological intervention. The results of the US study by Siguel and Lerman [32] did not suggest this kind of influence but the study from Australia, showing an association between the 9t and 10t isomers of C18:1, the predominant isomers found in partially hydrogenated vegetable oils, and severity of coronary artery disease, together with a direct association between linoleic acid and CAD [14,15], suggested that subjects with more severe disease had consumed more polyunsaturated margarines. A short-term indicator of intake, the fatty acid composition of platelets, was used and the results thus reflected recent dietary intake only. A change of diet during the preceding 10 years showed a significant correlation with linoleic acid in platelet phospholipids [15]. In the carefully conducted Dutch study where no case-control differences were observed, dietary changes due to diagnosed CHD probably did not affect the results, because in the Netherlands diet margarines that are supposed to be recommended for patients with CHD contain very low amounts of *trans* fatty acids [41].

The multicentre EURAMIC study is the largest of the case-control studies [5]. The results did not suggest an overall effect of C18:1 *trans* fatty acids on CHD risk. However, the findings indicated considerable differences between the participating countries, both in *trans* fatty acid intake and in the associations between intake and risk of CHD (Figure 8.1). The two Spanish centres with very low adipose tissue C18:1 *trans* fatty acids comprised 20-25 % of the subjects of the study. The lowest quartile of *trans* fatty acids included almost exclusively Spanish cases and controls and, consequently, the odds ratios reflected mainly comparison between Spain and the other countries. After exclusion of the Spanish participants, a suggestion of a positive association between C18:1 *trans* fatty acids and risk of CHD appeared, with highest relative risk in the third quartile of *trans* fatty acids (Table 8.3). In the absence of dietary intake data, no definite explanations could be given to the questions concerning the causes for differences between the countries [4]. The further analyses of the Norwegian and Finnish samples did not give specific answers, either. The higher proportions of all *trans* fatty acid classes in the Norwegian adipose tissue samples, with a significant case-control difference for C18:1 and total *trans* fatty acids suggested an effect by a high overall consumption of *trans* fatty acids. For Finland with much lower levels of *trans* fatty acids, there is a fair chance that adipose tissue *trans* fatty acids may be a surrogate risk indicator for concomitant saturated fatty acids [4]. Men in the south of England showed an inverse association between adipose tissue *trans* fatty acids and risk of fatal CHD in the study by Roberts *et al.* [30], but no association was evident in the Scottish participants of the EURAMIC Study on non-fatal CHD [5].

The two studies from Greece and Italy, showing some association between margarine intake and risk of CHD, did not include data on *trans* fatty acids [34,40]. In populations where the use of crude vegetable oils for cooking is customary, margarine consumption is an indicator of dietary habits that are associated with increased use of both saturated and *trans* fatty acids. Available information from the Mediterranean countries indicates very low intakes of *trans* fatty acids [5,8,20,23], although there may be differences between different areas and between urban and rural populations.

2. Cohort Studies

In both the Nurses' Health Study [44] and the Health Professionals Follow-up Study [7] and also in the US case-control study by Ascherio *et al.* [6] data derived from food frequency questionnaires that were based on the same food composition data [12,33] were used for the assessment of *trans* fatty acid intakes. All three studies showed essentially similar results, a significantly increased relative risk of CHD in the highest quintile of *trans* fatty acid intake. No continuous dose-response effect was evident and in the case-control study a J-shaped association with a significantly reduced relative risk of CHD in the middle quintile was observed [6] (Figure 8.3). The authors did not discuss the significance of this finding but it seems obvious that moderate intake of *trans* fatty acids was associated with some concomitant beneficial factors, and it is tempting to speculate that these were the *cis* unsaturated fatty acids.

The results from the ATBC study cohort in Finland, in men with considerably lower average intake of *trans* fatty acids, were essentially similar to those of the US studies with respect to *trans* fatty acids: the relative risk of CHD was moderately increased in the highest quintile of intake [27] (Figure 8.3). The intake of *trans* fatty acids in the highest quintile was quite high for the Finnish conditions, three-fold compared with the mean intake of the population. The high consumers were characterized by high consumption of

Multivariate RR

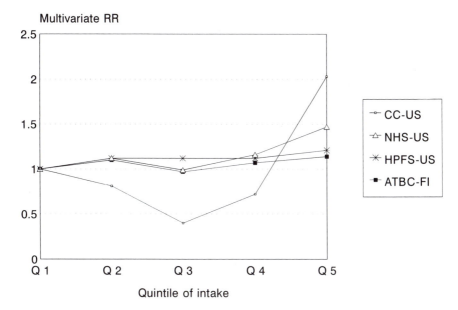

Quintile of intake

Fig. 8.3. Multivariate relative risk of coronary heart disease according to quintiles of *trans* fatty acid intake in four studies based on the assessment of intake by questionnaires. Studies: CC-US, the US case-control study by Ascherio *et al.* [6]; NHS-US, the Nurses' Health Study [44]; HPFS-US, the Health Professionals Follow-up Study [7]; ATBC-FI, the Finnish ATBC study [27]. For details, see Tables 8.2 and 8.4-6.

margarines and a surprisingly low butter intake. Although men with diagnosed CHD at baseline were excluded, the possibility cannot be excluded that the study subjects included men who had changed their dietary habits because of symptoms of CHD existing already before the start of the study. The health-conscious nurses and health professionals participating in the US studies were characterized by a low risk of CHD death, 16 per 100,000 person-years for the women and 97 per 100,000 person-years for the men, respectively [7,44]. On the other hand, the participants of the Finnish study were high-risk individuals with a 30-year history of smoking and high dietary fat intake. They showed a higher risk of CHD death, 483 per 100,000 person-years [27]. However, in all these studies with different risk levels, adjustment for other CHD risk factors only slightly attenuated the risk ratios for CHD in the highest quintiles of *trans* fatty acid intake.

The 25-year follow-up of the Seven Countries Study differed from the other cohort studies by the biochemical analysis of the food composites representing the average diets of the cohorts [22,23]. It is probable that the fatty acid composition of the most important foods had remained relatively unchanged between the 1960s and the 1980s, because the concern about the unfavourable effects on serum lipoproteins of *trans* fatty acids with subsequent changes in the composition of margarines in many countries was raised first in the 1990s, after the results of the dietary intervention study by Mensink and Katan were published [26]. The 25-year follow-up results confirmed the strong association between dietary saturated fatty acids and CHD mortality observed earlier during the study [21].

The association between dietary C18:1 *trans* fatty acids and CHD mortality was similar to that observed between saturated fatty acids and CHD mortality for 14 of the 16 cohorts, but the values for USA and Netherlands indicated a somewhat lower CHD mortality in relation to *trans* fatty acid intake than would have been expected according to the correlation found in the other countries (Figure 8.2). Intakes of *trans* fatty acids from animal and vegetable sources were not analysed separately in the study but it appears reasonable to assume that the correlation in the Mediterranean countries and Finland depended on *trans* fatty acids from animal products whereas in the USA and Netherlands *trans* fatty acid from vegetable fats were involved. Consumption of margarines with hydrogenated vegetable oils has been very low in the Mediterranean countries and a study conducted in 1991 in Crete still showed a low mean margarine intake of 2 g/d in adults [20]. The results of the 60 men representing the Finnish cohorts, studied in 1959, indicated a low intake of margarines (2.7-2.8% of energy) also, in contrast to the high contribution to total energy intake (34.7-39.7%) of butter and milk products [31]. The importance of foods of animal origin was reflected in a high correlation between the average intakes of individual saturated fatty acids, *trans* fatty acids and dietary cholesterol that prevented the study of the independent effects of individual fatty acids and cholesterol in multivariate models [23]. On the other hand, it is known that in the USA and Netherlands the majority of *trans* fatty acids are derived from partially hydrogenated vegetable oils [10,18].

The findings of the Seven Countries Study, presented in Figure 8.2, suggest that in the majority of the cohorts dietary C18:1 *trans* fatty acids acted as surrogate risk indicator for saturated fatty acids that are present in the same dairy and ruminant meat products in more than ten-fold amounts compared with *trans* fatty acids. The US and Dutch cohorts presented a slightly lower risk associated with dietary *trans* fatty acids from vegetable fats, which had been modified possibly by the concomitant *cis* unsaturated fatty acids. It should be emphasized that a major change in the dietary habits of the Finnish population has taken place between the start of the Seven Countries Study and the ATBC Cancer Prevention Study, with a 25 % reduction in the intake of saturated fatty acids, mainly due to reduced consumption of butter and fatty milk products, and a concomitant major reduction in mean serum cholesterol levels and age-adjusted CHD mortality [28]. Therefore, there are certain differences between the customary diets of the Finnish participants of the two studies.

3. *Effect of different trans isomers*

In an epidemiological analysis of studies conducted by 1995, it was concluded that case-control studies on specific *trans* fatty acids suggested an association between C16:1 *trans* (animal) fatty acids and CHD rather than between C18 *trans* isomers (commercial hydrogenation) and CHD [1]. This conclusion was based mainly on the two post-mortem studies of Thomas *et al.* and the study by Siguel and Lerman, all with certain problems both in the case-control design and analytical methods [32,37,38]. In the more recent analyses of the Dutch case-control study [41] and the further analyses of the Norwegian and Finnish samples of the EURAMIC study, no case-control differences in C16:1 *trans* fatty acids were found. Furthermore, as discussed earlier, it is not known how well C16:1 *trans* fatty acids in tissues reflect dietary intake of C16:1 *trans* fatty acids from animal foods, and thus it is impossible to differentiate between animal and vegetable sources of C16:1 *trans* fatty acids in human tissues.

The results of 356 women of the Nurses' Health Study who had not changed their margarine consumption during the preceding 10 years showed a significantly increased relative risk of CHD in the highest quintile of *trans* fatty acid intake from vegetable fats but not for the intake from animal fats [44]. The results were adjusted for confounders including dietary intakes of saturated, monounsaturated and polyunsaturated fatty acids. In animal fats the effect of the small amounts of *trans* fatty acids are expected to be small and less important as compared with the ten-fold higher amounts of saturated fatty acids. The impact of *trans* fatty acids from hydrogenated vegetable fats seems more important, particularly when adjusted for the confounding effect of linoleic acid. A similar result was found also in the case-control study of Ascherio *et al.* using a similar food frequency questionnaire for the assessment of *trans* fatty acid intake [6]. These results do not confirm, however, that the *trans* fatty acids from animal sources, dominated by 11*t* vaccenic acid, have different effects on CHD risk factors than the *trans* isomers from hydrogenated vegetable oils that comprise considerable amounts of 9*t* elaidic acid and 10*t* octadecenoate in addition to vaccenic acid. So far there are no studies on the specific effects of different individual *trans* isomers on serum lipoproteins or other risk factors of CHD. For appropriate studies on different C18:1 *trans* isomers, better methods for the separation and identification of these fatty acids will be needed.

Attempts at quantifying the long-chain C20-22 *trans* isomers have been done in few studies with negative findings concerning association with CHD [37,38]. The identification of these fatty acids that are found in hydrogenated fish oils is difficult and the results of the studies should be interpreted with caution. As it has been found that partially hydrogenated fish oils may have more adverse effects on CHD risk factors than partially hydrogenated vegetable oils [2], further studies on the effects of hydrogenated fish oils should be conducted in countries like Norway and Iceland where the consumption of foods containing these fats is still common.

According to present evidence, the effects of dietary *trans* fatty acids are most probably due to the most common C18:1 isomers that are derived mainly from partially hydrogenated vegetable oils in most populations. The effects of

C16:1 and C20-22 *trans* isomers are probably of less importance but should be elucidated further in future studies.

E. SUMMARY AND CONCLUSIONS

The results of the case-control studies, in many respects discordant, reflect the remarkable differences between different populations [5] and even within a single population [6] in the associations between dietary *trans* fatty acids and CHD. These differences are probably affected by both the amounts of *trans* fatty acids consumed and the different combinations of *trans* fatty acids with saturated and *cis* unsaturated fatty acids in foods. The prospective cohort studies suggest that a high consumption of *trans* fatty acids is associated with a moderately increased risk of CHD but they also suggest that moderate intakes are not associated with increased risk (Figure 8.3). The lack of a consistent dose-response relationship indicates that the findings cannot be explained by a simple causal relationship. High intake of *trans* fatty acids affects the ratio between LDL and HDL cholesterol in a way which is unfavorable compared with all other fatty acids [2,3,26]. When consumed in large amounts, the effects of *trans* fatty acids on risk of CHD are expected to be more or less comparable with those of C12-16 saturated fatty acids in the diet. At the levels of average intakes, *trans* fatty acids are a minor food component, accompanied by much higher amounts of saturated and *cis* unsaturated fatty acids. In margarines, deep-fried foods and many bakery products either saturated fatty acids or *trans* fatty acids are needed in order to achieve the desired texture and stability. In comparison with foods that are based on saturated animal fats like butter and tallow or vegetable fats like coconut oil, foods based on partially hydrogenated vegetable oils usually appear favorable with respect to the *cis* unsaturated fatty acid content [42].

Both biochemical and nutritional methods of assessing *trans* fatty acid intake have certain weak points. A major drawback of measuring the fatty acid composition of tissues is insensitivity to the dietary intake of saturated fatty acids, the quantitatively most important fatty acid class with respect to CHD risk. The reliability of semi-quantitative food frequency questionnaires is dependent on the quality of food composition data and they usually fail to reflect changes in the fatty acid composition of foods and loyalty to particular brands. It would be desirable to combine dietary survey methods with biochemical indicators of intake [4] but no such studies on *trans* fatty acids have been conducted so far. The food composition data should be based on updated analysis of representative food samples, and an indicator of long-term fatty acid intake, preferably the adipose tissue fatty acid composition, should be used as biomarker.

For the prevention of CHD, it is apparent that high intakes of both C12-16 saturated fatty acids and *trans* fatty acids should be avoided. The role of *trans* fatty acids should be considered in relation to the more abundant saturated and *cis* unsaturated fatty acid classes. Ultimately the effects of dietary *trans* fatty acids depend on how they affect the balance between the LDL-cholesterol-raising (C12-16 saturated plus *trans*) fatty acids and the neutral or hypocholesterolaemic *cis* unsaturated fatty acids in foods.

REFERENCES

1. Allison,D.B., *Am. J. Clin. Nutr.*, **62**, 670S-678S (1995).
2. Almendingen,K., Jordal,O., Kierulf,P., Sandstad,B. and Pedersen,J.I., *J. Lipid Res.*, **36**, 1370-1384 (1995).
3. Aro,A., Jauhiainen,M., Partanen,R., Salminen,I. and Mutanen,M., *Am. J. Clin. Nutr.* **66**, 1419-1426 (1997).
4. Aro,A., Kardinaal,A.F.M. and Kok,F.J., for the EURAMIC study group, *Lancet*, **345**, 1109-1110 (1995).
5. Aro,A., Kardinaal,A.F.M., Salminen,I., Kark,J.D., Riemersma,R.A., Delgado-Rodriguez,M., Gomez-Aracena,J., Huttunen,J.K., Kohlmeier,L., Martin,B.C., Martin-Moreno,J.M., Mazaev,V.P., Ringstad,J., Thamm,M., van 't Veer,P. and Kok,F.J., *Lancet*, **345**, 273-278 (1995).
6. Ascherio,A., Hennekens,C.H., Buring,J.E., Master,C., Stampfer,M.J. and Willett,W.C., *Circulation*, **89**, 94-101 (1994).
7. Ascherio,A., Rimm,E.B., Giovannucci,E.L., Spiegelman,D., Stampfer,M. and Willett,W.C., *BMJ*, **313**, 84-90 (1996).
8. Boatella,J., Rafecas,M. and Codony,R., *Eur. J. Clin. Nutr.*, **47**, S62-S65 (1993).
9. British Nutrition Foundation Task Force, *Trans fatty acids* (1995) (British Nutrition Foundation, London).
10. Brussaard,J.H., *Voeding*, **47**, 108-111 (1986).
11. Emken,E.A., in *Health Effects of Dietary Fatty Acids*, pp. 245-263 (1993) (edited by G.J. Nelson, American Oil Chemists' Society, Champaign, IL).
12. Enig,M.G., Pallansch,L.A., Sampugna,J. and Keeney,M., *J. Am Oil Chem. Soc.*, **60**, 1788-1794 (1983).
13. Expert Panel on Trans Fatty Acids and Coronary Heart Disease, *Am. J. Clin. Nutr.*, **62**, 655S-708S (1995)
14. Hodgson,J.M., Wahlqvist,M.L., Boxall,J.A. and Balazs,N.D., *Atherosclerosis*, **120**, 147-154 (1996).
15. Hodgson,J.M., Wahlqvist,M.L., Boxall,J.A. and Balazs,N.D., *Am. J. Clin. Nutr.* **58**, 228-234 (1993).
16. Hudgins,L.C., Hirsch,J. and Emken,E.A., *Am. J. Clin. Nutr.*, **53**, 474-482 (1991).
17. Hunter,D.J., Rimm,E.B., Sacks,F.M., Stampfer,M.J., Colditz,G.A., Litin,L.B. and Willett,W.C., *Am. J. Epidemiol.*, **135**, 418-427 (1992).
18. Hunter,J.E. and Applewhite,T.H., *Am. J. Clin. Nutr.*, **44**, 707-717 (1986).
19. Johansson,L., Rimestad,A.H. and Andersen,L.F., *Scand. J. Nutr.*, **38**, 62-66 (1994).
20. Kafatos,A., Chrysafidis,D. and Peraki,E., *Int. J. Food Sci. Nutr.*, **45**, 107-114 (1994).
21. Keys,A., *Circulation*, **41**(Suppl.1), 1-211 (1970).
22. Kromhout,D., Keys,A. Aravanis,C., Buzina,R., Fidanza,F., Giampaoli,S., Jansen,A., Menotti,A., Nedeljkovic,S., Pekkarinen,M., Simic,B.S. and Toshima,H., *Am. J. Clin. Nutr.*, **49**, 889-894 (1989).
23. Kromhout,D., Menotti,A., Bloemberg,B., Aravanis C., Blackburn,H., Buzina,R., Dontas,A.S., Fidanza,F., Giampaoli,S., Jansen,A., Karvonen,M., Katan,M., Nissinen,A., Nedeljkovic,S., Pekkanen,J., Pekkarinen,M., Punsar,S., Räsänen,L., Simic,B. and Toshima,H., *Prev. Med.*, **24**, 308-315 (1995).
24. London,S.J., Sacks,F.M., Caesar,J., Stampfer,M.J., Siguel,E. and Willett,W.C., *Am. J. Clin. Nutr.*, **54**, 340-345 (1991).
25. Mensink,R.P. and Hornstra,G., *Br. J. Nutr.*, **73**, 605-612 (1995).
26. Mensink,R.P. and Katan,M.B., *N. Engl. J. Med.*, **323**, 439-445 (1990).
27. Pietinen,P., Ascherio,A., Korhonen,P., Hartman,A.M., Willett,W.C., Albanes,D. and Virtamo,J., *Am. J. Epidemiol.*, **145**, 876-887 (1997).
28. Pietinen,P., Vartiainen,E., Seppänen,R., Aro,A. and Puska,P., *Prev. Med.*, **25**, 243-250 (1996).
29. Precht,D. and Molkentin,J., *Die Nahrung*, **39**, 343-374 (1995).
30. Roberts,T.L., Wood,D.A., Riemersma,R.A., Gallagher,P.J. and Lampe,F.C., *Lancet*, **345**, 278-282 (1995).

31. Roine,P., Pekkarinen,M. and Karvonen,M.J., *Voeding*, **25**, 383-393 (1964).
32. Siguel,E.N. and Lerman,R.H., *Am. J. Cardiol.*, **71**, 916-920 (1993).
33. Slover,H.T., Thompson,R.H., Davis,C.S. and Merola,G.V., *J. Am. Oil Chem. Soc.*, **62**, 775-786 (1985).
34. Tavani,A., Negri,E., D'Avanzo,B. and La Vecchia,C., *Eur. J. Clin. Nutr.*, **51**, 30-32 (1997).
35. The Alpha-Tocopherol Beta-Carotene Cancer Prevention Study Group, *N. Engl. J. Med.*, **330**, 1029-1035 (1994).
36. Thomas,L.H., Olpin,S.O., Scott,R.G. and Wilkins,M.P., *Hum. Nutr. Food Sci. Nutr.*, **41F**, 167-172 (1987).
37. Thomas,L.H. and Winter,J.A., *Hum. Nutr. Food Sci. Nutr.*, **41F**, 153-165 (1987).
38. Thomas,L.H., Winter,J.A. and Scott,R.G., *J. Epidemiol. Community Health*, **37**, (1983).
39. Troisi,R., Willett,W.C. and Weiss,S.T., *Am. J. Clin. Nutr.*, **56**, 1019-1024 (1992).
40. Tzonou,A., Kalandidi,A., Trichopoulou,A., Hsieh,C., Toupadaki,N., Willett,W. and Trichopoulos,D., *Epidemiology*, **4**, 511-516 (1993).
41. van de Vijver,L.P.L., van Poppel,G., van Houwelingen,A., Kruyssen,D.A.C.M. and Hornstra,G., *Atherosclerosis*, **126**, 155-161 (1996).
42. van Poppel,G., van Erp-Baart,M-A, Leth,T., Gevers,E., van Amelsvoort,J., Antoine,J., Kafatos,A. and Aro,A., *J. Food Composit. Anal.* (in press).
43. van Staveren,W.A., Deurenberg,P, Katan,M.B., de Groot,L.C. and Hoffmans,M.D., *Am. J. Epidemiol.* **123**, 455-463 (1986).
44. Willett,W.C., Stampfer,M.J., Manson,J.E., Colditz,G.A., Speizer,F.E., Rosner,B.A., Sampson,L.A. and Hennekens,C.H., *Lancet*, **341**, 581-585 (1993).

CHAPTER 9

CONJUGATED LINOLEIC ACID AND METABOLITES

S. Banni[a] and J.-C. Martin[b]

[a] *Dipartimento di Biologia Sperimentale sez. Patologia Sperimentale, Viale Regina Margherita, 45 09124 Cagliari, Italy.*
[b] *Unité de Nutrition Lipidique, INRA, 17 rue Sully, 21034 Dijon Cédex, France.*

A. INTRODUCTION

The presence of a conjugated diene (CD) structure in polyunsaturated fatty acids (PUFA) is not usual. They are formed during structural modification following autoxidation or partial hydrogenation, usually with formation of a *cis,trans* double bond system [1]. While autoxidation processes lead to formation of CD fatty acid hydro(pero)xides [2], during partial hydrogenation CD fatty acid non-hydro(pero)xides are formed [3]. Even though these two species have different origins and share only the double bond system, CD fatty acid non-hydro(pero)xides have often been assumed to be a product of lipid oxidation [4-7]. This is due to the simultaneous presence of these two molecular species containing CD in several biological systems, and the use of the CD structure for their detection. For many years both molecular species have been well known, but studied by very distinct groups of investigators.

PUFA hydroperoxides have been an important topic for all scientists involved in lipid peroxidation [2]. On the other hand, the formation and presence of CD fatty acid non-hydroperoxides were just a curiosity in the field of oil and dairy fat chemistry until about ten years ago.

In fact, the most common conjugated diene fatty acid, the so-called conjugated linoleic acid (CLA), a collective term describing a mixture of positional and geometric conjugated diene isomers of linoleic acid, has been recently receiving growing attention because of its several biological properties, and particularly as an anti-carcinogen and anti-atherogen in experimental models [8]. For this reason, many scientists from different branches of chemistry, biology, biochemistry and animal sciences are now involved in CLA research. As a consequence, an exponentially number of papers are currently being published in several journals rendering it difficult to keep track of the literature on CLA. In this chapter, we try to review the literature that has been published so far and the problems related to CLA research and analysis, aware that inevitably we may not have covered it all. Therefore we apologize in advance to readers and authors for any missing citations.

B. OCCURRENCE, CHEMISTRY AND NUTRITION OF CONJUGATED LINOLEIC ACID

1. Occurrence of CLA in Food

The origin of CLA in foods is due to partial hydrogenation either performed industrially [9] or biologically by anaerobic bacteria present in ruminants [10]. Studies by Shantha and coworkers [11], while not suggesting an increase in CLA content due to cooking, did demonstrate that CLA is relatively stable and not completely destroyed by cooking and storage. Banni *et al.* [12] demonstrated that CLA and its metabolites decreased steadily during oxidative stress and they are more prone to oxidation than the corresponding methylene-interrupted fatty acids.

i) Formation of CLA by industrial partial hydrogenation. Partial hydrogenation of fats proceeds by means of dissolved hydrogen in the presence of a catalyst, saturating some of the double bonds originally present in a liquid oil containing (poly)unsaturated fatty acids. This process changes the physical characteristics of the oils. By reducing the original degree of unsaturation of the oils, it also increases their stability and melting point (see also Chapter 2).

Catalysts are formally unchanged by the reaction and favour particular reaction mechanisms, depending on the prevailing conditions. The favoring of a certain mechanism represents the selective nature of a catalyst, in the present case the favored reactions catalysed by nickel mainly are: (i) the saturation of different double bonds present in the fatty acids; and (ii) the simultaneously geometric (*cis/trans*) and positional isomerization of double bonds (conjugation, shifting).

The process is essentially the same as the free radical-based double bond shifting and recombination as described for the autoxidation-caused conjugations; some isolated dienoic double bond systems may be transformed into conjugated dienoic isomers, whereas some single bonds may undergo *cis/trans* isomerization. More details of the mechanism of isomerization reactions are reviewed elsewhere [13].

ii) CLA formation by biohydrogenation. Biohydrogenation was first described by Reiser in 1951 [14]. Some years later, it was demonstrated that linoleic acid is hydrogenated by rumen microflora to stearic acid and during the process a mixture of geometrical and positional isomers of dienoic and monoenoic acids are formed. In 1964 [15], it was shown that similar products are produced by the flora of the caecum and colon of ruminant animals. In particular, *Butyrivibrio fibrisolvens*, a common rumen bacterium, was shown to hydrogenate linoleic acid to an octadecenoic acid, but not to the fully saturated derivative. Kepler *et al.* [16] reported that biohydrogenation of linoleic acid is not a one-step reduction of a dienoic acid to a monoenoic acid, but rather the system involves the production of a conjugated octadecadienoic acid, and two *trans*-monoenoic acids. Compared to the hydrogenation reaction, the formation of the conjugated dienoic acid is extremely rapid. Then the isomer that has been accumulated in the incubation mixture is converted to a *trans*-monoenoic acid. For this reason, it has been concluded that the production of the conjugated octadecadienoic acid from linoleic acid is the first reaction of the biohydrogenation pathway. The CD fatty acid produced from linoleic acid has been identified as an 9-*cis*,11-*trans*-octadecadienoic acid. The isomerization reaction, catalysed by linoleic acid isomerase [17], is unusual in several respects. It occurs in the middle of a long hydrocarbon chain remote from any activating functional groups and possesses no cofactor requirements. Previous studies of the isomerase have shown that it is particulate and that it exhibits maximum activity with the substrates linoleic and linolenic acids within a narrow concentration range [18]. From these studies it is apparent that at least three features of the substrate molecule are involved in binding substrate to the enzyme. These are the system of a substrate double bond, a hydrophobic interaction, and hydrogen bonding of the substrate carboxyl group.

iii) Dietary source of CLA. The principal dietary sources of CLA are animal products even though CLA is also present in plant oils and partially hydrogenated oils in low concentrations [19].

9-*cis*,11-*trans*-CLA accounts for less than 50% of the total CLA in vegetable oils, in contrast to the 80-90% range found in meat and dairy products. The presence in milk fat of fatty acids with conjugated unsaturation was established first by Booth *et al.* [20] who reported that when cows were turned out to pasture after winter, the fatty acids of milk fat showed greatly increased absorption in the ultra-violet region at 230 nm. The range of CLA in milk fat, determined by spectrophotometric methods in various studies, was summarized by Riel [21] and varies from 0.24 to 2.81% from winter to summer. Recently, Banni *et al.* [22] showed that the CLA content of Italian

TABLE 9.1.

Conjugated Fatty Acids, *Trans*-octadecenoic and linoleic acids in Italian milk and dairy products [22]

samples	18:1t mg/g of fat mean±S.D.	18:2 mg/g of fat mean±S.D.	CD18:2 mg/g of fat mean±S.D.	CD18:3 mg/g of fat mean±S.D.
pecorino	49.92 ± 28.87	17.93 ± 10.37	13.03 ± 5.88	1.01 ± 0.49
ricotta	91.04 ± 3.31	17.50 ± 1.38	24.19 ± 2.79	2.03 ± 0.48
parmesan	32.21 ± 8.62	17.67 ± 7.20	8.65 ± 0.63	0.53 ± 0.07
swiss cheese	37.02 ± 7.39	12.66 ± 0.94	14.23 ± 0.36	1.06 ± 0.26
sheep milk	85.74 ± 11.02	9.78 ± 0.47	29.68 ± 2.36	1.82 ± 0.08
sheep milk *	13.14 ± 1.45	24.40 ± 2.58	11.72 ± 1.29	0.75 ± 0.10
cow milk	19.22 ± 2.39	15.79 ± 0.23	7.10 ± 0.11	0.52 ± 0.03
yogurt	43.37 ± 5.94	25.12 ± 3.89	7.98 ± 1.21	0.16 ± 0.05

Footnotes:
*samples collected in the summer.
S.D.: standard deviation

cheeses and milk falls within the same range (Table 9.1), but highest values were detected in the winter season; this is due to the fact that in the south of Italy, where winters are mild and summers very dry, ruminants are at pasture during the winter rather than during the summer. In fact, variation of milk CLA is seasonal, highest values occurring when pastures are lush and rich in PUFAs. It seems that CLA increases at the expense of linoleic acid while *trans*-octadecenoic acid increases proportionally (Table 9.1) [22]. Interestingly, they detected low levels of conjugated linolenic acid also [22], presumably derived from biohydrogenation of linolenic acid per se [23].

The variation of CLA content in the raw material was not always taken into account in studies dealing with the influence of processing on CLA content in dairy products and is likely to explain discrepancies in the results, as observed for cheeses for instance [24-26]. Hence, only work starting with the same raw material shall be considered in comparing the process conditions. During processing, CLA content increased in cheeses while heating in air at 80°C to 90°C (3.8% of CLA increase) [27], or following addition of a whey protein concentrate (35% of CLA increase) [27], or of hydrogen donors as well as iron and dairy-based additives (16% - 59% of CLA increase). During aging or ripening of cheese, lipolysis by bacterial enzymes occurs which partially splits neutral fat into free fatty acids and glycerol [28]. Under such conditions, free fatty acids including CLA should become very vulnerable to further oxidation. This might indirectly reduce the CLA concentration in aged cheeses. In contrast, long ripening [25] did not modify the CLA concentrations in cheeses and dairy products [29-31]. It has been shown that lactalbumin and lactoglobulin, the predominant proteins in whey from milk [32], bind long-chain fatty acids [33]. Thus the stability of CLA during processing may be due to its binding to lactalbumin or lactoglobulin, which then protects against isomerization and oxidation. This might explain the relatively high CLA content found in ricotta cheese [22] (Table 9.1).

TABLE 9.2

CLA content in foods [24]

Foodstuff	number of samples	Total CLA (mg/g of fat)	c-9,t-11 %
Uncooked Meats and Seafood			
Round beef	4	2.9 ± 0.09	79
Fresh ground beef	4	4.3 ± 0.13	85
Veal	2	2.7 ± 0.24	84
Lamb	4	5.6 ± 0.29	92
Pork	2	0.6 ± 0.06	82
Chicken	2	0.9 ± 0.02	84
Fresh ground turkey	2	2.5 ± 0.04	76
Egg yolk	2	0.6 ± 0.05	82
Salmon	4	0.3 ± 0.05	n.d.
Lake trout	3	0.5 ± 0.05	n.d.
Sea scallops	2	0.3 ± 0.05	n.d.
Shrimp	2	0.6 ± 0.10	n.d.
Mussels	2	0.4 ± 0.04	n.d.
Natural and Processed Cheeses			
Natural cheese:			
Romano	2	2.9 ± 0.22	92
Parmesan	4	3.0 ± 0.21	90
Sharp cheddar	3	3.6 ± 0.18	93
Cream	3	3.8 ± 0.08	88
Medium cheddar	4	4.1 ± 0.14	80
Colby	3	6.1 ± 0.14	92
Mozzarella	4	4.9 ± 0.20	95
Cottage	3	4.5 ± 0.13	83
Ricota	3	5.6 ± 0.44	84
Brick	2	7.1 ± 0.08	91
Natural Muenstre	2	6.6 ± 0.02	93
Reduced fat Swiss	2	6.7 ± 0.56	90
Blue	2	5.7 ± 0.18	90
Processed cheese:			
American processed	3	5.0 ± 0.13	93
Cheez whiz	4	5.0 ± 0.07	92
Velveeta	2	5.2 ± 0.03	86
Old English spread	2	4.5 ± 0.21	88
Oils and Fats			
Commercial oil or fat:			
Safflower	2	0.7 ± 0.14	44
Sunflower	2	0.4 ± 0.02	38
Peanut	2	0.2 ± 0.01	46
Canola	2	0.5 ± 0.07	44
Vegetable	2	0.3 ± 0.02	41
Corn	2	0.2 ± 0.03	39
Coconut	2	0.1 ± 0.01	47
Olive	2	0.2 ± 0.01	47
Beef Tallow	2	2.6 ± 0.01	84

TABLE 9.3

CLA content in foods [24]

Foodstuff	number of samples	Total CLA (mg/g of fat)	c-9,t-11 %
Processed Foods			
Beef frank	2	3.3 ± 0.13	83
Turkey frank	2	1.6 ± 0.08	70
Beef smoked sausage	2	3.8 ± 0.07	84
Smoked turkey	2	2.4 ± 0.01	62
Smoked bratwurst	3	2.4 ± 0.05	77
Smoked bacon:	3	2.6 ± 0.12	75
Peanut butter	2	0.2 ± 0.01	n.d.
Canned Foods			
Potted meat	2	3.0 ± 0.08	72
Corned beef	2	6.6 ± 0.08	85
Chicken breast	2	0.4 ± 0.06	71
Vienna sausage	2	1.5 ± 0.06	76
Pork and beans	2	1.2 ± 0.06	69
Baked beans	2	0.7 ± 0.04	56
Pink salmon	2	0.8 ± 0.03	n.d.
Crabmeat	2	0.5 ± 0.05	n.d.
Tuna	2	< 0.1	n.d.
Clams	2	0.2 ± 0.03	n.d.
Anchovy fillets	2	0.4 ± 0.05	n.d.
Other Dairy Products			
Homogenized milk	3	5.5 ± 0.30	92
Condensed milk	3	7.0 ± 0.29	82
Cultured buttermilk	3	5.4 ± 0.16	89
Butter	4	4.7 ± 0.36	88
Butter fat	4	6.1 ± 0.21	89
Sour cream	3	4.6 ± 0.46	90
Ice cream	3	3.6 ± 0.10	86
Nonfat frozen dairy dessert	2	0.6 ± 0.02	n.d.
Lowfat yogurt	4	4.4 ± 0.21	86
Custard style yogurt	4	4.8 ± 0.16	83
Plain yogurt	2	4.8 ± 0.26	84
Nonfat yogurt	2	1.7 ± 0.10	83
Frozen yogurt	2	2.8 ± 0.20	85
Milk chocolate	2	3.5 ± 0.06	76
Double chocolate	2	3.1 ± 0.48	71
Vanilla	2	3.8 ± 0.10	84
Extracted in laboratory:			
Garlic	2	< 0.1	n.d.
Corn	2	0.2 ± 0.01	n.d.

Also, no influence of different starting bacterial cultures in cheeses was detected [31]. Thus the CLA values of the various dairy products and different varieties of natural cheeses probably do not represent different types of products or processing conditions but rather reflect the fluctuating levels of

CLA in the raw materials [21]. Interestingly, the peak distribution of CLA in dairy products is not substantially affected by the process conditions [29-31]. This is probably due to the primary rumen bacterial origin of CLA found in the final dairy products.

It is possible to increase CLA content in the milk by elevating dietary intakes of unsaturated fatty acids, the substrate for CLA synthesis. Feeding cows on pasture with supplements of crushed full fat rapeseeds and soybeans, rich in 18:1 and 18:2 fatty acids, respectively, yielded lower proportions of saturated fatty acids and higher levels of 18:1 fatty acids in milk fat [34]. Furthermore, an increase of 65% in milk fat CLA levels was demonstrated by supplementing pasture diets with full rapeseeds over unsupplemented controls [35,36]. Increases of up to 50% in CLA content of milk were also obtained by dietary restriction of the animals [37]. In addition, the levels of 18:1 were higher in milk from cows on pasture compared to that produced from cows kept indoors, because of increased 18:2 and 18:3 intakes [34]. More recently, it has been confirmed that CLA levels are higher in milk produced from pasture-fed cows than from cows fed diets containing only a proportion of their diet as pasture, the remainder being composed of conserved forages [38]. Other parameters that may influence the levels of CLA in fat have received less attention. A positive association between milk fat CLA levels and lactation number has been reported [36,39]. Cows with seven or more lactations produced significantly more CLA in their milk than cows with one to six lactations [39] and cows with lactation numbers greater than four produced more CLA in the milk than cows with two to four lactation numbers [36]. The stage of lactation of the animals, which ranged from 12 to 193 days throughout the study, had no effect on milk fat CLA levels [36].

Chin et al. [24] have provided a general survey of the CLA content of various representative foods normally consumed by humans in the USA. As expected, meats from ruminants contained considerably more CLA than meats from non-ruminants (Tables 9.2 and 9.3). The CLA content for beef ranged from 2.9 to 4.3 mg CLA/g of fat. Among ruminants, lamb was the highest (5.6 mg/g of fat). Pork and chicken were far lower in CLA content. The exception among non-ruminants was turkey, which contained 2.5 mg CLA/g of fat. The CLA content in seafood ranged from 0.3 to 0.6 mg CLA/g of fat.

The method of cooking has been reported to modify the CLA content in food in some circumstances. Only grilling [25] but not broiling, frying, baking or microwaving [11] seemed to increase the CLA concentration in beef meat (from 2.05 to 9.3 mg/g of fat in uncooked and grilled beef, respectively). Differences in the temperatures between grilling and the other ways of cooking could probably afford an explanation to these observations. However, it should be taken into account that during cooking, highly unsaturated fatty acids are destroyed concentrating all the other fatty acids more resistant to high temperatures. Moreover, Shantha et al. [40] did not observe any changes in the CLA concentration of cooked beef during storage (up to 7 days at 4°C in air) either in the presence of hydrogen donors or antioxidants or with no additives at all. This pointed out that during storage, formation of CLA through autoxidation is probably not significant.

An important question is whether all CLA found in the tissues of non-ruminants is a consequence of dietary intake, or perhaps it is due to the conversion of linoleic acid to the 9-*cis*,11-*trans* CLA isomer by bacterial flora in the intestines.

The presence of CLA in plant oils was thought to be due to oxidation and/or bleaching effects. However, oil extracted from corn oil [24] was shown to contain amounts of CLA comparable to that in commercial oil. It was claimed that CLA is naturally present in plant oils. This would not be really surprising since fatty acids with conjugated double bonds system are well-known in the vegetable kingdom [41,42] and must arise from specific enzyme systems. It was also shown that plant oils have more 10-*trans*,12-*cis* and less 9-*cis*,11-*trans* CLA isomer than animal fats.

Infant formulae, either soy or milk protein-based, was found to contain only small amounts of CLA; this is to be expected since the major fat source is plant oils. The total CLA concentrations of infant foods that contained beef, lamb or turkey were similar to those of the comparable raw materials. Interestingly, infant food with veal had a CLA concentration that was three fold higher than that of the unprocessed starting material.

Very recently another source of CLA has become available as dietary supplements. Yurawecz *et al.* [43] examined eight commercial products alleged to contain CLA. Five of the products were either available as dietary supplements or may be soon marketed as such. Of the remaining three products, one was purchased from a specialty chemical company, another was obtained from a university laboratory and the last was standard castor oil. By area percent, the content of the 9-*cis*,11-*trans* isomer varied from 0 to 40 in the eight products, leading to the conclusion that not all samples labeled as containing CLA actually do. It has recently been recognized that commercial CLA preparations, which are produced by alkaline isomerization of linoleate-rich oils, can contain a much wider range of isomers than is admitted by the manufacturers [48].

Of the vast number of naturally occurring substances that have been demonstrated to possess anticarcinogenic and antiatherogenic activity in experimental models, all but a handful of them are of plant origin. CLA is unique because it is present in food from animal sources. CLA is widely present in the human diet, but since its concentration varies according to the type of dietary regimen in ruminants, increasing with natural pasture, it is likely that the concentration of CLA in the human diet was higher in the past than nowadays and sufficient to exert its biological activities. It could be argued that a new policy should be envisaged in order to promote those practices that can increase the CLA level in foods.

2. *Occurrence and Metabolism of CLA in Human and Animal Tissues*

i) Animal studies. As already indicated, for many years the detection of CD

TABLE 9.4.

Conjugated fatty acids of liver and adipose tissues of 1 month-old lambs [22]

samples	CD 18:2 mg/g of fat mean±S.D.	CD 18:3 mg/g of fat mean±S.D.	CD 20:3 mg/g of fat mean±S.D.	CD 20:4 mg/g of fat mean±S.D.	t18:1 mg/g of fat mean±S.D.
liver	12.57±1.00	1.83±0.31	1.03±0.21	0.46±0.13	29.83±13.84
adipose tissue	16.90±2.96	1.75±0.42	0.56±0.31	n.d.*	79.27±24.46

Footnotes:
*not detected.
S.D.: standard deviation

fatty acids in animal tissues had been taken as a reliable marker of lipid peroxidation. In a series of papers [44-47], Banni *et al.* characterized different types of CD fatty acids present in tissues of rats fed a semipurified diet containing 10% partially-hydrogenated fats; 0.4% in the latter were CD fatty acids. The scope of the studies was to investigate whether the diet did indeed trigger a peroxidative process in liver lipids [4]. By HPLC with a diode-array detector, separating CD fatty-acid hydroperoxides from non-hydroperoxide CD fatty acids, they demonstrated that only non-hydroperoxide CD fatty acids of dietary origin were present in the liver lipids of the rats [45,47]. Further analyses by HPLC, with both a diode-array and a MS detector in line, showed that the CD fatty acids in the partially-hydrogenated fats were CLA isomers [19]. In similar analyses of the rat liver lipids, two additional CD fatty acids besides CLA were detected, and were characterized as CD 18:3 and CD 20:3 [46]. This finding was taken as an indication that elongation and desaturation of CLA occur in rat liver, and do not affect the CD structure. However, no CD 20:4, the expected end-product of such processes, was detected. In a subsequent study [22], CLA, CD 18:3, CD 20:3 and CD 20:4 were all found to be present in liver phospholipids (phospholipids) of lambs (Table 9.4), animals which are naturally overexposed to CLA due to its abundance in their milk-diet, and its generation by intestinal flora.

These findings were recently extended by Sébédio *et al.* [48], while studying the chain-elongation and desaturation, in rat liver, of a commercial CLA fed (180 mg/day) as triacylglycerols. By using complementary high-performance liquid chromatography (HPLC) and gas chromatography-mass spectrometry techniques, they identified three higher conjugated metabolites of CLA, namely, $20:3\Delta8,12,14$, $20:4\Delta5,8,12,14$ and $20:4\Delta5,8,11,13$. Formation of $20:4\Delta5,8,12,14$ *in vivo* by $\Delta5$-desaturation of chemically prepared $20:3\Delta8,12,14$ in rat had been reported earlier [49]. In Sébédio's study, the identification of $20:3\Delta8,12,14$ indicated that Δ-6 desaturation of CLA ($18:2\Delta10,12$ isomer) also occurred *in vivo*, as observed *in vitro* with rat liver microsomes [50]. Hence, in Sébédio's work, $20:3\Delta8,12,14$, and $20:4\Delta5,8,12,14$ undoubtedly arose from the $18:2\Delta10,12$ CLA isomer, whereas $20:4\Delta5,8,11,13$ occurred from the $18:2\Delta9,11$ CLA isomer through the well-known desaturation and elongation pathways. No other long-chain derivatives that could arise from the other CLA isomers initially present in the mixture ($18:2\Delta8,10$, $18:2\Delta11,13$) were detected. In addition, these authors found higher quantity of $20:4\Delta5,8,12,14$ than $20:4\Delta5,8,11,13$ in the total liver lipids. Altogether, this suggests that only two isomers of CLA can be efficiently chain-elongated and desaturated, and interestingly that the $18:2\Delta10,12$ CLA isomer appears to be a better substrate than the $18:2\Delta9,11$ CLA isomer, which is the major isomer found in natural products. Alternatively, differences in the incorporation into tissue lipids between the two conjugated 20 carbon isomers could explain the finding.

In another study [50], female SENCAR mice were fed for six weeks semi-purified diets containing 0.0, 0.5, 1.0, or 1.5% CLA by weight (Diets A,B,C and D, respectively). Mice fed Diets B, C and D exhibited lower body-weights and elevated amounts of extractable liver total-lipids. Fatty acid analyses by GC revealed that dietary CLA was incorporated into liver triacylglycerols and phospholipids at the expense of linoleate; in triacylglycerols, the content of oleate increased whereas that of arachidonate decreased. A positive association was observed between greater intakes of dietary CLA and linoleate reduction in phospholipids. In an *in vitro* assay, CLA was desaturated to an unidentified 18:3 product, to an extent similar to that of linoleate conversion into gamma-linolenate (9.9% and 13.6%, respectively).

Analyses of neutral lipids and phospholipids of the rat mammary fat pad indicated that the accumulation of CLA in this tissue was dose dependent from 0.5% to 2% [53]. Furthermore, the CLA concentration was 10 times higher in neutral lipids than in phospholipids, and the incorporation of CLA into either fraction was not affected by the availability of linoleic acid and did not appear to displace linoleic acid or arachidonic acid in the mammary tissue.

One distinct peculiarity of CD fatty acids in general is their sluggish or non-incorporation into rat liver and other tissue phospholipids [44,45,51,52], with the 9*cis*,11*trans* isomer been so far the only CLA isomer detected in phospholipids [51,52].

However, Fogerty *et al.* [54] found veal tissue phospholipids to be more rich in 'natural' CLA (9*cis*,11*trans* isomer) than those of adult-beef tissues (43-67% of total CLA was found in phospholipids in veal compared to 1 to 8% in beef). This finding suggests that CLA may be metabolized differently at different ages, and raises the question of whether its biological importance may change during the life span of animals. In conclusion, more work is needed to establish whether the diet content and/or the feeding period affect the incorporation of CLA into tissue phospholipids and its conversion to CD 20:4. CD 20:4 in phospholipids could compete with regular arachidonic-acid in the formation of eicosanoids, and this competition could be a basis for some of the biological activities of CLA [49,55].

ii) Human studies. Since the discovery of the beneficial effects of CLA in experimental animals, several studies have been carried out on its occurrence, absorption, deposition and metabolism in humans. Actually, the first reports of its occurrence in blood plasma and other body's fluids appeared in the

80's, in a series of papers by Dormandy [5,56-59], who subsequently observed the presence of CLA in the phospholipids of all fractions of serum lipoproteins [60,61]. This author attributed such a presence to free-radical attacks on the lipids [5], but there is no evidence to date that oxidative stresses can generate CLA [12,62,63]. On the other hand, dietary CLA is absorbed and assimilated by humans [54,64], and in samples of human adipose-tissue, Banni et al. [65] found a very good correlation between the levels of CLA and those of *trans* fatty acids, the other unusual fatty acids present in partially hydrogenated fats. An identical correlation was found in a variety of dairy products, the main dietary-sources of CLA [22]. It is highly probable, therefore, that most of the CLA detectable in human tissues is of dietary origin, even though there can be endogenous sources also in humans. Indeed, bacteria capable of generating it have been isolated from patients with anaerobic vaginosis [57] or inflammatory lung conditions [69]. Also, it has been shown by several investigators [66-68] that 18:1Δ11*t*, the major *trans* monoenoic fatty acid in dairy products, can be Δ9-desaturated by rat liver-microsomes to the 18:2Δ9c,11*t* CLA-isomer. This metabolic pathway, if active in human liver, could be another potential endogenous source. High levels of CLA have been observed in patients with certain clinical entities [5,6,56,58,70-72]. Lucchi et al. detected high level of CD fatty acids in blood plasma and adipose tissue lipids of patients with chronic renal failure [73]. Separate analyses of CD fatty acid hydroperoxides and of non-hydroperoxide CD fatty acids revealed increased levels of the latter, but not of the former [74], ruling out the origin of the detected CDs from lipoperoxidation processes. Analyses of CLA and its metabolites [65] showed that the CD fatty acids were CLA, and revealed decreased levels of CD 20:3 [70,72]. These data suggest that the increased levels of CD fatty acids, observed in these patients, may result from an impaired metabolism of CLA, rather than from endogenous synthesis. In conclusion, CLA has been detected in human adipose tissue at concentrations about one third of those found in adipose tissue of rats fed for 2 weeks 0.04% CLA [65], a dietary level already shown to be too low to exert anticarcinogenic effects in the rat [75]. A 300 g wt rat fed 0.1% CLA, having those effects, consumes about 0.015 g CLA per day [52]. By extrapolation, the daily dietary intake of a 70 Kg human would have to be 3.5 g to be effective, assuming that CLA might prove to have beneficial effects also in humans, and at dietary dose-responses comparable to those effective in the rat. This amount is considerably higher than the estimated current intake in the United States of several hundred mg/person/day [25]. The length of intake might however be an accruing factor, since it may influence the metabolism of CLA, as indicated by a higher CLA-metabolites/CLA ratio found in human adipose tissue than in the same tissue of rats fed CLA for 2 weeks [65]. Any attempted supplementation of the dietary intake would have to take into account also the possibility that abnormalities in the metabolism of CLA may be present in patients with several clinical entities.

3. Biological Effects of CLA

Several biological properties have been attributed to CLA. However, while some of them have been fairly-well documented, others are still open to question.

i) Antioxidant activity. Studies in vitro by Ha et al. [51] produced results that were interpreted as indicating that CLA is an effective antioxidant, more potent than α-tocopherol, and almost as effective as butylated hydroxyl toluene (BHT). Ip et al. [52] reported that feeding CLA results in decreased lipoperoxidation (as measured by the TBA test) in the mammary gland but not the liver of rats. However, Van Den Berg et al. [76] tested whether CLA could protect membranes composed of 1-palmitoyl-2-linoleoyl phosphatidylcholine (PLPC) from oxidative modification under conditions of metal ion-dependent or independent oxidative stress. Progress of oxidation was determined by measurement of conjugated diene formation and by analysis of fatty acids. When oxidation of PLPC (1.0 mM) was initiated using the lipid-soluble 2,2'-azobis(2,4-dimethylvaleronitrile) or the water-soluble 2,2'-azobis(2-amidinopropane) hydrochloride, α-tocopherol and BHT at 0.75 μM efficiently inhibited PLPC oxidation, as evident from a clear lag phase. In contrast, 0.75 μM CLA did not have any significant effect on PLPC oxidation. Inhibition of PLPC oxidation by higher concentrations of CLA appeared to be due to competition, not to an antioxidant effect. When oxidation of PLPC was initiated by hydrogen peroxide/Fe^{2+} (500 μM/0.05-20 μM), both α-tocopherol (1 μM) and ethylene glycol-bis(aminoethyl ether) tetraacetic acid (50 μM) efficiently inhibited PLPC oxidation. However, CLA (1-50 μM) did not show a clear protective effect under any of the conditions tested. They concluded that CLA, under those test conditions, did not act as an efficient radical scavenger in any way comparable to vitamin E or BHT. CLA also does not appear to be converted into a metal chelator under metal ion-dependent oxidative stress. In another study in vitro, Banni et al. [12] challenged CLA, incorporated in vivo into rat-liver triacylglycerols and phospholipids, with two different prooxidants, Fe-ADP and tButylHP. An improved methodology was used for the analysis of CD fatty acids containing or not containing an hydroperoxide group. The results obtained indicated that CLA is not endowed with antioxidant activity and is not generated by free-radical attacks on PUFAs, and that CD fatty acids are more susceptible to oxidation than their parent non-conjugated fatty acids. In a recent study in vitro [77], oxidation of CLA was shown to generate several furan derivatives, identified as 8,11-epoxy-8,10-octadecadienoic; 9,12-epoxy-9,11-octadecadienoic; 10,13-epoxy-10,12-octadecadienoic; and 11,14-epoxy-11,13-octadecadienoic. CD fatty acids should therefore be considered as a possible source of furan fatty acids. The furan derivatives were found to protect cultured fibroblasts from hydrogen peroxide-induced cytotoxicity [78]; however, it is not known whether the derivative can be generated also in vivo.

ii) Effect on eicosanoid production. It has been shown that dietary CLA reduced *ex vivo* bone PGE$_2$ production in rats [79]. Weanling rats were given a diet containing 70g/Kg of added fat for 42 d. The fat treatments included two levels of CLA (0 and 10g/Kg) and soybean oil (SBO) or menhaden oil+safflower oil (MSO) following a 2×2 factorial design. Rat growth was not influenced by the diets. CLA was incorporated into all tissues analysed except brain. The 9*cis*,11*trans* and 10*trans*,12*cis* were the primary CLA isomers present in liver and bone. In liver, CLA decreased the concentrations of 16:1, 18:1, 18:2n-6, and total n-6 PUFAs but increased those for 22 carbon n-3 PUFA, and total saturated fatty acids. Rats given the MSO treatment had significantly lowered *ex vivo* PGE$_2$ production in liver homogenate and bone organ culture (tibia and femur) compared to SBO. CLA lowered *ex vivo* PGE$_2$ production in bone. Based on the fatty acid composition data in liver, the authors speculate that CLA inhibits Δ-9 desaturase activity. This study showed that CLA reduced *ex vivo* PGE$_2$ biosynthesis in bone and altered fatty acid composition of tissues in rats fed dietary n-6 and n-3 PUFAs. Furthermore, the decrease in n-6 PUFA and accumulation of 22 carbon n-3 PUFA suggest that CLA may selectively promote the elongation of n-3 PUFA and increase the degradation of n-6 PUFA. Additionally, Sugano *et al.* [80] demonstrated that rats fed a synthetic mixture of CLA isomers as free fatty acids for two weeks (1% in diet) had decreased PGE$_2$ content in serum and to a lesser extent in spleen, with no alteration in the tissue lipid content of the parent PGE$_2$ fatty acid, arachidonic acid. This effect could be mediated through competition of the conversion of arachidonic acid to PGE$_2$ by higher metabolites of CLA at the cyclooxygenase level. Such an effect was demonstrated *in vitro* by Nugteren [49] who showed that 20:3Δ8*c*,12*t*,14*c* and 20:4Δ5*c*,8*c*,12*t*,14*c*, the long-chain derivatives of 18:2Δ10*t*,12*c* CLA isomer, were very strong inhibitors of prostaglandin biosynthesis. On the other hand, octadecatrienoic acid-comprising CD systems are also powerful inhibitors of prostaglandin synthesis, and amongst them, those from the n-6 family with ZEE and ZEZ double bonds configuration (such as 18:3Δ8*c*,10*t*,12*t* and 18:3Δ8*c*,10*t*,12*c*, respectively) were the most potent [55]. The structural similarity between those conjugated octadecatrienoic acids with CLA (especially the 18:2Δ10*t*,12*c* isomer) would also suggest that CLA themselves could be inhibitors of eicosanoid synthesis. Furthermore, a possible effect of CLA on lipoxygenase could be also suggested by an old report of Holman [81] who determined a direct inhibition of soybean lipoxygenase by CLA, and more recently by Funk *et al.* [82] who found a similar effect but using thia-analogues of the 18:2Δ9*c*,11*t* and 18:2Δ9*c*,11*c* CLA isomers.

iii) Bacteriostatic activity. The bacteriostatic activity of CLA was evaluated in brain and heart infusion broth and in milk against *Listeria monocytogenes* [83], which is a human food-borne pathogen that can grow even at refrigeration temperatures. Although 5 to 10 fold less potent than lauric acid, linolenic acid or monolaurin, CLA as a potassium salt was bactericidal in brain heart infusion broth at 50 to 200 μg/mL and for up to 6 days. On the other hand, only potassium CLA and monolaurin were potent in inhibiting bacterial growth in whole and skim milk at 4°C. At 25°C, potassium CLA prolonged the bacteriostatic lag-phase (30h over control, at 300µg/mL). This antibacterial activity of CLA in foodstuff requires further confirmations with other microorganisms, but offers another very appealing application for CLA in food. Peculiarly, sorbic acid, a very well-known bacteriostatic agent widely used in the food industry, also contains a CD structure.

iv) Anticarcinogenic activity. Research from several laboratories has shown that CLA expresses powerful activity in cancer protection in a number of animal models. CLA is not the only fatty acid known to inhibit carcinogenesis. Eicosapentaenoic and docosahexaenoic acids, which are representative of the n-3 PUFA, also fit this category [84]. However, CLA differs from the fish oil fatty acids in two distinct aspects as far as their efficacies are concerned. Whereas fish oil is usually required at levels of about 10%, CLA at levels of 1% or less seems to be sufficient to produce a significant cancer protective effect [75]. Additionally, there are a number of papers which have indicated that an optimal ratio of fish oil to linoleate in the diet is critical in achieving maximal tumor inhibition [84-86]. Whereas the potency of CLA in cancer prevention is largely dissociated from the quantity and type of dietary fats consumed by the host [87].

The first studies on the anticarcinogenic activity of CLA were conducted by Pariza's group. Pariza *et al.* [88], in studying the effects of frying under carefully controlled temperatures (143°C, 191°C and 210°C) on the production of mutagens in hamburger, found evidence of inhibition of mutagenesis in extracts of hamburger heated to 191°C. Using a modified Ames test [89] they showed that inhibition of mutagenesis was present when using the liver S_9 fraction from normal but not Aroclor 1254-treated rats. A later paper [90] indicated that the modulator of mutagenesis had been purified partially and was now active against isoquinoline mutagenicity when using liver S_9 fraction from normal rats or those in which it had been induced by phenobarbital or Aroclor 1254. When 2-aminofluorene was the mutagen, the modulator lowered mutagenic activity when uninduced liver S_9 was used but not when the S_9 fraction had been induced by phenobarbital or Aroclor 1254.

Before the structure of the mutagenesis modulator was established, Pariza and Hargraves [91] tested it for its effect on epidermal tumors in mice. The tumors were induced by application of 7,12-dimethylbenz(a)anthracene (DMBA). Using the assay of Slaga and Boutwell [92], the mutagenesis modulator extracted from beef was tested for antiproliferative properties. Female SENCAR or CD-1 mice were used. The mice were treated with the modulator, followed immediately by an initiating dose of DMBA. One to two weeks later, the tumor promoter TPA (12-O-tetradecanoylphorbol-13-acetate) was applied twice weekly for the duration of the study. In all, three experiments were carried out. In the first (using SENCAR mice), after 20 weeks 30.2 ± 3.4 papillomas/mouse developed in the controls vs. 13.4 ± 2.4 in the inhibitor-treated mice. The second study also used SENCAR mice for 19 weeks and papillomas yields per mouse were 9.6 ± 0.8 in controls versus 4.5 ± 0.9 in those treated with the modulator. The third study involved the use of CD1 mice and after 12 weeks yields of papillomas per mouse in the control and test mice were 10.0 ± 1.7 and 3.2 ± 1.1, respectively. In the meantime this mutagenesis modulator had been identified as CLA [93].

Ha et al. [51] showed that CLA inhibits the initiation of mouse forestomach tumorigenesis by benzo(a)pyrene. Four and two days prior to per os treatment with benzo(a)pyrene, female ICR mice were given 0.1 mL of CLA in olive oil. Three days later the cycle was repeated for a total of four times. At thirty weeks of age, the mice were killed. Mice treated with CLA developed only about half as many neoplasms/animal as mice in the control groups treated with linoleic acid.

Studies on the anti-carcinogenic activity of CLA in the DMBA induced-mammary tumor model have been conducted by Ip's group. Their experimental design contrasted with the acute dosing of CLA as described above, and was intended to simulate human intake of CLA [94]. They reported that feeding CLA at levels of 0.5, 1, 1.5% continuously, reduced the total number of mammary adenocarcinomas by 32, 56, and 60%, respectively [52]. Animals were given 10 mg of DMBA (high dose) in this experiment. In general they observed that the protection was dose dependent at levels of 1% CLA or below, but no further beneficial effect was evident at levels above 1%.

In an attempt to extend the CLA efficacy curve below 0.5%, the same group of investigators performed another experiment similar to the previous one but with two modifications: 1) animals were given 5 mg of DMBA (low dose) and 2) the sample size was increased from 30 to 50, in order to ensure adequate statistical power because of the reduced number of tumors produced by the low dose of carcinogen [75]. Rats were fed the CLA-containing diets at 0.05, 0.1, 0.25 and 0.5% starting 2 weeks before DMBA administration and continuing for 9 months. Tumors took a longer time to develop with the low dose of DMBA. There was clearly a dose-dependent effect on mammary cancer inhibition. Total mammary tumor yield was reduced by 22, 36, 50, and 58% with 0.05, 0.1, 0.25, 0.5% CLA diets, respectively. Intergroup comparison showed that as little as 0.1% was sufficient to cause a significant reduction in the number of tumors.

Similar results were obtained using a different carcinogen, methylnitrosourea (MNU). Unlike DMBA which requires metabolic activation MNU is a direct alkylating agent. They described two distinct activities of CLA in mammary cancer prevention with the use of MNU model [95]. First exposure to CLA during the early post-weaning and peripubertal period only (21-55 days of age) is sufficient to block subsequent tumorigenesis induced by a single dose of MNU given at 56 days of age [95]. This observation suggests that CLA is able to effect certain changes in the developing mammary gland and render it less susceptible to neoplastic transformation later in life [95]. Secondly, CLA is active in suppressing tumor promotion/progression when it is fed only after carcinogen administration [95]. However, this mode of action is different from the first in that once the mammary cells have been initiated by a carcinogen, a continuous intake of CLA is necessary to achieve maximal inhibition.

In another study Ip *et al.* [53] evaluated how changes in the concentration of CLA in mammary tissue as a function of CLA exposure and withdrawal were correlated with the rate of occurrence of mammary carcinoma. The results showed that the rate of decay of CLA in neutral lipid following CLA withdrawal seemed to match more closely the rate of emergence of new tumors. Furthermore it seemed that an uninterrupted supply of CLA was necessary to achieve tumor inhibition. As long as there is an abundant source of CLA present in the target organ, tumor appearance will be blocked or delayed.

Ip's group [53] also studied whether CLA might selectively inhibit the clonal expansion of DMBA-initiated cells carrying either the wild type or codon 61 mutated Ha-ras gene. Previous work from Thompson's laboratory has shown that high dietary levels of linoleic acid preferentially increased the number of wild type Ha-ras tumors, in the rat MNU model [96]. In chemical carcinogenesis, specific ras mutations are induced and are believed to be involved in early stages of tumor development [97-100]. Generally, ras mutation is considered to be permissive but not sufficient for carcinogenesis. Thus, the ras genotype was used as a marker to identify a subpopulation of neoplastically transformed cells that might be differentially modulated by CLA intervention. Their study indicated that CLA inhibited mammary carcinogenesis irrespective of the presence or absence of ras mutation [53]. Thus, this characteristic of CLA in chemoprevention is apparently not dependent on specific genomic mutation induced at the time of initiation.

In another study Ip and co-workers [87] showed that mammary cancer prevention by CLA is independent of the level of fat in the diet. Because CLA is an isomer of linoleic acid, another study [101] was undertaken in order to answer the question regarding whether the effect of CLA is due to a displacement of linoleic acid in cells. In fact, linoleic acid has been found consistently to enhance mammary tumorigenesis in rodents over a wide concentration range [102-104]. The study was designed to examine the dose response to CLA (at 0.5%, 1%, 1.5%, and 2%) in rats fed a 2% or a 12% linoleate diet (both basal diets contained 20% total fat by weight) [101]. The mammary carcinogenesis results showed that the efficacy of tumor suppression by CLA was not affected by linoleate intake. With either linoleate diet, no further protection was evident with levels of CLA over 1% [101]. Another very recent study was designed by Ip *et al.* [105] to characterize certain morphological and biochemical changes of the mammary gland induced by 1% dietary CLA fed for 50 days, that might potentially render it less susceptible to cancer induction. Their results showed that CLA treatment did not affect total fat deposition in the mammary tissue or the extent of epithelial invasion into the surrounding fat pad, but it was able to cause a 20% reduction in the density of the ductal-lobular tree as determined by digitized image analysis of the whole mounts. This was accompanied by a suppression of bromodeoxyuridine labeling in the terminal end buds and lobuloalveolar buds. The recovery of desaturation and elongation products of CLA, namely CD 18:3 and CD 20:3, in the mammary gland confirmed their earlier suggestion [22,53,101] that the metabolism of CLA might be critical to risk modulation. The significance of the above findings was confirmed by the data which show that inhibition of mammary tumors, when CLA was started at weaning and continued for 6 months until the end of the experiments, was essentially of the same magnitude as that produced by 1 month of CLA feeding from weaning. The observation is consistent with the hypothesis that exposure to CLA during the time of mammary gland maturation may modify the developmental potential of a subset of target cells that are normally susceptible to carcinogen-induced transformation.

Using a two-stage model of skin carcinogenesis in mice, Belury *et al.* investigated whether CLA affects the promotion stage [106,107]. The mice were fed a basal diet during the initiation stage, and the same diet containing 1.5% CLA during promotion with tetradecanoylphorbol-13-acetate. Control mice were fed throughout only the basal diet. The tumor yields were an average of 6.2 tumors per mouse in controls, and 4.3 in mice fed CLA ($p < 0.05$) [106]. A decrease in body weight and development of a fatty liver were observed in the latter mice, and incorporation of CLA into liver triacylglycerols and phospholipids was found to be associated with increased amounts of oleic acid and reduced levels of linoleic and arachidonic acids [107]. A decrease in body-weight, liver steatosis and alterations in the fatty acid composition of hepatic lipids are effects typically displayed by hypolipidemic agents in rodents [108]. For this reason, the authors suggested that CLA may share with those agents not only the noted effects, but also their biological activities as peroxisome proliferators, compounds that are non-genotoxic hepatocarcinogens in rodents. To evaluate this possibility, a set of biochemical parameters known to be stimulated by peroxisome proliferators in rodent liver [107] were studied in mice fed 0.0, 0.5, 1.0 or 1.5% dietary CLA. In mice fed CLA, the liver level of acyl-CoA oxidase mRNA was increased about 6-, 9- and 9-fold, respectively, and maximal increases in the steady-state mRNA levels of fatty acid-binding protein and cytochrome P4504A1 were present in mice fed 1% CLA. Western blot analyses revealed 2.5, 3 and 3 fold increases in acyl-CoA oxidase protein. In mice fed 1.0 and 1.5% CLA, the activity of hepatic ornithine-decarboxylase, which is often taken as an index of cell proliferation, was 10-fold greater than in mice fed 0.0 or 0.5% CLA. The authors concluded therefore that CLA elicits many of the liver responses typical of peroxisome proliferators and, as such, if CLA induction of peroxisome proliferation occurs through activation of peroxisome proliferator-activated receptor-α, the biological activity of this chemoprotective compound may occur through a tissue specific, receptor-mediated mechanism to alter fat metabolism, cell proliferation and/or differentiation and apoptosis. It is worthy of note, though, that CLA was not found to affect the body weight and liver lipid-content of female rats [52,53,75,87,95,101].

In an other report Liew *et al.* [109] showed that CLA protects against 2-amino-3-methylimidazo[4,5-f]quinoline-induced colon carcinogenesis in the F344 rat. CLA was administrated to male Fisher 344 rats by gavage on alternating days in weeks 1-4, while 2-amino-3-methylimidazo[4,5-f]quinoline (IQ) was given by gavage every other day in weeks 3 and 4 (100mg/Kg body weight). Rats were killed 6 h after the final carcinogen dose in order to quantify IQ-DNA adducts or after week 16 in order to score colonic aberrant crypt foci. CLA inhibited by about 70% the development of IQ-induced aberrant foci, which are putative preneoplastic lesions. Consistent with the inhibition of aberrant crypt foci, CLA also protected against IQ-DNA adduct formation in the colon by 50%. These results extended previous studies in which CLA inhibited IQ-DNA adducts in the liver, lung, kidney and large intestine of CDFI mice [113].

CLA has also been shown to inhibit the growth of human cancer cells *in vitro*. At a concentration of 1.78×10^{-5} M, CLA inhibited the growth of M21-HPB, HT-29 and MCF-7 cells by 18, 47 and 54%, respectively. At a concentration of 3.57×10^{-5} M, CLA inhibited M21-HPB cell growth by 33%, inhibited growth of MCF7 cells totally and had no further effect on HT-29 cells [114,115]. However, DesBordes and Lea [116] found that CLA was less inhibitory than linoleic acid even when they examined the same cell line, namely the MCF7 breast cancer cells. There were differences in the incubation conditions and length of exposure between these two studies, but it appears that some caution is necessary before concluding that CLA has notable growth inhibitory effect against cancer cells [116].

v) Antiatherogenic activity. Lee *et al.* [117] tested the influence of CLA at 0.5% on atherogenesis in rabbits fed a semi-purified diet containing 25% casein, 19.5% sucrose, 19.3% corn starch, 12% hydrogenated coconut oil, 2% corn oil, 15% cellulose, 5.3% mineral mix, 1% vitamin mix and 0.1% cholesterol. The rabbits were bled at monthly intervals and the study was terminated at 22 weeks. At about 12 weeks, total and LDL cholesterol and triacylglycerols were markedly lower in the CLA-fed group. For the 12-22 week period, LDL cholesterol and the LDL/HDL cholesterol ratio were significantly different $p < 0.05$. The aortic surface of the CLA-fed group showed 22% fewer lesions and their aortas contained 30% less cholesterol. Lipid deposition and connective tissue development determined histologically were less severe in the CLA-fed rabbits.

Kowala *et al.* [118] have shown that a commercial diet augmented with 10% coconut oil and 0.12% cholesterol leads to aortic fatty streak formation in hamsters. Using this model it has been shown that CLA treatment lowers plasma cholesterol and triacylglycerol levels. When CLA was fed at levels of 0.06, 0.11 or 1.1% of energy, aortic fatty streaking was reduced by 19, 26 and 30%, respectively. Linoleic acid fed at a level of 1.1% of energy reduced fatty streaks by 25% [119].

vi) Catabolic/anabolic activity. Another very interesting nutritional property of CLA lies in the anabolic and catabolic response of the body seemingly brought about by this lipid in different physiological situations. The anabolic response of the body was first observed in animals challenged with an endotoxin injection (Lipopolysaccharide of *E. coli*) and experiencing cachexia (immune-induced weight loss) [120,121]. For instance, three week-old mice fed for 2 weeks with a diet containing 0.5% of CLA had a 3 fold less weight loss over 24 h than control animals after the endotoxin trial [120]. This immune-induced growth depression was ascribed to a likely decrease in the fractional synthesis rate of muscle proteins, mediated through interleukine-1 [122]. A similar amount of fish oil in the diet (0.5%) did not exert such an effect. However, others [123] found that high doses of marine oil (8% in the diet) were able to prevent anorexia induced by interleukine-1 in the rat. The effect was thought to be mediated through a decrease of PGE_2 availability, a cyclooxygenase product of arachidonic acid that is decreased during fish oil feeding. As CLA lowered arachidonic acid content also in animal tissues (muscle [120] and abdominal fat pads [121]) during the immune-challenged experiments, the postulated mechanism of prevention of the immune-induced catabolic effect was presumed to be through PGE_2 synthesis. However, in the studies indicated above, the effect of CLA on the body weight of the animals was monitored for only 24 hrs, and in the context of a non-physiological condition. It was therefore not known if CLA could sustain such an effect in other conditions and for a longer period of time. In addition, the weight changes observed were not ascribed to either the fat body mass or the lean body mass, although the latter was more likely. Growth performances of young rats at day 10 of lactation were slightly improved (8.6%) when their mothers were supplemented during gestation and lactation with 0.5% CLA in the diet. A 0.25% dose was inefficient, as well as a 0.5% dose during lactation only [124]. This result suggests that CLA supplementation during fetal life seems necessary to observe an effect during lactation. After weaning, dietary supplementation with 0.25% or 0.5% of CLA allowed detection of a small but significant enhancement of body weight (5.6 or 8.1g of weight gain in 5 weeks for the 0.25% or the 0.5% CLA diet for the males, respectively; 7.1 or 9.2g of weight gain in 7 weeks for the females in the same conditions), together with a better feed efficiency for the CLA groups. Therefore, CLA had a moderate effect on both the weight gain and feed efficiency of growing rats, irrespective of the gender. Again, the partitioning between the fat and the lean body mass was not known.

Pariza *et al.* [125] observed that a 0.5% CLA supplementation to young mice for 4 to 8 weeks reduced the fat body mass by 70 to 57% in females and males, respectively, by 23% in rats and 22% in chicks. Such an effect was not limited to young animals but was noticed for adult mice fed for 4 weeks with a diet containing 20% of lipids comprising 0.5% of CLA. The CLA group exhibited a 46% lowering of the fat body mass and a 9% increase of the lean body mass, leading to a decrease of body weight for a similar food intake to control group. Albright *et al.* [126] reported that this effect required a continuous supply of CLA. Otherwise, CLA withdrawals from diet led to a 19 to 20.5% decrease of the whole protein content in the body. Park and coworker [127] found also that the lower fat body mass induced by CLA in mice was associated with an increase of the enzyme related to fatty acid β-oxidation (carnitine palmitoyl transferase) both in fat pad and skeletal muscle. A lower fat deposition was found to be involved in the reduction of the fat body mass, as CLA treatment significantly reduced heparin-releasable lipoprotein lipase activity (–66%) and increased the release of free glycerol (+ 22%) in the culture medium of adipocytes.

In conclusion, CLA elicited an opposite effect on the fat and the lean body mass, *i.e.* a catabolic response for the former and an anabolic response for the latter, thereby modifying the fat over lean mass balance. This opposite effect could explain the moderate impact of CLA on the weight changes. Seemingly, CLA are more prone to decrease the fat mass than to increase the lean mass. Although an effect mediated through PGE_2 metabolism was suggested at least for the immune-induced growth depression, the exact mechanism of action is widely unknown and may be related to the physiological conditions.

4. Conclusions

The biochemical mechanism whereby CLA elicits the biological responses is still largely unknown. This is due in part to an ignorance of which compound is actually active: is it CLA itself, or one of its metabolically derived molecule such as conjugated arachidonic acid [22,48] or furan-derived molecules [77]? Nonetheless, some indications of the biological action of CLA might be given by data already published: *i)* CLA modulates numerous biological responses in which eicosanoids can also be involved (from cancer to the immunological responses and atherosclerosis) *ii)* the primary products of the cyclo- and lipo-oxygenation enzymes also possess a conjugated dienoic unit [128], and *iii)* because CLA can be chain-elongated and desaturated to eicosanoid fatty acid precursors (namely conjugated diene dihomo-γ-linolenic acid and conjugated diene arachidonic acid, [22,46,48]), it is possible that CLA could exert its biological effect through interference with the metabolism of the eicosanoids, either at the oxygenation enzyme level or at the eicosanoid site of action level.

Another important issue is to establish in which position CLA and its metabolites are incorporated into phospholipids and triacylglycerols. To serve as a substrate for phospholipase A2, PUFAs have to be in position 2 of phospholipids. Our preliminary results indicate that CLA and its metabolites are better incorporated into position 1 of phospholipids as might also be predicted by their structure, closer to a monounsaturated fatty acid than a PUFA.

The position at which CLA and its metabolites are incorporated into triacylglycerols will also be relevant, since positional isomerism of triacylglycerols affects atherogenesis [129].

Another major drawback to interpretation of the data is the different lipid metabolism changes elicited by CLA according to different experimental conditions applied and different animal species used.

An intriguing observation is related to the insect pheromones, some of which have been identified as conjugated linoleic alcohol [130] and its acetate [131], for example. These types of compounds act at a very low concentration and likely through a receptor-mediated mechanism. These agents have a structure similar to that of CLA. It would therefore be of great interest to determine whether CLA also acts via such a mechanism, as has been suggested [107]. Another unresolved issue relates to the identification of which CLA isomer is biologically potent, or whether one isomer is more potent than the others. It is therefore necessary to conduct studies with pure individual isomers and to develop analytical technologies in identifying isomer-specific metabolites. Finally, most of the studies published thus far report the beneficial aspect of CLA on health, and underline the high potencies of CLA to regulate key function of cells. The toxicological aspect of these very active molecules should not be ignored and is worthy of further consideration.

As suggested by Ip *et al.* [132], there is no compelling need to adopt an aggressive approach to chemoprevention in the general population, since a great majority of individuals is in reasonably good health, and morbidity from cancer is only one of many maladies to which they will succumb in the aging process. Therefore, the agents selected should have no toxicity. Because of the intrinsic requirement for a wide distribution in this situation, an expeditious way of delivering these protective agents is through the food system. In view of the impossible task of persuading the public to eat only those foods that are presumably good for their health and the need of providing consumers with a variety of food choices, the time may have come to enrich our foods with known preventive agents, so that their beneficial effects can be realized fully over the life span of an individual. As reviewed in this paper, it has been demonstrated that CLA can be increased in the starting material by different dietary regimens in ruminants, and the beneficial properties of CLA have been associated with very low levels of CLA in the diet. Therefore, further studies are needed to determine the optimal levels for human consumption.

C. ANALYSIS

1. Introduction.

We will focus only on the methods developed to analyse CLA and their long-chain desaturated and elongated derivatives, *i.e.* fatty acids containing only one conjugated double bond system (CLA) or in combination with methylene-interrupted double bonds. These will be referred to below as conjugated fatty acids (CFA).

One of the key step in the biochemical study of CFA is the analytical methodology. The presence of a conjugated bond within the aliphatic chain makes these molecules very unstable to the usual techniques developed for

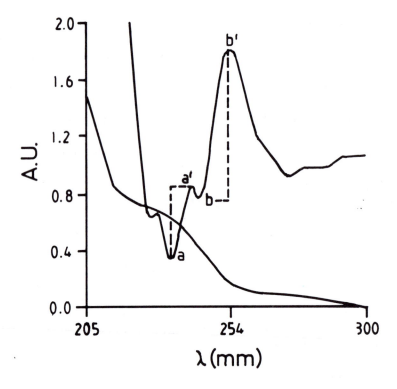

Fig. 9.1. Conventional (lower curve) and second-derivative (upper curve) UV absorption spectra of a sample of partially peroxidized cod liver oil containing fatty acids with conjugated double bonds. Abscissa, wavelength in nm; ordinate, absorption units. From Banni *et al.* [3]. (Reproduced with permission from the authors and from Richelieu press).

normal fatty acid analysis, and could thereby lead to artifactual results. The lack of individual isomer standards of CFA from commercial sources adds difficulties also in their identification in samples from processed or natural products. Hence, distinctive methods are required for a proper analysis of this class of compounds.

1. Spectroscopic Analysis

i) Ultra-violet analysis. The usual unsaturated fatty acids containing methylene-interrupted double bonds can be detected spectrophotometrically in the ultra-violet (UV) region around 200-210 nm, whereas those containing conjugated systems absorb at around 230-235 nm [42]. However, CFA are usually present in low amount in lipid extracts. Therefore, the residual absorption from the maximum in the region 200-210 nm from the bulk of the lipids can interfere with the sharp absorption spectra of CFA around 234 nm, unless further purified. Corongiu *et al.* [133] have designed a mathematical technique to overcome this drawback. By taking the differential of the first derivative spectrum, thus calculating a second derivative, two distinct peaks

TABLE 9.5

Infrared absorption maxima of geometrical isomers of CLA

octadecadienoic ester isomers	infrared absorption maxima	
	=C-H Strecht	=C-H out-of-plane
methylene interruped *cc*	3018 cm⁻¹ᵃ	729 cm⁻¹ᵃ
conjugated *cc*		none in the region 730 cm⁻¹ᵇ
methylene-interrupted *ct* (*tc*)		971 cm⁻¹ᵃ; 730 cm⁻¹ᵃ
conjugated *ct* (*tc*)	3028 cm⁻¹ᵇ; 3009 cm⁻¹ᵇ	978 cm⁻¹ to 987 cm⁻¹ᵇ,ᶜ
		946 cm⁻¹ to 950 cm⁻¹ᵇ,ᶜ
		731 cm⁻¹ᵇ
methylene interrupted *tt*		972 cm⁻¹ᵃ,ᵇ
conjugated *tt*	3023 cm⁻¹ᵇ; 3005 cm⁻¹ᵇ	981 cm⁻¹ to 990 cm⁻¹ᵇ,ᶜ

ᵃ: only from Mossoba et al, 1993 [136]
ᵇ: from Mossoba et al, 1991 [137], Yurawecz et al, 1994 [150]
ᵃ,ᵇ,ᶜ: from Hopkins 1972 [42], Mossoba et al, 1991 [137], Berdeaux *et al.*, 1997 [157], Lavillonnière *et al.*, 1997 [29]
All data were from GC/FTIR, except [42] which were from IR spectroscopy alone.

with minima at around 233 and 242 nm are extracted from the shoulder (Figure 9.1). In addition, second derivative spectroscopy affords a more direct and accurate mean to quantify CFA, since the linear relationship between sample absorption and concentration, as governed by the Lambert-Beer law, is unaffected by differentiation. The CFA content of a given sample can thus be easily determined, provided a standard curve is prepared with suitable reference compounds [3,134]. Nevertheless, this method does not allow to distinguish between the CFA and the hydroxy and peroxy CFA present and produced by oxidative enzymatic systems or during autoxidation, unless a separation of these two classes of compounds is achieved by HPLC (see below).

ii) Infrared analysis. Infrared studies of fatty acids included those with conjugated double bond systems was done many years ago [42,135, AOCS Cd 14-61]. Characteristic maxima of absorption are featured in the infrared spectra for *cis* and *trans* fatty acids, due to =C-H stretch (in the region 3000-3040 cm⁻¹) and =C-H out-of-plane deformation (in the region 720-980 cm⁻¹) [136]. However, these kinds of study required the isolation of pure conjugated fatty acids from natural sources or their chemical preparation, which is not a trivial exercise. The advent of powerful instrumental separation techniques and the use of computerised treatment of the signal together with IR detection, such as gas chromatography coupled with a Fourier transformed infrared detector, enabled a better application of IR spectroscopy to the study of the geometrical isomers of unsaturated fatty acids, including CFA [136,137] (see below for practical applications).

Specific IR spectra also allowed to distinguish between the different geometrical isomers of dienoic fatty acids (*cis,cis-*, *cis,trans-/trans,cis-*, *trans,trans-*) [136,137]. Compared to their counterparts with methylene-interrupted double bonds, conjugated octadecadienoic acids give rise to specific and distinct infrared maxima, as depicted in Table 9.5. Usually, the =C-H out-of-plane deformation (in the region 720-990 cm^{-1}) alone permits precise determination of the geometry of conjugated double bonds [136]. In contrast to the other positional isomers, conjugated *cis,cis*-octadecadienoic acids give negligible spectra in the region 900-1000cm^{-1} [42]. Applications of IR spectroscopy to GC analysis are discussed below.

3. Chromatographic Analysis

i) Thin-layer chromatography. This method is generally used as a preparative means to carry out further detailed analysis on purified CFA. The procedure is primarily based on silver nitrate-impregnated thin-layer chromatography (TLC) silica gel G plates [138-140]. CFA form less stable polar complexes than do their counterparts with methylene-interrupted double bonds [141], therefore the former migrate ahead of the latter. As a result, CLA migrate with the *cis*-monoenoic acid [138,139]. Solvents such as toluene, mixtures of hexane:diethyl ether [18,139], benzene:hexane and benzene:diethyl ether [142], or benzene:petroleum ether:diethyl ether [140] were used. Although a complete separation of the geometrical isomers of CFA cannot be achieved by this procedure, a relative enrichment of either *cis,trans-*, *cis,cis-* or *trans,trans*-isomers might be obtained by carefully selecting portions of the spot containing the CFA, on the grounds that during chromatography the *trans,trans* isomers form less stable complex with silver ions than do the *cis,trans* isomers and then the *cis,cis* isomers. Repeated chromatography gave fractions enriched in the different CFA geometrical isomers [49].

Reversed-phase thin-layer chromatography was also utilized to separate Δ6-desaturated metabolites of CLA [50] or CLA produced by isomerisation of linoleic acid by *Butyrivibrio fibrisolvens* [142,143]. In this case, CLA migrated according to their partition numbers (PN) (PN = Number carbons - 2 × number of double bonds; PN = 14 for CLA).

ii) High-performance liquid chromatography. HPLC techniques using different types of columns in combination with UV, refractometric or mass-spectrometry detectors have been used for the separation and identification of CFA. The most widespread method employed C-18 reversed-phase columns with small particle size (5 μm), such as ODS Hypersil®, Spherisorb ODS2®, Ultrasphere ODS®, or Lichrosorb RP18® columns. Monitoring is carried out by UV spectrophotometry, either directly at 234 nm [59,144] or by second-derivative spectroscopy [19,22,46]. The mobile phase is usually made up of a mixture of acetonitrile:water:acetic acid, and CFA can be chromatographed either in the free form [46,59,144] or as methyl esters [19,46]. It should be noted that HPLC affords both the separation and selective quantification of non-hydroxy-CFA from hydroxy-CFA [12,45,47] which may be present simultaneously in samples. Furthermore, reversed-phase columns enabled the separation of CLA into two peaks, absorbing in second derivative spectroscopy

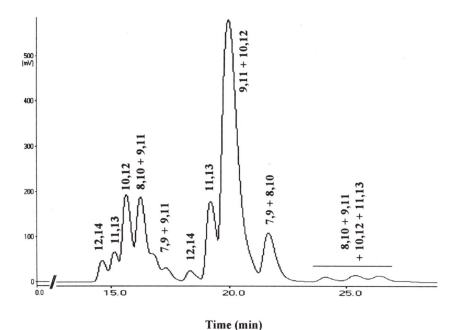

Time (min)

Fig. 9.2. Silver-nitrate high-performance liquid chromatography of methyl esters of CLA obtained from alkali-isomerized linoleic acid. Identification of the positional isomers in each collected fraction was made by gas chromatography-mass spectrometry after conversion of the methyl esters into dimethyloxazoline derivatives. The geometry of the double bonds is not specified. Column: Nucleosil 5SA loaded with silver nitrate, 25 cm × 4.8 mm i.d.; mobile phase: 100% hexane at 1 mL/min; detection in UV at 234 nm. Reproduced with kind permission (P. Juanéda, personnal communication, unpublished results).

with minima at 236 nm and 245 nm, and 232 nm and 242 nm, respectively [22], but no attempt has been made to further characterised these peaks. However, Banni *et al.* [145] showed that the *cis,trans*-CLA hydroperoxide spectrum has two minima at 236 and 245 nm and the *trans,trans* form at 233 and 242 nm. Since the hydroperoxide group should not interfere with the CD absorption, the two peaks with minima at 236 nm and 245 nm, and 232 nm and 242 nm, could be ascribed to *cis,trans*- and *trans,trans*-CLA isomers, respectively. Using a methanol:acetate buffer 0.5% as the mobile phase, Smith *et al.* [146] also obtained two peaks from a mixture of CLA methyl ester standards, the first eluting one containing the *trans,trans* isomers and the last eluting peak the *cis,cis*-, *cis,trans*- and *trans,cis*-isomers. HPLC fitted with an on-line electrospray-mass spectrometer together with UV detection provided a powerful tool to unambiguously identify CLA and their elongated and desaturated metabolites, UV detection confirming the conjugated double bond system whereas mass-spectrometric analysis confirmed the molecular weight of the molecule [19,46].

Apart for identification and quantitative analysis, reversed-phase HPLC has been used as a preparative mean to obtain fractions enriched in CFA [25,29,48,93]. Then, CFA detected by UV spectrophotometry at 234 nm eluted

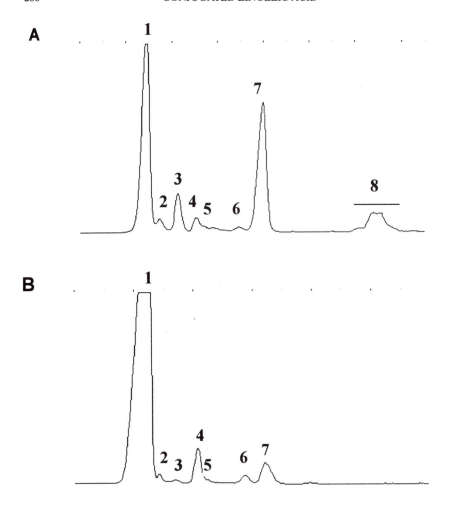

Fig. 9.3. Partial gas liquid chromatograms of conjugated linoleic acid methyl ester isomers from the same lipid extract of cheese. Panel A: chromatogram obtained after *trans*esterification with 14% boron trifluoride in methanol:benzene (3:2; v/v) at 90°C for 1.5 h. Panel B: chromatogram obtained after *trans*esterification with sodium methoxide (2N in methanol) for 10 min at room temperature. Peak identification: (1) 9c11t+8c10t; (2) 9t11c+10c12t; (3) 10t12c+11?,13?; (4) 8c10c+9c11c; (5) 10c12c+11c13c; (6) 11t13t; (7) 8t10t+9t11t+10t12t; (8) unidentified (only after boron trifluoride). Conditions: Fused silica column (50 m x 0.32 mm i.d.) coated with 70% cyanopropylsiloxane (BPX 70); injector and detector temperature set at 250°C; programmed mode: 60°C for 6 min, then 8°C/min up to 170°C and held for 35 min. Reproduced with kind permission (F. Lavillonnière, personnal communication, unpublished results).

according to partition number, similarly to fatty acids with methylene-interrupted double bonds.

Some attempts have been made to use silver-nitrate HPLC. Scholfield [147] used a 2 mm × 61 cm column packed with Amberlite™ XE-284 resin treated with silver nitrate and methanol as a mobile phase. Monitoring was made by refractometry. Synthetic geometrical isomers of 9,11-CLA, analysed as methyl esters, were thoroughly separated between 2.8 and 14.5 min. 18:2Δ9t,11t eluted first, then 18:2Δ9c,11t, and finally 18:2Δ9c,11c. No further practical application was made with this procedure. Juanéda (unpublished results) attempted to fractionate a more complete mixture of CLA (prepared from alkali-isomerized linoleic acid) on a bonded benzenesulphonate column impregnated with silver nitrate and hexane:acetonitrile as the mobile phase. Detection was made at 234 nm. Figure 9.2 depicts a typical separation obtained in the authors' laboratory. With the mass-spectrometric identification method used, none of the HPLC peak was found to be pure but this could arise from the procedure to obtain the dimethyloxazoline derivatives (discussed below). Consequently, only the identification of the major isomer(s) present in each peak is reported. In addition, although the mixture was fractionated into 13 peaks, critical pairs still remained and a complete separation of all of the individual isomers was not achieved by this method. Nevertheless, so far no other chromatographic technique has allowed the fractionation of a CLA mixture to such an extent, and some overlaps among isomers may be a minor inconvenience. It is now required that the retention times of the geometrical isomers be determined for a complete evaluation of the procedure. At the present time, silver-nitrate HPLC seems to be a very promising tool to quantify the individual isomers of CLA in a mixture.

iii) Gas liquid chromatography. Because of the high resolution of polar capillary columns, gas chromatography (GC) is the method of choice to analyse and quantify the different isomers of CLA in foodstuffs [24,25,29], or in biological samples [48,50,51,80,140]. It is a powerful means of molecular identification when used in combination with different kinds of detector, including Fourier transform infrared (FTIR) and mass spectrometry (MS) detectors.

For quantification purposes, analyses are usually performed with fatty acid methyl ester (FAME) derivatives with flame ionisation detection. Great care should be exercised in the preparation of the derivatives of CFA, and this is a critical step in the analysis. For instance, although it has long been known that CLA can be easily isomerized by acidic catalyst reagents, such as boron trifluoride (BF$_3$), used to prepare FAME [18], this was not always taken into account in the following studies. Thus, although BF3 did not modify to a great extent the total recovery of CLA [31], it dramatically changed the profile of the isomers to be analysed and especially increased the proportion of the di*trans* isomers as well as producing unidentified compounds eluting after the CLA peaks (Figure 9.3). In addition, the hydroxy conjugated fatty acids found in biological samples have been shown to be converted to non-hydroxy CFA when using the acidic catalyst [148] and thereby could lead to over-estimation of the content of CFA. This has shed doubt on the results obtained with the methods utilized to derivatize the usual fatty acids, and these should obviously not be employed for CFA. Such results must be interpreted with caution, unless appropriate control have been made to check that no isomerisation has occurred. Thorough examinations of the methods used to derivatize CFA into methyl esters were made recently [31,149]. It was shown that base-catalyst

TABLE 9.6

Separation of different CLA isomers by gas chromatography with various column types.

isomers[a]	column characteristics	samples	authors
c9t11, t10c12, t11t13, t9t11/t10t12	polar: cyanopropylsiloxane coatings in glass capillary column (Silar 10C type), 50m × 0.25mm	mixture of CLA isomers standards	Scholfield, 1981 [147]
all ct + tc, all cc, all tt	non-polar: dimethyl polysiloxane coatings in fused silica capillary column (OV-1 type), 25m × 0.32mm	mixture of CLA isomers standards	Smith, 1991 [146]
c9t11/t9c11, c10t12, t10c12, c11c13, c9c11, c10c12, t9t11/t10t12	polar: polyethylene glycol coatings in fused silica capillary column Supelcowax 10(DBWax types), (60m × 032 mm	home-made c9c11, c9c11, t9t11, t10c12 isomers; CLA from dairy products and grilled-ground-beef	Ha, 1989 [25]
c9t11/c8t10, t9c11/c10t12, t10c12/, ?11?13, c8c10/c9c11, c10c12/c11c13, t11t13, t9t11/t8t10/t10t12	polar: 70% cyanopropylsiloxane (BP × 70 type), 50m × 0.32mm	home made c9t11, c9c11, t9t11; CLA from cheese-fat	Lavillonnière, 1997 [29]

[a]: the octadecadienoic acid isomers are denoted by both the geometry (cis or trans) and the position of the double bond.

reagents such as sodium methoxide and trimethylguanidine in methanol can be used safely for transesterification operations. It should be noted that although trimethylguanidine was reported to esterify free fatty [150], this was not observed with CLA [149]. Some problems may therefore arise with samples containing substantial amounts of CFA in the free form. A total conversion of the esterified and non-esterified CFA may be achieved by a two-step procedure, employing basic hydrolysis of the esterified acids, followed by methylation of the free fatty acids by BF_3 in methanol under mild conditions (room temperature for 30 min), as described by Werner et al. [31].

The separation of seven different peaks of CLA isomers, analysed as methyl esters and extracted from cheese-fat [25, 29] or grilled-ground beef [25] has been achieved by using polar capillary columns (Figure 9.3 and Table 9.6). In such studies [25,29,48,146], the complete identification of the isomers was accomplished by a combination of chemical and/or spectroscopic techniques with GC (see explanation below), as well as by reference to appropriate standards [25,29,146] (Table 9.6). Nevertheless, attempts made to separate and to quantify each of the CLA isomers by GC have not been entirely successful to date, and critical pairs that are not resolved with the different column coatings still remain. Finally, due to the high susceptibility of isomerization of the conjugated double bond system, it might be prudent to estimate the various injection techniques (injector type, injection temperature) on the CFA isomer profile.

4. Structural Analysis

Until recently, the conventional procedure to determine both the double bond position and geometry of CFA was based on the analysis of the products of chemically degraded molecules [49,139,142,151-153]. Briefly, CFA were first isolated as methyl esters by preparative silver nitrate-TLC, and then subjected to partial hydrazine reduction. The resulting esters were then isolated by another preparative silver nitrate-TLC and the *cis* and *trans* monoenes collected separately. These monoenes were representative of the parent dienes with regard to the original double bond position and geometry. Each *cis* and *trans* monoene isomers was then cleaved at the double bond by periodate-permanganate or ozonolysis into di- and monocarboxylic acids. Finally, the identity of each fragment was determined by GC. This allowed assignation of the position and geometry of each double bond. Although time-consuming, this multi-step procedure afforded the identification of the major isomer in milk, *i.e.* 18:2Δ9*c*,11*t* [152], and bile [139]. However, it required large amount of purified compounds (several milligrams), so minor isomers are hardly analysable. More sensitive and accurate methods based on GC analysis have been developed in the last decade to overcome these difficulties.

i) *Gas chromatography/mass spectrometry.* This is the method of choice to locate double bond in the fatty acyl chain, including CFA. Because the usual electron impact ionization in mass spectrometry of underivatised olefins show rearrangements and extensive fragmentation [154], it is necessary to stabilise the double bonds system during the ionisation process by an appropriate modification of the original molecule. This can be achieved either by adding a stabilising group at the carbonyl end of the molecule, or by reacting the double bond with appropriate reagents. These procedures aim at yielding characteristic diagnostic ions of high intensities that prevent misinterpretation of the spectra. However, there are several constraints to keep in mind when using this methodology. First, as stated above, it is not possible to resolve all the CLA isomers present in commercial mixture or in biological samples with the capillary columns available, and the kind of derivatives used to carry out mass spectrometry analysis may also lower this resolution. As a consequence, overlaps among isomers can complicate interpretation of the spectra, especially when dealing with minor isomers. Secondly, as observed when preparing methyl esters, methods used to obtain suitable derivatives for mass spectrometry may possibly induce isomerisation and/or loss of compounds, and thereby be misleading in location of the conjugated double bonds in the original molecule. Thirdly, the utilisation of at least two different procedures for structural studies might be wise to cross-check that no isomerization has occurred and to minimise difficulties in interpretation of the spectra.

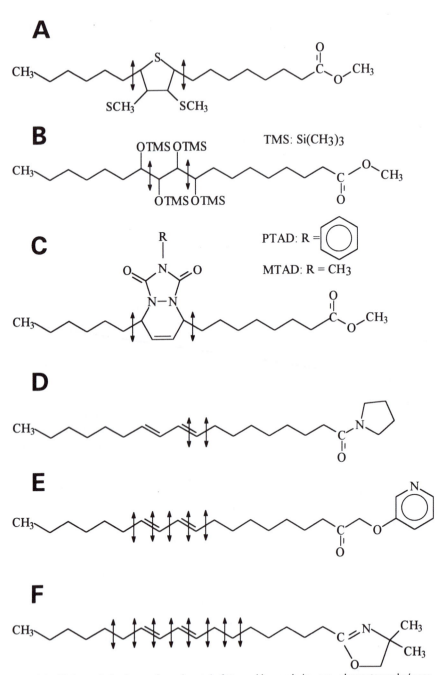

Fig. 9.4. Various derivatives of conjugated fatty acids used in gas chromatography/mass spectrometry analysis. Dimethyldisulfide (A); tetrakis-trimethylsilyl ether (B); Diels-Alder adducts of 4-phenyl-1,2,4-triazoline-3,5-dione (PTAD) and of 4-methyl-1,2,4-triazoline-3,5-dione (MTAD) (C); pyrrolilide esters (D); picolinyl esters (E); dimethyloxazoline (F). Arrows indicate fragmentation that gives rise to diagnostic ions.

There are several ways to react the double bonds with reagents in order to produce characteristic fragments that help to locate the position of the original double bonds. But only a limited number of procedure might be useful when dealing with CFA. Among them, Diels-Alder adducts of conjugated fatty acids appear to be particularly efficient to detect the presence [25] or position of conjugated double bonds [48,146,155-158]. Two kinds of electrophilic reagent can be utilised, namely 4-phenyl-1,2,4-triazoline-3,5-dione (PTAD) and 4-methyl-1,2,4-triazoline-3,5-dione (MTAD). The reaction is highly specific for conjugated double bonds and is almost instantaneous when carried out at 0°C or at room temperature [155-158], but different rates have been monitored when operating at -20°C according to the geometry of the double bonds (in decreasing order *EE > ZE=EZ > ZZ* isomers) [158], thus giving potential for differentiation between the geometrical isomers. Both kinds of adducts (MTAD and PTAD) give very simple and informative mass spectra, including a molecular ion and two intense diagnostic ions formed by α-cleavage of the side chains around the bicyclic ring structure (Figure 9.4). MTAD adducts may be preferred over PTAD for lower volatility reasons in GC analysis (lower Kovats indices). Non-polar capillary columns are required for the analysis of Diels-Alder adducts, which do not permit a separation of positional isomers. Nonetheless, because of the intense diagnostic ions, the spectra of a whole mixture of alkali-isomerized linoleate can be easily interpreted and the major isomers of CLA identified [146,158]. On the other hand, identification of minor isomers can be troublesome. Moreover, Diels-Alder reagents only form adducts with conjugated double bonds and therefore the derivatives cannot be informative for the positional determination of non-conjugated double bonds, when dealing with polyenoic fatty acids containing both conjugated and non-conjugated double bonds. In any case this method is useful to complement other procedure used for a complete structural determination. Alternatively, complexion of the conjugated double bond with maleic anhydride to form adducts has also been employed [42,154]. This reagent only reacts with di*trans* conjugated double bonds and is only informative of this geometric configuration, but it does not require sophisticated instrumentation.

Other attempts have been made over the past to derivatize double bonds in conjugated systems, such as dimethyl disulfide adducts [159] or trimethylsilyl ether derivatives of the hydroxylated double bonds [160,161] (Figure 9.4). Only the latter has demonstrated its usefulness and accuracy for CLA, but the PTAD and certainly MTAD adduct method should be preferred, because of the simplicity and high specificity for conjugated double bond system.

Derivatization of the carboxylic function of unsaturated fatty acid with suitable groups, such as nitrogen-containing substituents, stabilises the charge and minimises double bond ionisation and migration in the mass spectrometer. In contrast to methyl ester derivatives, these substituents bring about simple radical-induced cleavage along the chain, resulting in a decreased abundance of low-mass ions and an increase in a series of ions from the cleavage of each C-C bond [162]. The cleavages yield ions separated by 14 amu and this even-

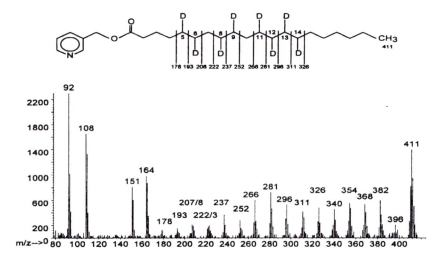

Fig. 9.5. Mass spectrum of picolinyl esters of deuterated 20:4Δ5,8,11,13 eicosatetraenoic acid. From Sébédio *et al.* [48] (Reproduced with kind permission of the authors and of Elsevier publications).

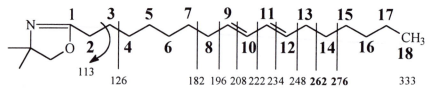

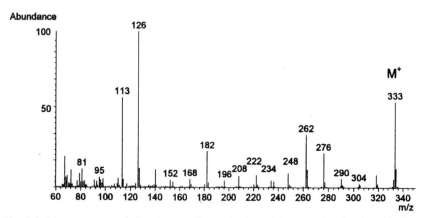

Fig. 9.6. Mass spectrum of dimethyloxazoline derivative of 9,11-octadecadienoic acid. From Berdeaux *et al.* [157] (Reproduced with kind permission of the authors and of AOCS press)

mass sequence is interrupted for C=C bonds or other structural features. Pyrrolidide [59,163], picolinyl [164] and dimethyloxazoline derivatives [29,48,165] have found applications in the elucidation of the double bond position of CFA. The mass spectra of pyrrolidide derivatives of CLA produced key fragments that enabled the location of the double bond closest to the carboxyl group, but the remote double bond is not indicated by the observed fragmentation pattern [163]. This derivative is therefore of limited value. Picolinyl esters of CLA gave spectra which were less easily interpreted in terms of double bonds location [164], and the spectra of methylene-interrupted double bonds from polyenoic acids with picolinyl esters are sometimes difficult to interpret [162]. On the other hand, Sébédio *et al.* [48] used picolinyl esters to examine the double bond position of chain-elongated and desaturated metabolites of CLA after deuteration of the double bonds. The mass spectra of the resulting saturated molecules had ions of high intensity and separated by 14 amu, resulting from the cleavage of each C-C bond. This even-mass sequence was interrupted when carbons were linked to deuterium at the position of the original double bonds, yielding ions separated by 15 amu. This enabled clear identification of 20:4Δ5,8,11,13, a conjugated isomer of arachidonic acid (Figure 9.5).

Recently, dimethyloxazoline (DMOX) derivatives have been successfully used to identify isomers of CLA [29,165] or of their long-chain metabolites [48]. DMOX derivatives of CFA gave the most distinctive spectra [165]; for instance, for octadeca-9,11-dienoic acid, the even-mass homologous series *m/z* = 126 + 14 is interrupted in the region of the double bonds, giving a gap of 12 amu (Figure 9.6). Two other ions of high intensity, separated by 80 amu and resulting from an allylic cleavage of the C-C bond flanking the conjugated double system (*i.e.* between the n-1 and n-2 carbons ahead of the first double bond and the n+1 and n+2 carbons of the last conjugated double bond) also support the structure assignments. Interestingly, another prominent ion (*m/z* 276) is shown 14 amu away from the last ion indicative of the allylic cleavage (*i.e.* at *m/z* 262). Hence, aside from the ions separated by gaps of 12 amu resulting from cleavages at the double bonds, additional diagnostic ions help to support the structural elucidation. For higher metabolites of CLA such as 20:3Δ8,12,14 and 20:4Δ5,8,12,14, specific diagnostic ions such as those separated by a gap of 12 amu and those resulting from the allylic cleavage and separated by 80 amu were evident [48] (Figure 9.7).

Although both DMOX derivatives and picolinyl esters can be analysed efficiently by highly polar capillary columns, DMOX derivatives are more volatile and their elution pattern is very similar to those of FAME, considered as a reference. This prevents, to some extent, overlaps of isomers facilitating interpretation of the spectra. However, the method of preparation is rather vigorous, and could lead to geometrical isomerisation.

Chemical ionisation performed with isobutane could afford an alternative way to electron impact ionisation to study the positional isomers of CFA. This method has been assessed for a variety of olefins containing one conjugated double bond system, such as conjugated diene aldehydes, alcohols, formates, and hydrocarbons, but did not provide satisfactory results for all olefinic isomers [166]. Nonetheless, based on this work, Ha *et al.* [25] examined the structure of CLA derivatised into methyl esters, previously extracted from cheese-fat and grilled-ground beef. Several different positional isomers of CLA

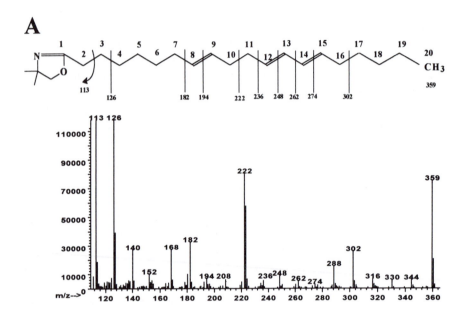

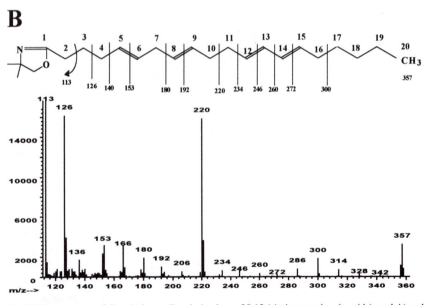

Fig. 9.7. Mass spectra of dimethyloxazoline derivatives of 8,12,14-eicosatetrienoic acid (panel A) and of 5,8,12,14-eicosatetraenoic acid (panel B) [48] (Reproduced with kind permission of the authors and of Elsevier publications).

were thereby identified with characteristic diagnostic ions, representative of cleavages at the level of the double bonds. However, no further control has been made with other methods for structural determination. In addition, octadeca-8,10-dienoic acids found by Parodi *et al.* [152] and Lavillonnière *et al.* [29] in similar products (dairy products) were not found. Hence, GC/MS in chemical ionisation mode makes this method equivocal unless further validations can be made.

ii) Gas chromatography/Fourier transformed infrared spectroscopy (GC/FTIR). This is an easy method for characterisation of the double bond geometry in the absence of pure standards of individual CLA isomers or of their related long-chain desaturated and elongated metabolites. GC/FTIR allows on-line measurement of IR spectra of the analytes eluted stepwise from the capillary column of the GC instrument. The principles and application of this technique to unsaturated fatty acid analysis has been reviewed elsewhere [135,136]. The application of GC/FTIR is recent and the examples with conjugated fatty acids studies are rare (Table 9.4).

The sequence of elution of the different geometrical isomers of CLA by gas chromatography has been readily determined by using this method [25,29,157]. Especially with the high polar capillary columns used (cyanopropylsiloxane), the pattern of elution of the geometrical isomers of CLA differed from that of their methylene-interrupted double bond homologues: the order is *cis,trans-/trans,cis-, cis,cis-, trans,trans-*CLA (Figure 9.3) and *trans,trans-, cis,trans-, trans,cis-, cis,cis-* for the geometrical isomers of linoleic acid. This method has been also applied to determine the geometry of the double bonds of higher CLA metabolites (Sébédio, personal communication).

5. Practical Considerations and Future Directions

i) General considerations. First of all, since CFA must be hydrolysed from triacylglycerols, phospholipids or cholesterol esters and/or derivatised before any analysis can be undertaken, great care should be exercised in these operations due to the high sensitivity of the conjugated double bond to chemical treatment, and heat and oxidative deterioration. Saponification should be conducted at room temperature preferably [12,19,29] and transesterification carried out with a basic catalyst such as sodium methoxide in methanol [149].

ii) Quantitative analysis. For quantitative analysis, because it relied only on the absorbing properties of the conjugated double bond system, reversed-phase HPLC with second derivative-UV detection affords a simple, direct, specific and efficient mean of measurement of CFA from samples of processed or biological origin, without regard to the different isomers present. The major pitfall is that identification only relies on UV absorption characteristics and HPLC mobility, although confirmation of the molecular identity can be obtained with LC/MS. After such validation is made, the method can be successfully applied to quantify either CLA in dairy foods and conjugated dihomo-γ-linolenic and arachidonic acids in biological samples [19,22,46].

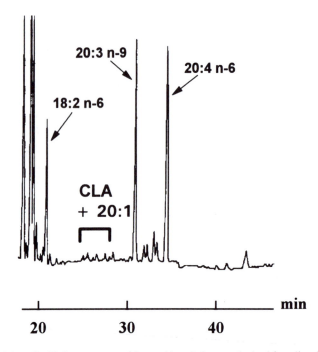

Fig. 9.8. Partial gas liquid chromatogram of fatty acid methyl esters obtained from liver lipids of rats fed with a mixture of conjugated linoleic acid isomers [48]. (Reproduced with kind permission of the authors and of Elsevier publications).

Numerous clinical studies have been undertaken to assess the CLA content in human tissues using this method .

Although more efficient in terms of resolution than reversed-phase HPLC and UV spectroscopy when dealing with CFA isomers, GC analysis presents more difficulties and requires the utilisation of several techniques in combination, according to the accuracy and type of measurement required. In contrast to a UV detector, the response of the flame ionisation detector in GC analysis is not specific. Because of the generally low concentrations of some of the CFA isomers in most of the products, the relatively low signal of CFA might be in the order of magnitude of the background noise and thereby not quantifiable (Figure 9.8). In addition, overlaps of the CLA peaks with other fatty acids such as C20:0 and C20:1 present in some samples might complicate identification and quantification. These problems might not be considered when examining only the major 'natural' isomer (e.g. $18:2\Delta9c,11t$) in samples containing high amount of CLA such as dairy products (1 to 2% of total fatty acids as CLA, near 80% of total CLA as $18:2\Delta9c,11t$) [24], as this elutes in a region of the chromatogram free of contaminants. Most of the GC studies focusing only on the contents of the 'natural' CLA in foodstuff or biological samples adopted this protocol [24,40,54,167]. Alternatively, CFA might be first partially purified by using a reversed-phase HPLC step prior to GC analysis [25,29,48] (Figure 9.9). The same procedure may be applied to other CFA.

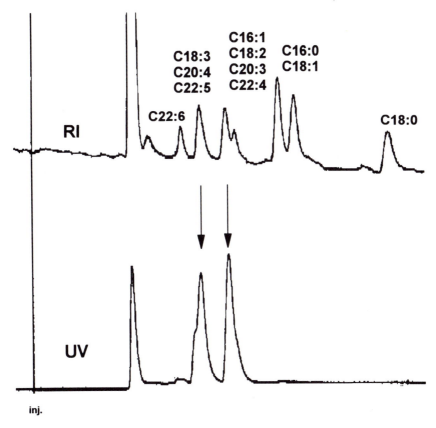

Fig. 9.9. Reversed-phase high-performance liquid chromatography of fatty acid methyl esters obtained from liver lipids of rats fed with a mixture of conjugated linoleic acid isomers. Detection was made both by refractometry (upper panel) and in ultra-violet (lower panel). Reproduced with kind permission (P. Juanéda, personnal communication).

This additional step facilitates CFA analysis substantially, by concentrating CFA in the mixture, allowing thus an accurate GC analyses [25,29]. Then, identification of the CLA peaks by GC might be achieved by comparison of their retention times with pure standards of a CLA mixture commercially available. Unfortunately, standards for other CFA, such as those occurring from the metabolic conversion of CLA, are not available yet. However, UV monitoring may indicate which HPLC fractions contain CFA (Figure 9.9). Then, the conjugated nature of unknown fatty acids present in any of the fractions might be determined by GC by observing the disappearance of peaks in the fatty acid profile before and after reacting the fatty acid ester mixture with PTAD or MTAD (the adducts formed from the CFA methyl ester do not elute from a polar column). Independently of a structural study, GC/MS can additionally support the identification of the CFA esters, simply by measuring the molecular ion.

The separation and quantification of all of the 18:2Δ9,11 geometrical

isomers has been achieved by GC on a high polar capillary column (50 m ×
0.33 mm, BPX 70™ (SGE Ltd), coated with 70% cyanopropylsiloxane
(0.25 μm film thickness)) [157]. However, the issue becomes more complicated
when dealing with complex mixture of both geometrical and positional isomers.
As a result, to date, the quantification of *all* of the individual isomers of CLA
present in natural or processed samples has not been fully attained.
Improvement to the silver ion-HPLC technique in combination with high
resolution GC columns might help to overcome the problem.

iii) Structural study. It is necessary that a combination of the chromatographic
techniques described in the previous section are used for an extensive
structural elucidation of the isomers present in samples. The CFA in natural
samples should be purified as much as possible, for example by reversed-phase
HPLC, though additional fractionation of the CFA fraction might be achieved
by silver ion TLC or HPLC. Double bonds location can be obtained by
GC/MS in EI mode by using at least two kinds of derivative, such as DMOX
(preferentially prepared from free acids) and MTAD adducts, in order to cross-
check results. Also, picolinyl esters of deuterated CFA (prepared from FFA)
can be used but only with highly purified molecules [48].

The double bond geometry can be established by GC/FTIR. An additional
confirmation can be achieved by a combination of the chemical analysis
depicted above and GC/MS; the positional isomers from each of the *cis* and
trans monoenes fractions obtained after hydrazine reduction can be determined
using DMOX derivatives [29]. This set of techniques used in combination
affords better analytical reliability and minimises the limitations present in any
of the single methods. Similar protocols were applied to identify the CLA
isomers in cheese-fat [25,29], grilled-ground beef [25], bile [139] and long-
chain metabolites in rats fed with a mixture of commercial CLA [48].
However, these kinds of procedure are not simple and other possibilities could
be explored, such as nuclear magnetic resonance (NMR) which has been
applied very recently to chemically-prepared CLA from castor oil [168].

LIST OF ABBREVIATIONS

amu, atom mass unit; BF3, boron trifluoride; BHT, butylated hydroxy
toluene; CD, conjugated dienes; CFA, conjugated fatty acids; CLA, conjugated
linoleic acid; DMBA, 7,12-dimethylbenz(a)anthracene; DMOX,
dimethyloxazoline; FAME, fatty acid methyl ester; Fe-ADP, iron-adenosine
diphosphate; FTIR, Fourier transform infra-red; GC, gas chromatography;
HDL, high density lipoprotein; HPLC, high-performance liquid
chromatography; IQ, 2-amino-3-methylimidazo[4,5-f]quinoline; IR, infra-red;
LC/MS, liquid chromatography/mass spectrometer; LDL, low density
lipoprotein; MNU, N-methylnitrosourea; mRNA, messenger ribonucleic acid;
MS, mass spectrometer; MSO, menhaden plus safflower oil; MTAD, 4-methyl-
1,2,4-triazoline-3,5-dione; NMR, nuclear magnetic resonance; PGE2,
prostaglandin E2; PLPC, L-α-Phosphatidylcholine, β-palmitoyl-γ-linoleoyl;

PTAD, 4-phenyl-1,2,4-triazoline-3,5-dione; PUFA, polyunsaturated fatty acids; SBO, soybean oil; tButylHP, tert-butyl-hydroperoxide; TLC, thin-layer chromatography; TPA, 12-O-tetradecanoylphorbol-13-acetate; UV, ultra-violet.

REFERENCES

1. Banni,S., Carta,G., Contini,M.S., Angioni,E., Deiana,M., Dessì,M.A., Melis,M.P. and Corongiu,F.P., in *Nutrition, Lipids, Health, And Disease*, pp. 218-224 (1995) (edited by A.S.H. Ong, E. Niki, and L. Packer, AOCS Press, Champaign, Illinois).
2. Porter,N.A., Caldwell,S.E. and Mills,K.A., *Lipids*, **30**, 277-290 (1995).
3. Banni,S., Dessì,M.A., Melis,M.P. and Corongiu,F.P., in *Free Radicals and Antioxidants in Nutrition*, pp. 347-364 (1993) (edited by F.P. Corongiu, S. Banni, M.A. Dessì and C. Rice-Evans, Richelieu Press, London UK).
4. Ghoshal,A.K. and Farber,E., *Lab. Invest.*, **64**, 255-260 (1993).
5. Dormandy,T.L. and Wickens,D.G., *Chem. Phys. Lipids*, **45**, 353-364 (1987).
6. Situnayake,R.D., Crump,B.J., Thurnham,D.I., Davies,J.A., Gearty,J. and Davis,M. *Gut*, **31**, 1311-1317 (1990).
7. Guyan,P.M., Uden,S. and Braganza,J.M., *Free Rad. Biol. Med.*, **8**, 347-354 (1990).
8. Kritchevsky,D., *Mal. Oil Sci. Technol.*, **4**, 47-51 (1995).
9. Dutton,H.J.,. In *Geometrical and Positional Fatty Acid Isomers*, pp. 1-16 (1979) (edited by E.A. Emken, and H.J. Dutton, American Oil Chemists' Society, Champaign).
10. Huges,P.E., Hunter,W.J. and Tove,S.B., *J. Biol. Chem.*, **257**, 3643-3649 (1982).
11. Shantha,N.C., Crum,A.D. and Decker,E.A., *J. Agr. Food Chem.*, **42**, 1757-1760 (1994).
12. Banni,S., Angioni,E., Contini,M.S., Carta,G., Casu,V., Iengo,G.A., Melis,M.P., Deiana,M., Dessì,M.A. and Corongiu,F.P., *J. Am. Oil Chem. Soc.*, **75**, 261-268 (1998).
13. Hoffmann,G., in *The Chemistry and Technology of Edible Oils and Fats and their High Fat Products*, pp. 203-245 (1989) (edited by G. Hoffmann, Academic Press, New York).
14. Reiser,R., *Fed. Proc.*, **10**, 236 (1951).
15. Ward,P.V.F., Scott,T.W., and Dawson,R.W.C., *Biochem. J.*, **92**, 60-68 (1964).
16. Kepler,R.C., Hirons,K.P., McNeill,J.J. and Tove,S.B., *J. Biol. Chem.*, **241**, 1350-1354 (1966).
17. Kepler,R.C., Tucker,P.W. and Tove,S.B., *J. Biol. Chem.*, **245**, 3612(1970).
18. Kepler,R.C. and Tove,S.B., *J. Biol. Chem.*, **242**, 5686-5692 (1967).
19. Banni,S., Day,B.W., Evans,R.W., Corongiu,F.P. and Lombardi,B., *J. Am. Oil Chem. Soc.*, **71**, 1321-1325 (1994).
20. Booth, R.G., Kon, S.K., Dann, W.J., and Moore, T. *Biochem. J.* **29**, 133-137 (1935).
21. Riel, R.R. *J. Dairy Sci.* **46**, 102-106 (1963).
22. Banni,S., Carta,G., Contini,M.S., Angioni,E., Deiana,M., Dessì,M.A., Melis,M.P. and Corongiu,F.P., *J. Nutr. Biochem.*, **7**, 150-155 (1996).
23. Fujimoto,K., Kimoto,H., Shishikura,M., Endo,Y. and Ogimoto,K., *Biosci. Biotech. Biochem.*, **57**, 1026-1027 (1993).
24. Chin,S.F., Liu,W., Storkson,J.M., Ha,Y.L. and Pariza,M.W., *J. Food Comp. Anal.*, **5**, 185-197 (1992).
25. Ha,Y.L., Grimm,N.K. and Pariza,M.W., *J. Agric. Food Chem.*, **37**, 75-81 (1989).
26. Lin,H., Boylston,T.D., Chang,M.J., Luedecke,L.O. and Shultz,T.D., *J Dairy Sci.*, **78**, 2358-2365 (1995).
27. Shantha,N.C., Decker,E.A. and Ustunol,Z., *J. Am. Oil Chem. Soc.*, **69**, 425-428 (1992).
28. Kosikowski,F. in *Cheese and Fermented Milk Foods*, pp. 91-108 (1982) (edited by F. Kosikowski, Brooktondale, New York).
29. Lavillonnière,F., Martin,J.C., Bougnoux,P. and Sebedio,J.L. *J. Am. Oil Chem. Soc.*, **75**, 343-352 (1998).
30. Shantha,N.C., Ram,L.N., O'Leary,J., Hicks,C.L. and Decker,E.A., *J. Food Res.*, **60**, 695-720 (1995).
31. Werner,S.A., Luedecke,L.O. and Shultz,T.D., *J. Agric. Food Chem.*, **40**, 1817-1821 (1992).
32. McDermott,R.L., *Food Technol.*, **41**, 91-103 (1987).
33. Perez,M.D., DeVillegas,C.D., Sanchez,L., Aranda,P., Ena,J.M. and Calvo,M., *J. Biochem.*, **106**, 1094-1097 (1989).
34. Murphy,J.J., Connolly,J.F. and McNeill,G.P., *Livest. Prod. Sci.*, **44**, 13-25 (1995).
35. Lawless,F., Murphy,J., Kjellmer,G., Connolly,J.F., Devery,R., Aherne,S., O'Shea,M. and Stanton,C., *Irish J. Agric. Food Res.*, **35**, 208A (1996).

36. Stanton,C., Lawless,F., Kjellmer,G., Harrington,D., Devery,R., Connolly,J.F. and Murphy,J., *J. Food Sci.*, in press.
37. Jiang,J., Bjoerck,L., Fonden,R. and Emanuelson,M., *J. Dairy Sci.*, **79**, 438-445 (1996).
38. Dhiman,T.R., Anamd,G.R., Satter,L.D. and Pariza,M.W., *J. Dairy Sci.*, **79**, 137(1996).
39. Lal,D. and Narayanan,K.M., Indian *J. Dairy Sci.*, **37**, 225-229 (1984).
40. Shantha,N.C. and Decker,E.A., *J. Food Lipids*, **2**, 57-64 (1995).
41. Badami,R.C. and Patil,K.B., *Prog. Lipid Res.*, **19**, 119-153 (1981).
42. Hopkins,G.Y., in *Topics in Lipid Chemistry*, pp. 37-87 (1972) (edited by F.D. Gunstone, Logos Press, London).
43. Yurawecz,M.P., Sehat,N., Mossoba,M.M., Calvey,E., Roach,J.A.G., and Ku,Y. 88th Annual Meeting of AOCS, Seattle,WA (1997) (Abstract).
44. Banni,S., Basford,R.E., Corongiu,F.P. and Lombardi,B., *Adv. Biosci.*, **76**, 187-201 (1989).
45. Banni,S., Evans,R.W., Salgo,M.G., Corongiu,F.P. and Lombardi,B., *Carcinogenesis*, **11**, 2053-2057 (1990).
46. Banni,S., Day,B.W., Evans,R.W., Corongiu,F.P. and Lombardi,B., *J. Nutr. Biochem.*, **6**, 281-289 (1995).
47. Banni,S., Evans,R.W., Salgo,M.G., Corongiu,F.P. and Lombardi,B., *Carcinogenesis*, **11**, 2047-2051 (1990).
48. Sébédio,J.L., Juanéda,P., Dobson,G., Ramilison,I., Martin,J.C., Chardigny,J.M. and Christie,W.W., *Biochim. Biophys. Acta*, **1345**, 5-10 (1997).
49. Nugteren,D.H., *Biochim. Biophys. Acta*, **210**, 171-176 (1970).
50. Belury,M.A. and Kempasteczko,A., *Lipids*, **32**, 199-204 (1997).
51. Ha,Y.L., Storkson,J. and Pariza,M.W., *Cancer Res.*, **50**, 1097-1101 (1990).
52. Ip,C., Chin,S.F., Scimeca,J.A. and Pariza,M.W., *Cancer Res.*, **51**, 6118-6124 (1991).
53. Ip,C., Jiang,C., Thompson,H.J. and Scimeca,J.A., *Carcinogenesis*, **18**, 755-759 (1997).
54. Fogerty,A.C., Ford,G.L. and Svoronos,D., *Nutr. Rep. Int.*, **38**, 937-944 (1988).
55. Nugteren,D.H. and Christ-Hazelhof,E., *Prostaglandins*, **33**, 403-417 (1987).
56. Braganza,J.M., Wickens,D.G., Cawood,P. and Dormandy,T.L., *Lancet*, **1983**, 375 (1983).
57. Fairbank,J., Hollingworth,A., Griffin,J., Ridgway,E., Wickens,D., Singer,A. and Dormandy,T., *Clin Chim Acta*, **186**, 53-58 (1989).
58. Fairbank,J., Hollingworth,A., Griffin,J., Ridgway,E., Wickens,D.G., Singer,A. and Dormandy,T.L., *Lancet*, **2**, 329-330 (1988).
59. Iversen,S.A., Cawood,P., Madigan,M.J., Lawson,A.M. and Dormandy,T.L., *Fed. Eur. Biol. Soc.*, **171**, 320-324 (1984).
60. Harrison,K., Cawood,P., Iversen,S.A. and Dormandy,T.L., *Life Chem. Rep.*, **3**, 41-44 (1985).
61. Iversen,S.A., Cawood,P., Madigan,M.J., Lawson,M. and Dormandy,L., *Life Chem. Rep.*, **3**, 45-48 (1985).
62. Halliwell,B. and Chirico,S., *Am. J. Clin. Nutr.*, **57**, 715S-725S (1993).
63. Thompson,S. and Smith,M.T., *Chem. Biol. Interact.*, **55**, 357-366 (1985).
64. Britton,M., Fong,C., Wickens,D.G. and Yudkin,J., *Clin. Sci.*, **83**, 97-101 (1992).
65. Banni,S., Lucchi,L., Ip,C., Angioni,E., Carta,G., Contini,M.S., Deiana,M., Dessì,M.A., Melis,M.P., Rapanà,R. and Corongiu,F.P., *Biochim. Biophys. Acta*, submitted.
66. Mahfouz,M.M., Valicenti,A.J. and Holman,R.T., *Lipids*, **15**, 306-314 (1980).
67. Pollard,M.R., Gunstone,F.D., James,A.T. and Morris,L.J., *Lipids*, **15**, 306-314 (1980).
68. Riisom,T. and Holman,R.T., *Lipids*, **16**, 647-654 (1981).
69. Jack,C.I.A., Jackson,M.J., Ridgway,E. and Hind,C.R.K., *Clin. Sci.*, **81**, 17(1991).
70. Banni,S., Angioni,E., Carta,G., Casu,V., Deiana,M., Dessì,M.A., Lucchi,L., Melis,M.P., Rosa,A., Scrugli,S., Sicbaldi,D., Solla,E. and Corongiu,F.P., 88th AOCS Annual Meeting 94-95 (1997) (Abstract).
71. Erskine,K.J., Iversen,S.A. and Davies,R., *Lancet*, **1**, 554-555 (1985).
72. Lucchi,L., Banni,S., Botti,B., Rota,C., Bergamini,S., Cappelli,G., Corongiu,F.P., Angioni,M.P. Melis,E., Perrone,S., Tomasi,A. and Lusvarghi,E., *Kidney Int.*, submitted.
73. Lucchi,L., Banni,S., Botti,B., Cappelli,G., Medici,G., Melis,M.P., Tomasi,A., Vannini,V. and Lusvarghi,E., *Nephron*, **65**, 401-409 (1993).
74. Banni,S., Lucchi,L., Baraldi,A., Botti,B., Cappelli,G., Corongiu,F.P., Dessì,M.A., Tomasi,A. and Lusvarghi,E., *Nephron*, **72**, 177-183 (1996).
75. Ip,C., Singh,M., Thompson,H.J. and Scimeca,J.A., *Cancer Res.*, **54**, 1212-1215 (1994).
76. van den Berg,J.J., Cook,N.E. and Tribble,D.L., *Lipids*, **30**, 599-605 (1995).
77. Yurawecz,M.P., Hood,J.K., Mossoba,M.M., Roach,J.A. and Ku,Y. *Lipids*, **30**, 595-598 (1995).
78. Wamer,W., Yurawecz,M.P., Wei,R., Sehat,N. and Ku,Y., *FASEB J.*, **10**, A272 (1996).

79. Li,Y., Allen,K.G.D. and Watkins,B.A., *FASEB J.*, **11**, A165 (1997).
80. Sugano,M., Tsujita,A., Yamasaki,M., Yamada,K., Ikeda,I. and Kritchevsky,D., *Nutr. Biochem.*, **8**, 38-43 (1997).
81. Holman,R.T., *Arch. Biochem. Biophys.*, **15**, 403-413 (1947).
82. Funk,M.O. and Alteneder,A.W., *Biochem. Biophys. Res. Commun.*, **114**, 937-943 (1983).
83. Wang,L.H. and Johnson,E.A., *Appl. Environ. Microbiol.*, **58**, 624-629 (1992).
84. Cave,W.T., *FASEB J.*, **5**, 2160-2166 (1991).
85. Cohen,L.A., Chen-Backlund,J.Y., Sepkovic,D.W. and Sugie,S., *Lipids*, **28**, 449-456 (1993).
86. Rose,D.P., Rayburn,J., Hatala,M.A. and Connolly,J.M., *Nutr. Cancer*, **22**, 131-141 (1994).
87. Ip,C., Briggs,S.P., Haegele,A.D., Thompson,H.J., Storkson,J. and Scimeca,J.A., *Carcinogenesis*, **17**, 1045-1050 (1996).
88. Pariza,M.W., Ashoor,S.H., Chu,F.S. and Lund,D.B., *Cancer Lett.*, **7**, 63-69 (1979).
89. Ames,B.M., McCann,J. and Yamasaki,E., *Mutation Res.*, **31**, 347-364 (1975).
90. Pariza,M.W., Loretz,L.J., Storkson,J. and Holland,N.C., *Cancer Res.*, **43**, 2444S-2446S (1983).
91. Pariza,M.W. and Hargraves,W.A., *Carcinogenesis*, **6**, 591-593 (1985).
92. Slaga,T.J. and Boutwell,R.K., *Cancer Res.*, **37**, 128-133 (1977).
93. Ha,Y.L., Grimm,N.K. and Pariza,M.W., *Carcinogenesis*, **8**, 1881-1887 (1987).
94. Ip,C., Scimeca,J.A. and Thompson,H.J., *Cancer*, **74**, 1050-1054 (1994).
95. Ip,C., Scimeca,J.A. and Thompson,H.J, *Nutr. Cancer*, **24**, 241-247 (1995).
96. Lu,J., Jiang,C., Forntaine,S. and Thompson,H.J., *Nutr. Cancer*, **23**, 283-290 (1995).
97. Sukumar,A., Notario,V., Martin-Zanca,D. and Barbacid,M., *Nature*, **306**, 658-661 (1983).
98. Zarbl,H., Sukumar,S., Arthur,A.V., Martin-Zanca,D. and Barbacid,M., *Nature*, **315**, 382-385 (1985).
99. Kumar,R., Sukumar,S. and Barbacid,M., *Science*, **248**, 1101-1104 (1990).
100. Schneider,B.L. and Bowden,G.T., *Carcinogenesis*, **6**, 1-4 (1992).
101. Ip,C. and Scimeca,J.A., *Nutr. Cancer*, **27**, 131-135 (1997).
102. Ip,C., Carter,C.A. and Ip,M.M., *Cancer Res.*, **45**, 1997-2001 (1985).
103. Fischer,S.M., Conti,C.J., Locniskar,M., Belury,M.A., Maldve,R.E., Lee,M.L., Leyton,J., Slaga,T.J. and Bechtel,D.H., *Cancer Res.*, **52**, 662-666 (1992).
104. Welsch,C.W., *Cancer Res.*, **52**, 2040S-2048S (1992).
105. Thompson,H., Zhu,Z., Banni,S., and Darcy,K., Loftus,T. and Ip,C., *Cancer Res.*, (In Press).
106. Belury,M.A., Nickel,K.P., Bird,C.E. and Wu,Y., *Nutr. Cancer*, **26**, 149-157 (1996).
107. Belury,M.A., Moya-Camarena,S.Y., Liu,K.L. and Vanden Heuvel,J.P., *J. Nutr. Biochem.*, **8**, 579-584 (1997).
108. Pennachiotti,G.L., Rotstein,N.P. and Aveldano,M.I., *Lipids*, **31**, 179-185 (1996).
109. Liew,C., Schut,H.A.J., Chin,S.F., Pariza,M.W. and Dashwood,R.H., *Carcinogenesis*, **16**, 3037-3043 (1995).
110. Shivapurkar,N., Tang,Z., Ferreira,A., Nasim,S., Garett,C. and Alabaster,O., *Carcinogenesis*, **15**, 775-778 (1994).
111. Pretlow,T.P., O'Riordan,M.A., Somich,G.A., Amini,S.B. and Pretlow,T.G., *Carcinogenesis*, **13**, 1509-1512 (1992).
112. Lam,L.K.T. and Zhang,J., *Carcinogenesis*, **12**, 2311-2315 (1991).
113. Zu,H.X. and Schut,H.A.J., *Food Chem. Toxicol.*, **30**, 9-16 (1992).
114. Shultz,T.D., Chew,B.P. and Seaman,W.R., *Anticancer Res.*, **12**, 2143-2146 (1992).
115. Shultz,T.D., Chew,B.P., Seaman,W.R. and Luedecke,L.O., *Cancer Letters*, **63**, 125-133 (1992).
116. Des Bordes,C. and Lea,M., *Anticancer Res.*, **15**, 2017-2022 (1995).
117. Lee, K.N., Kritchevsky, D. and Pariza, M.W., *Atherosclerosis*, **108**, 19-25 (1994).
118. Kowala, M.C., Nunnari, J.J., Durham, S.K. and Nicolosi, R.J., *Atherosclerosis*, **91**, 35-49 (1991).
119. Nicolosi, R.J., Courtemanche, K.V., Laitinen, L., Scimeca, J.A. and Huth, P.J., *Circulation*, **88**, 2458 (1993).
120. Miller, C.C., Park, Y., Pariza, M.W. and Cook, M.E., *Biochem. Biophys. Res. Commun.*, **198**, 1107-1112 (1994).
121. Cook, M.E., Miller, C.C., Park, Y. and Pariza, M.W., *Poultry Sci.*, **72**, 1301-1305 (1993).
122. Klasing, K.C., Laurin, D.E., Peng, R.K. and Fry, D.M., *J. Nutr.*, **117**, 1629-1637 (1987).
123. Hellerstein,M.K., Meydany,S.N., Meydany,M., Wu,K. and Dinarello,C.A., *J. Clin. Invest.*, **84**, 228-235 (1989).
124. Chin,S.F., Storkson,J.M., Albright,K.J., Cook,M.E. and Pariza,M.W., *J. Nutr.*, **124**, 2344-2349 (1994).

125. Pariza,M.W., Park,Y., Cook,M.E., Albright,K.J. and Liu,W., *FASEB J.*, **4**, A3227 (1996).
126. Albright,K.J., Liu,W., Storkson,J.M., Hentges,E., Lofgren,P., Scimeca,J.A., Cook,M.E. and Pariza,M.W., *J. Anim. Sci.*, **74**, 152 (1996).
127. Park,Y., Albright,K.J., Liu,W., Storkson,J.M., Cook,M.E. and Pariza,M.W., *Lipids*, **32**, 853-858 1997.
128. Smith,W.L., Borgeat,P. and Fitzpatrick,F.A., in *Biochemistry of Lipids, Lipoproteins and Membranes*, pp. 297-325 (1991) (edited by D.E. Vance and J. Vance, Elsevier, Amsterdam).
129. Kritchevsky,D., *J Nutr. Biochem.*, **6**, 172-178 (1995).
130. Butenandt,A., Hecker,E., Hopp,M. and Koch,W., *Annalen*, **658**, 39-64 (1962).
131. Roelofs,W., Kochansky,J., Cardè,R., Arn,H. and Rauscher,S., *Bull. Soc. Entomol. Suisse*, **46**, 71-73 (1973).
132. Ip,C., Lisk,D.J. and Scimeca,J.A., *Cancer Res.*, **54**, 1957s-1959s (1994).
133. Corongiu,F.P. and Milia,A., *Chem. Biol. Interac.*, **44**, 289-297 (1983).
134. Corongiu,F.P. and Banni,S., *Meth. Enzymol.*, **233**, 303-310 (1994).
135. Firestone,D. and Sheppard,A., in *Advances in Lipid Methodology - One*, pp. 273-322 (1992) (edited by W. W. Christie, Oily Press, Ayr).
136. Mossoba,M., *Inform*, **4**, 854-859 (1993).
137. Mossoba,M.M., MacDonald,R.E., Armstrong,D.J. and Page,S.W., *J. Chromatogr. Sci.*, **29**, 324-330 (1991).
138. Strocchi,A., Borgatti,A.R., Pallota,U. and Viviani,R., *Ind. Agrar.*, **6**, 644 (1968).
139. Christie,W.W., *Biochim. Biophys. Acta*, **316**, 204-211 (1973).
140. Ackman,R.G., Eaton,C.A., Sipos,J.C. and Crewe,N.F., *Can. Inst. Food Sci. Technol. J.*, **14**, 103-1074 (1981).
141. Nikolova-Damyanova,B., in *Advances in Lipid Methodology - One*, pp. 181-237 (1992) (edited by W. W. Christie, Oily Press, Ayr).
142. Kemp,P. and Lander,D.J., *J. Gen. Microbiol.*, **130**, 527-533 (1984).
143. Polan,C.E., MacNeill,J.J. and Tove,S.B., *J. Bacteriol.*, **88**, 1056-1064 (1964).
144. Cawood,P., Wickens,D.G., Iversen,S.A., Braganza,J.M. and Dormandy,T.L., *Fed. Eur. Biol. Soc.*, **162**, 239-243 (1983).
145. Banni,S., Contini,M.S., Angioni,E., Deiana,M., Dessi,M.A., Melis,M.P., Carta,G. and Corongiu,F.P., *Free Radic. Res.*, **25**, 43-53 (1996).
146. Smith,G.N., Taj,M. and Braganza,J.M., *Free Rad. Biol. Med.*, **10**, 13-21 (1991).
147. Scholfield,C.R., *J. Am. Oil Chem. Soc.*, **57**, 331-334 (1980).
148. Yurawecz,M.P., Hood,J.K., Roach,J.A.G., Mossoba,M.M., Daniels, .H., Ku,Y.K., Pariza,M.W. and Chin,S.F., *J. Am. Oil Chem. Soc.*, **71**, 1149-1155 (1994).
149. Shantha,N.C., Decker,E.A. and Hennig,B., *J. AOAC Int.*, **76**, 644-649 (1993).
150. Schuchardt,U. and Lopes,O.S., *J. Am. Oil Chem. Soc.*, **65**, 1940-1941 (1988).
151. Kemp,P., White,R.W. and Lander,D.J., *J. Gen. Microbiol.*, **90**, 100-114 (1975).
152. Parodi,P.W., *J. Dairy Sci.*, **60**, 1550-1553 (1977).
153. Sébédio,J.L., Grandgirard,A. and Prevost,J., *J. Am. Oil Chem. Soc.*, **65**, 362-366 (1988).
154. Spitzer,V., Marx,F., Maia,J.G.S. and Pfeilsticker,K., *J. Am. Oil Chem. Soc.*, **68**, 183-189 (1991).
155. Young,D.C., Vouros,P., Decosta,B. and Holick,M.F., *Anal. Biochem.*, **59**, 1954-1957 (1987).
156. Young,D.C., Vouros,P. and Holick,M.F., *J. Chromatogr.*, **522**, 295-302 (1990).
157. Berdeaux,O., Christie,W.W., Gunstone,F.D. and Sébédio,J.L., *J. Am. Oil Chem. Soc.*, **74**, 1011-1015 (1997).
158. Dobson,G., *J. Am. Oil Chem. Soc.*, **75**, 137-142 (1998).
159. Vincenti,M., Guglielmetti,G., Cassani,G. and Tonini,C., *Anal. Biochem.*, **59**, 694-699 (1987).
160. Janssen,G. and Parmentier, ., *Biomed. Mass Spectrom.*, **5**, 439-443 (1978).
161. Janssen,G., Verhulst,A. and Parmentier,G., *Biomed. Environ. Mass Spectrom.*, **15**, 1-6 (1988).
162. Dobson,G. and Christie,W.W., *Trends Anal. Chem.*, **15**, 130-137 (1996).
163. Andersson,B.A., Christie,W.W. and Holman,R.T., *Lipids*, **10**, 215-219 (1975).
164. Harvey, D.J., in *Advances in Lipid Methodology - One*, pp. 19-80 (1992) (edited by W. W. Christie, Oily Press, Ayr).
165. Spitzer,V., Marx,F. and Pfeilsticker,K., *J. Am. Oil Chem. Soc.*, **71**, 873-876 (1994).
166. Doolittle,R.E., Tumlinson,J.H. and Proveaux,A., *Anal. Biochem.*, **57**, 1625-1630 (1985).
167. Shantha,N.C., Ram,L.N., O'Leary,L., Hicks,C.L. and Decker,E.A., *J. Food Sci.*, **60**, 695-697 (1995).
168. Lie Ken Jie,M.S.F., Khysar Pasha,M., Shahim Alam,M., *Lipids*, **32**, 1041-1044 (1997)

APPENDIX 1

NATURAL TRANS *FATTY ACIDS*

R.G. Ackman

Canadian Institute of Fisheries Technology, DalTech, Dalhousie University, P.O. Box 1000, Halifax, Nova Scotia, B3J 2X4, Canada

A list of linear fatty acids including *trans* ethylenic bonds compiled from *The Lipid Handbook*, 1st edition, editors, F.D. Gunstone, J.L. Harwood and F.B. Padley. Chapman and Hall, London, 1986. Melting points, trivial names, and origins are given if of interest, otherwise they may be synthetic. *Trans* fatty acids occur in the seed oils of many plants, some very common, others rare. In preparing this list it has been necessary to accept the listed plant names. Both plants and oils and fats can have different names in different places. To permit cross referencing among botanical names and seed fats as sources of fatty acids the reader should, if necessary, consult another tabulation by K. Aitzetmuller, "Vegetable Oils of the World: Names of Oils and Fats and their Botanical Source", *Fat Sci. Technol.* **97**, 539-544 (1995).

Formula	Name	Melting point (°C)
$C_4H_5BrO_2$	2-bromo-2*t*-butenoic acid	92
$C_{16}H_{29}BrO_2$	16-bromo-9*t*-hexadecenoic acid	43
$C_4H_6O_2$	2*t*-butenoic acid Crotonic acid	72
$C_{10}H_6O_4$	2*t*,8*t*-decadiene-4,6-diynedioic acid from *Polyporus anthracophilus*	200 (decomp)
$C_{10}H_8O_2$	2*t*,8*t*-decadiene-4,6-diynoic acid from *Matricaria tenuifolia*, *Polyporus anthracophilus*	173 (decomp)
$C_{10}H_{16}O_2$	2*t*,4*t*-decadienoic acid 2*t*,4*c*-decadienoic acid Stillingic acid (from *Stillingia sebifera*) also as pear (Bartlett) ester	49-50 -

$C_{10}H_{16}O_2$	2t,6t-decadienoic acid	-
	2t,6c-decadienoic acid	-
	2c,6t-decadienoic acid	-
$C_{10}H_{16}O_2$	4t,6t-decadienoic acid	39
$C_{10}H_{12}O_2$	2t,6t-decadien-4-ynoic acid from *Aster schreberi*	-
$C_{10}H_{14}O_2$	2t,6c,8t-decatrienoic acid from *Heliopsis longipes* etc. 2t,6t,8t-decatrienoic acid synthetic	- -
$C_{10}H_{16}O_4$	2t-decendioic acid *Penicillium notatum* metabolite	172
$C_{10}H_8O_4$	2t-decene-4,6-diynedioic acid from *Polyporus anthracopilus*	-
$C_{10}H_{10}O_2$	2t-decen-4,6-diynoic acid from *Bellis perennis* etc.	124
$C_{10}H_6O_2$	2t-decene-4,6,8-triynoic acid from *Matracaria inodora*	-
$C_{10}H_6O_2$	8t-decene-2,4,6-triynoic acid	-
$C_{10}H_{18}O_2$	3t-decenoic acid *cis* is insect pheromone	18
$C_{16}H_{20}Br_2O_2$	14,16-dibromo-7t,13,15c-hexadecatrien- 5-ynoic acid from sponge *Xestospongia muta*	-
$C_{20}H_{32}O_6$	5,15-dihydroperoxy-6t,8c,11c,13t- icosatetraenoic acid 8,15-dihydroperoxy-5c,9t,11c,13t- icosatetraenoic acid metabolites of arachidonic acid	- -
$C_{10}H_{10}O_4$	2,10-dihydroxy-4t-decene-6,8-diynoic acid *Poria sinuosa* metabolite	163
$C_{22}H_{42}O_2$	11t-docosenoic acid Cetelaidic acid (from hydrogenated fish oil)	-
$C_{22}H_{42}O_2$	13t-docosenoic acid Brassidic acid (isomer of natural *cis*)	61
$C_{12}H_{20}O_2$	2t,4t-dodecadienoic acid from *Anacyclus pyrethrum*	50

$C_{12}H_{20}O_2$	2*t*,6*c*-dodecadienoic acid from Bartlett pear (as methyl or ethyl ester)	-
$C_{12}H_{20}O_2$	2*t*,8*t*-dodecadienoic acid 2*t*,8*c*-dodecadienoic acid 2*c*,8*t*-dodecadienoic acid	35 - -
$C_{12}H_{20}O_2$	7*t*,9*c*-dodecadienoic acid	-
$C_{12}H_{20}O_2$	8*t*,10*t*-dodecadienoic acid 8*t*,10*c*-dodecadienoic acid	- -
$C_{12}H_{16}O_2$	2*t*,6*c*,8*t*,10*t*-dodecatetraenoic acid from *Echinacea angustifolia* root etc. (also isomerized to all-*trans* form)	95 -
$C_{12}H_{20}O_4$	2*t*-dodecenedioic acid plant (Traumatic) acid	167
$C_{12}H_{22}O_2$	2*t*-dodecenoic acid ester in pear volatiles	13-18
$C_{12}H_{22}O_2$	7*t*-dodecenoic acid from human depot fat	-
$C_{12}H_{22}O_2$	9*t*-dodecenoic acid from human depot fat	-
$C_{12}H_{18}O_2$	8*t*-dodecen-10-ynoic acid	-
$C_{18}H_{33}FO_2$	18-fluoro-9*t*-octadecenoic acid cis isomer is toxic principle in *Dichapetalum toxicarium*	53-54
$C_{17}H_{30}O_2$	8*t*,10*t*-heptadecadienoic acid	-
$C_{17}H_{30}O_2$	10*t*,12*c*-heptadecadienoic acid	-
$C_{17}H_{26}O_2$	10*t*,16-heptadecadien-8-ynoic acid from *Acanthosyris spinescens*	-
$C_{17}H_{20}O_2$	7*t*,9*t*,15*t*-heptadecatriene-11,13-diynoic acid from *Baccharis trinervis* as methyl ester	-
$C_{17}H_{30}O_4$	8*t*-heptadecenedioic acid Civetic acid	94-95
$C_{17}H_{32}O_2$	2*t*-heptadecenoic acid	57
$C_{17}H_{32}O_2$	9*t*-heptadecenoic acid	38
$C_{17}H_{28}O_2$	10*t*-heptadecen-8-ynoic acid from *Pyrularia pubera*	33-34

$C_{16}H_{16}O_2$	6*t*,8*t*-hexadecadiene-10,12,14-triynoic acid from *Chrysocoma* spp. as methyl ester	-
$C_{16}H_{28}O_2$	2*t*,4*t*-hexadecadienoic acid from *Piper guineense* as isobutylamide (2*t*,4*c* isomer synthetic)	-
$C_{16}H_{28}O_2$	6*t*,8*c*-hexadecadienoic acid	-
$C_{16}H_{28}O_2$	10*t*,12*c*-hexadecadienoic acid	-
$C_{16}H_{24}O_2$	2*t*,6*c*,8*c*,12*t*-hexadecatetraenoic acid 2*t*,6*c*,8*c*,12*c*-hexadecatetraenoic acid	- -
$C_{16}H_{20}O_2$	6*t*,8*t*,12*t*,14*t*-hexadecatetraen-10-ynoic acid from *Chrysocoma* spp. as methyl ester	-
$C_{16}H_{18}O_2$	6*t*,8*t*,14*t*-hexadecatrien-10,12-diynoic acid 6*t*,8*t*,14*c*-hexadecatrien-10,12-diynoic acid both from *Chrysocoma* spp. as methyl esters	- -
$C_{16}H_{22}O_2$	6*t*,8*t*,12*t*-hexadecatrien-10-ynoic acid from *Chrysocoma tenuifolia* as methyl esters	-
$C_{16}H_{30}O_2$	2*t*-hexadecenoic acid	47
$C_{16}H_{30}O_2$	3*t*-hexadecenoic acid from leaves of spinach, red clover and many other plants	53-54
$C_{16}H_{30}O_2$	6*t*-hexadecenoic acid from *Picramnia sellowii*, parsley, human fat	-
$C_{16}H_{30}O_2$	9*t*-hexadecenoic acid palmitelaidic (hide beetle pheromone) acid	32-33
$C_{10}H_8O_3$	10-hydroxy-2*t*,8*t*-decadiene-4,6-diynoic acid from *Fistulina hepatica*	-
$C_{10}H_{10}O_3$	10-hydroxy-2*t*-decene-4,6-diynoic acid from *Polyporus anthracophilus*	154
$C_{10}H_{10}O_3$	10-hydroxy-8*t*-decene-4,6-diynoic acid from *Fistulina pallida*	-
$C_{10}H_{18}O_2$	9-hydroxy-2*t*-decenoic acid from royal jelly	-
$C_{10}H_{20}O_3$	10-hydroxy-2*t*-undecenoic acid from royal jelly	52/64
$C_{17}H_{26}O_3$	7-hydroxy-10*t*,16-heptadecadien-8-ynoic acid from *Acanthosyris spinescens*	-

$C_{17}H_{20}O_3$	3-hydroxy-7t,9t,15t-heptadecatriene-11,13-diynoic acid from *Baccharis trinervis*	42-43
$C_{17}H_{28}O_2$	7-hydroxy-10t-heptadecen 8-ynoic acid from *Acanthosyris spinescens*	-
$C_{16}H_{30}O_3$	16-hydroxy-9t-hexadecenoic acid	55/70
$C_{20}H_{36}O_3$	11-hydroxy-12t,14c-icosadienoic acid made by sheep vesicular glands from 20:2n-6	48
$C_{20}H_{30}O_3$	9-hydroxy-2c,5c,7t,11c,14c-icosapentaenoic acid antimicrobial from *Laurencia hybrida*	-
$C_{20}H_{32}O_3$	n-hydroxy icosatetraenoic acids various HETE, metabolites of arachidonic acid with bonds in positions 4-15	
$C_{20}H_{38}O_3$	14-hydroxy-11t-icosenoic acid *Lesquerella* spp. provides natural *cis* isomer	-
$C_9H_8O_3$	9-hydroxy-7t-nonene-3,5-diynoic acid from *Poria selecta*	-
$C_{18}H_{24}O_3$	18-hydroxy-9c,16t-octadecadiene-12,14-diynoic acid comparable to isano oil fatty acids	-
$C_{18}H_{32}O_3$	9-hydroxy-10t,12t-octadecadienoic acid β-Dimorphecolic acid, from *Dimorphotheca aurantiaca* the α form, -10t,12c-, is found in *Xeranthemum annum*	-
$C_{18}H_{32}O_3$	13-hydroxy-9c,11t-octadecadienoic acid Coriolic acid, *Coriaria nepalensis* the -9t,11t-form is found in *Absinthium* oils	-
$C_{18}H_{28}O_3$	8-hydroxy-11t,17-octadecadien-9-ynoic acid from *Pyrularia pubera* and *Acanthosyris spinescens*	-
$C_{18}H_{30}O_3$	18-hydroxy-9t,11t,13t-octadecatrienoic acid β-Kamolenic acid, α-Kamolenic acid from *Mallotus phillipinensis* seed has a -9c- bond	(β) 89 (α) 78
$C_{18}H_{34}O_3$	5-hydroxy-2t-octadecenoic acid	76-77
$C_{18}H_{34}O_3$	9-hydroxy-10t-octadecenoic acid product of oxidation of oleic acid	-

$C_{18}H_{34}O_3$	9-hydroxy-12t-octadecenoic acid Probably isomerized from *cis* form found in *Strophanthus* spp. seed fat	57-59
$C_{18}H_{34}O_3$	12-hydroxy-9t-octadecenoic acid Ricinelaidic acid is made from natural ricinoleic acid of castor oil	51-52
$C_{18}H_{30}O_3$	8-hydroxy-11t-octadecen-9-ynoic acid 8-Hydroxyximenynic acid	29-34
$C_{18}H_{30}O_3$	9-hydroxy-10t-octadecen-12-ynoic acid Helenynolic acid from *Helichrysum bracteatum*	-
$C_{20}H_{30}O_2$	2t,5c,8c,11c,14c-icosapentaenoic acid	-
$C_{20}H_{34}O_2$	8c,12t,14c-icosatrienoic acid	8
$C_{20}H_{38}O_2$	9t-icosenoic acid Gadelaidic acid (isomerized gadoleic acid)	54
$C_{17}H_{32}O_2$	14-methyl-8t-hexadecenoic acid pheromone for kaphra beetle *Trogoderma glabrum*	-
$C_9H_{16}O_2$	2t-nonenoic acid from tobacco, also honeybee pheromone	-
$C_9H_{16}O_2$	3t-nonenoic acid possibly in plant flavours	-
$C_9H_{16}O_2$	4t-nonenoic acid 6t-nonenoic acid	- -
$C_{18}H_{24}O_2$	9c,16t-octadecadiene-12,14-diynoic acid	-
$C_{18}H_{32}O_2$	5c,9t-octadecadienoic acid metabolic product of *trans*-9-monomer	-
$C_{18}H_{32}O_2$	5,12-octadecadienoic acid four possible isomers described	-
$C_{18}H_{32}O_2$	6t,8t-octadecadienoic acid	52
$C_{18}H_{32}O_2$	6t,9t-octadecadienoic acid followed by 6,10- 6,11- 6,12- all *trans* isomers, and 7,12-, 8,10-, 8,12-, 9,11- and 10,12-all-*trans* isomers	14-16
$C_{18}H_{32}O_2$	9,12-octadecadienoic acid Linolelaidic acid, from *Chilopsis linearis* all four isomers described	28-29
$C_{18}H_{28}O_2$	11t,13t-octadicadien-9-ynoic acid from root lipids of *Ximenia americana*	45-46

$C_{18}H_{28}O_2$	3*t*,9*c*,12*c*,15*c*-octadecatetraenoic acid from *Tecoma stans* seed oil	-
$C_{18}H_{28}O_2$	9*t*,11*t*,13*t*,15*t*-octadecatetraenoic acid β-Parinaric acid	95-96
	9*c*,11*t*,13*t*,15*c*-octadecatetraenoic acid α-Parinaric acid from *Parinarium laurinum* kernel oil	85-86
$C_{18}H_{30}O_2$	2*t*,9*c*,12*c*-octadecatrienoic acid pollen attractant for honey bees	-
$C_{18}H_{30}O_2$	5,6,16*t*-octadecatrienoic acid Lamenallenic acid (*Laminium purpureum* oil)	-
$C_{18}H_{30}O_2$	5*t*,9*c*,12*c*-octadecatrienoic acid Columbinic (Ranunculeic) acid from Gymnospermae seed lipids	-
$C_{18}H_{30}O_2$	8*t*,10*t*,12*t*-octadecatrienoic acid isomerized from natural forms	77-78
	8*t*,10*t*,12*c*-octadecatrienoic acid Calendic acid from *Calendula officinalis* seed oil	40
	8*c*,10*t*,12*c*-octadecatrienoic acid Jacaric acid from *Jacaranda mimosifolia*	43-44
$C_{18}H_{30}O_2$	9*t*,11*t*,13*t*-octadecatrienoic acid β-Eleostearic acid	72
	9*c*,11*t*,13*t*-octadecatrienoic acid α-Eleostearic acid from tung oil	49
	9*t*,11*t*,13*c*-octadecatrienoic acid Catalpic acid from *Catalpa* ovata	31-32
	9*c*,11*t*,13*c*-octadecatrienoic acid Punicic (Trichosanic) acid, from pomegranate oil	44-45
$C_{18}H_{30}O_2$	9*t*,12*t*,15*t*-octadecatrienoic acid Linolenelaidic acid (all eight isomers of α-Linolenic acid described)	29-30
$C_{18}H_{30}O_2$	10*t*,12*t*,14*t*-octadecatrienoic acid Pseudoeleostearic acid (isomerization product from α-linolenic acid)	77

Note - 18:1 *trans* fatty acids listed are mostly not natural but can be isolated from hydrogenated fats

$C_{18}H_{34}O_2$	2*t*-octadecenoic acid	58
$C_{18}H_{34}O_2$	3*t*-octadecenoic acid	64-65
$C_{18}H_{34}O_2$	4*t*-octadecenoic acid	58-59

$C_{18}H_{34}O_2$	5t-octadecenoic acid from *Thalictrum polycarpum* seed oil	47
$C_{18}H_{34}O_2$	6t-octadecenoic acid Petroselaidic acid	53 or 54-59
$C_{18}H_{34}O_2$	7t-octadecenoic acid	44
$C_{18}H_{34}O_2$	8t-octadecenoic acid	51-53
$C_{18}H_{34}O_2$	9t-octadecenoic acid Elaidic acid	44-45
$C_{18}H_{34}O_2$	10t-octadecenoic acid	52-52
$C_{18}H_{34}O_2$	11t-octadecenoic acid Vaccenic acid	43-44
$C_{18}H_{34}O_2$	12t-octadecenoic acid	40 or 52-53
$C_{18}H_{34}O_2$	15t-octadecenoic acid	59-61
$C_{18}H_{34}O_2$	16t-octadecenoic acid	65-66
$C_{18}H_{30}O_2$	9t-octadecen-12-ynoic acid *cis* form is Crepenynic acid from *Crepis foetida* or *Afzelia cuanzensis*	128 10-11
$C_{18}H_{30}O_2$	11t-octadecen-9-ynoic acid Ximenynic or santalbic acid from seed lipids of Santalaceae or Olacaceae	39-40
$C_8H_{12}O_2$	5t,7-octadienoic acid	-
$C_8H_{14}O_2$	3t-octenoic acid	-
$C_{10}H_{16}O_3$	9-oxo-2t-decenoic acid Queen (bee) substance	54-55
$C_{18}H_{26}O_3$	4-oxo-9c,11t,13t,15c-octadecatetraenoic acid from *Chrysobalanus icaco* seed oil	-
$C_{15}H_{26}O_2$	10t,12c-pentadecadienoic acid	-
$C_{24}H_{46}O_2$	15t-tetracosenoic acid	66-67
$C_{14}H_{24}O_2$	2t,4t-tetradecadienoic acid	57-58
$C_{14}H_{24}O_2$	3t,5c-tetradecadienoic acid Megatomic acid (beetle sex attractant)	-
$C_{14}H_{24}O_2$	10t,12c-tetradecadienoic acid	-

$C_{14}H_{14}O_4$	all-*trans*-form of 2,4,6,8,10,12-tetradecahexaene dioic acid Corticrocin, fungal pigment from *Corticium croceum*	
$C_{14}H_{26}O_2$	2t-tetradecenoic acid	33 or 50-53
$C_{14}H_{26}O_2$	9t-tetradecenoic acid Myristelaidic acid	18
$C_{13}H_{22}O_2$	3t,5t-tridecadienoic acid	-
$C_{13}H_{24}O_2$	2-tridecenoic acid	-
$C_{27}H_{52}O_2$	2,4,6-trimethyl-2t-tetracosenoic acid Mycolipenic acid (tuburcle bacilli)	28
$C_{11}H_{10}O_2$	2t,8t-undecadiene-4,6-diynoic acid 2t,8c-undecadiene-4,6-diynoic acid	143-145 85-88
$C_{11}H_{18}O_2$	7t,9c-undecadienoic acid	-
$C_{11}H_{20}O_2$	2- to 9-undecenoic acids are listed	-

Omitted from this list are certain fatty acids that are metabolites of polyunsaturated fatty acids. Two have been retained as typical examples, but most other hydroxy acids associated with biosynthesis of prostaglandins are left out. Cyclopentyl, epoxy and fatty acids with similar structures are also omitted. Many fatty acids of plant origin exist with ethylenic bonds both *cis* and *trans* forms, but the higher melting point of the *trans* isomer favours their isolation, especially through crystallization, and a *cis* isomer should always be suspected in undefined cases. For obvious reason plant attributions are usually to a seed oil.

INDEX